SOLAR ENERGY

Solar Energy

An Introduction

Michael E. Mackay

University of Delaware

OXFORD
UNIVERSITY PRESS

Great Clarendon Street, Oxford, OX2 6DP,
United Kingdom

Oxford University Press is a department of the University of Oxford.
It furthers the University's objective of excellence in research, scholarship,
and education by publishing worldwide. Oxford is a registered trade mark of
Oxford University Press in the UK and in certain other countries

First Edition published in 2015
Impression: 1

Published in the United States of America by Oxford University Press
198 Madison Avenue, New York, NY 10016, United States of America

British Library Cataloguing in Publication Data
Data available

Library of Congress Control Number: 2015932259

ISBN 978–0–19–965210–5 (hbk.)
ISBN 978–0–19–965211–2 (pbk.)

Printed and bound by
CPI Group (UK) Ltd, Croydon, CR0 4YY

To my cousin, Vance, who helped me more than he knows.

Preface

Solar Energy is the ultimate energy source for humankind. In a very subtle way the Sun gives energy to atoms which digest it by processes we have only recently understood. It is truly amazing that materials have been developed to absorb that energy and produce electricity or imbibe it with negligible release to become incredibly hot. The challenge now is use all that has been learned and developed to harness the electricity and heat in a rational, economically and ecologically sound manner. The purpose in writing this book is to present disparate, solar-based technologies so the reader can generalize the information and make a holistic decision when using this renewable energy source.

Bringing these different technologies together was not an easy task and this book is the result of teaching a course of the same name for four consecutive years to upper level undergraduate and beginning level graduate students at the University of Delaware. Students from various disciplines have taken the course including: chemical engineering, electrical engineering, materials science and engineering, mechanical engineering, physics and chemistry. The scientists and electrical engineers thrive at the beginning and struggle at the end when heat transfer calculations are involved. The opposite occurs for the others, especially the mechanical engineers who are very familiar with heat transfer calculations. In spite of this I can say all have enjoyed it and learned new principles.

Educating students with such different backgrounds places a greater burden on the instructor and he or she should be sympathetic with the students when they are not comfortable with the subject matter. I tell them they really do not know what they will be doing five years from now as can be seen in Fig. 1. I show this figure to the students in lecture and they suddenly realize that they will have a number of jobs throughout their lives. It becomes more real to them. So, in a course based on this book they will learn new information that is unfamiliar to them, similar to what they will experience in a new job. However, it is mostly the learning process that they develop when in these difficult situations that will aid their career. In other words, they will always be in situations where they have to learn something new, something that does not match their background or training, and they must adapt, learn and solve the problem. This should be made clear to them at the very beginning.

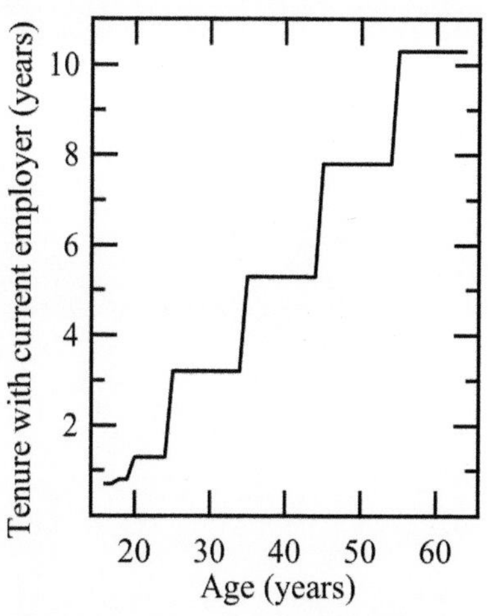

Fig. 1 Amount of time an employee has with their present employer as a function of age. Source: United States Department of Labor, Bureau of Labor Statistics.

Due to its nature *Solar Energy* is truly a capstone course. The student must learn principles from chemistry, physics and engineering. However, I am a firm believer in a student being able to *generalize* information rather than merely knowing facts about a topic. This is the second time I have used this word, generalize, and by it I mean that the student should be able to take information and apply it to their problem at hand, which may only be tangentially related to what they have learned here. It is hoped that this pledge is fulfilled.

This book is roughly divided into two parts: one that is concerned with photovoltaics and the other with thermal processes. The part concerning photovoltaics follows the principles used by many researchers and authors and I have merely tried to put everything into a consistent nomenclature with the second part. Furthermore, an effort to explain the assumptions, models and theories found in this area has been made by keeping in mind the various students' backgrounds.

In the second part a strong deviation from the literature has been done, especially in the design of a solar thermal flat plate energy collector. I have gone back to the basic heat transfer calculations which, although unwieldy, are now possible to manipulate. The challenge is solving coupled algebraic equations that have the temperature T to different powers: T^4 and T^1, representative of radiative and conductive or convective heat transfer, respectively. In addition, the solution to the coupled equations requires an iterative process. Present day computers and equation solvers are readily available to the student and can be used to easily solve these equations, which was not possible for the average student to do only 20 years ago. So, it is believed the generality introduced by this fundamental approach will allow the student a true understanding of the factors that affect solar thermal processes without the expense of burdening them with algebraically complex, iterative calculations. Overall, knowledge of the design procedure details should allow the scientist or engineer the ability to make new and novel devices.

Also, in this second part, I have considered the so-called solar tower which channels heated air to a central chimney to turn turbines and create electricity. This is typically not considered in introductory textbooks. Although this technology is not used commercially, it is a natural extension of the solar chimney, so its inclusion seemed to be warranted. In addition, the design of solar thermal energy power plants (Rankine cycle) has been included. As it turns out these power plants have much in common with the flat plate collector, so, some care has been taken to relate the two processes.

Teaching a class from this book can be done in many ways. For those who want a more semiconductor physics bent to their course then Chapter 4 will be emphasized and expanded by the lecturer. On the other hand, this may not be a priority and Chapter 4 can be deemphasized. A list of topics I include when teaching the class is given in the table at the end of this *Preface*. Even though I do not use the entire book it is a good way to present the entirety of Solar Energy to the students. The information not discussed was placed in the textbook for reference and to allow students that wish to know the details a way to find them.

Finally, some of the information to solve the *Exercises* at the end of each chapter may not be available in this book. This was done on purpose to expand the scope of the textbook with the ubiquitousness of the World Wide Web. An example is the very first Exercise, (1.1), where the student is to find the lifetime of CO_2 in the atmosphere. I mention in Chapter 1 it may last hundreds of years, however, there are better estimates than this in the scientific literature that the student can readily find. This is actually a more complicated exercise than expected. A given CO_2 molecule may be in the atmosphere for a much shorter time than this after being absorbed by the ocean, for example. Yet, it is not destroyed *per se* and will eventually go back to the atmosphere! This is the type of openendedness I have included in some of the exercises.

The production of this book has benefited from discussions with many individuals and I would like to first thank my former colleague Professor Phillip M. Duxbury with whom I worked on polymer-based solar cells as well as other research areas. Our collaboration is treasured by me and his insights as a physicist only made me grow academically from my engineering roots. My past mentors, Professors Jerold M. Schultz, Michael E. Paulaitis, Anthony J. McHugh and David V. Boger all shaped me to where I could tackle an effort such as this. I would also like to thank Professor Abraham (Bramie) Lenhoff for valuable discussions on the design of the flat plate solar energy collector which crystallized my thoughts on this topic and Professor James Kolodzey for clarifying semiconductor physics for me (any error is due to me). Finally, I would like to thank my former students who have aided me along the way, without them I would not have learned what I have.

The challenges that lay ahead for humankind are interesting and the future will certainly be different to now. The 19th century saw the beginnings of industrialization while the 20th had some of the most amazing advances in engineering, science and medicine, and simultaneously began our dependence on fossil fuels. The last century was truly incredible and although some call those who participated in it from the 1930s onward as *The Greatest Generation* I don't see them in that light. Anyone could have done what they did, and in fact anyone did, they just happened to be alive at that time. Now is the time for those alive *now* to have *great* achievements. I hope this monograph will inspire someone to this end.

I often think of this quote when trying to understand why a person wants to become richer, a corporation to become larger or a university to expand:

"Growth for the sake of growth is the ideology of the cancer cell."

Edward Abbey

Are humans the equivalent of cancer cells? There appears no way the population rise of the Earth will stop until the planet can support no more people. What will be the Earth's condition at that time? I cannot predict the future, yet, believe utilization of Solar Energy is one way to mitigate the negative effect we will have on the Earth.

Michael E. Mackay
Newark, Delaware
April 2015

Suggested topics to be covered in the course *Solar Energy*.

Chapter	Comments
1	**Why solar energy is important.** Covering this chapter will show why Solar Energy utilization is important, and gives the student an introduction to the scale of energy usage on Earth.
2	**Solar radiation.** The ability to determine how much solar radiation strikes a device is presented and is an emphasis of this chapter. This material is presented differently to other textbooks by considering an instantaneous radiative power (Watts) and the last section is written to rectify my choice. The reason I decided to present the information in this way is because a solar powered device will demand a rate of energy (*i.e.* power in Watts or Joules per second) to function, while location of that device is dictated by how much energy (Joules) is present at a given location. Since the material in this monograph is centered on device design one needs to have the design equations written in terms of power.
3	**Basic principles.** The basic principles to be used in this textbook are covered in this chapter. Many students will be familiar with much within these topics, such as the First Law of Thermodynamics for a flowing system and this will seem redundant to them. Other students may never have seen it before and it is for this reason that this, and other topics, were included.
4	**Electrons in solids.** The main results of this chapter are eqns (4.22), (4.23) and (4.24) which can merely be presented and subsequently justified if need be. One should keep in mind that this is a chapter necessary to understand semiconductor physics and if it is de-emphasized or skipped the student may not know how the devices actually function.
5	**Light absorption.** This is an important chapter as I find that students don't appreciate it is the electrons that interact with solar radiation to produce the absorption spectrum. In addition, if I give them the absorption coefficient versus wavelength they are challenged to find the band gap energy. This is a very good exercise to give to them and they learn much from it.
6	**The photovoltaic device.** In some cases the lecturer may not want to cover the details of what occurs in a solar cell. I wrote Section 6.1 with this in mind which can be covered as well as Sections 6.3 and 6.4. One should also present Figs. 6.4 and 6.5 since they represent the heart of a solar cell. Two good projects for students to do is to

Suggested topics to be covered in the course *Solar Energy* (cont.)

Chapter	Comments
	give them synthetic current–voltage data, knowing the two resistances, with random noise of 0.1% up to 10% and have them determine the various parameters discussed in the chapter. Another is to design a Silicon-based solar cell, a more difficult project, and calculate the light absorption as a function of depth as well as the current–voltage relation.
7	**The solar chimney and tower.** Depending on the emphasis of the lecturer, this chapter can be included if wanted. One consideration is that the Bernoulli equation must be presented, which is the basic relation in fluid mechanics. If the student has never seen the Bernoulli equation I give an example on how pressure and velocity have an inverse relation along a streamline (careful in using the airplane wing as an example, see H. Babinsky, "How do wings work?," Phys. Ed. **38** (2003) 497–503). I include this chapter when teaching this class as it is relatively simple, since the heat transfer is easy to consider, and prepares the student for the much more difficult heat transfer calculations in the next chapter.
8	**The flat plate solar energy collector.** As mentioned above, the presentation of the information in this chapter is a strong deviation from the literature with an emphasis on heat transfer calculations. I prefer to give the rationale of how to design this type of solar hot water heater and give them a design problem to do in class in groups. I like to have two students work together with one usually the better in use of the equation solver.
9	**Solar thermal energy generated electricity.** This may or may not be presented to the students. This topic is not usually presented in detail for a course of this type and certainly discussion of the Rankine power cycle is not normally given. Although this is not a power systems monograph it is important that students know where most of their electricity is from, a Rankine cycle. This discussion was included so the student knows that the Sun is ultimately going to drive a Rankine cycle. This topic is better introduced and then to let the students perform a design problem, just like in the previous chapter.

Contents

Why solar energy is important

The Sun produces an incredible amount of energy each and every day. There is enough to supply all the energy needs for everyone in the World in one hour to emphasize the magnitude. In spite of this, solar energy is still a fledgling technology compared to others that use oil, gas and coal, the so-called fossil fuels. Why?

The challenge lies in harvesting the energy with efficient and cost effective devices. One must take solar radiation, or insolation,[1] that is made of electromagnetic waves and convert them to useful heat or electricity. Both processes require a material that can absorb a photon's energy by placing an electron into a higher energy level. Heat is produced by the electron falling back to a lower energy level to make surrounding atoms vibrate through the energy exchange. More vibration corresponds to a higher temperature. This energy is then exchanged to an intermediary providing useful heat. Electricity generation demands that the electron not fall to a lower energy level, instead the electron and its oppositely charged partner, the hole, move to different electrodes and their potential (voltage) difference is used to power a device.

[1] The phrase *solar insolation* is redundant as insolation implies radiant energy from the Sun. The word comes from the Latin *insolare*.

It is the efficiency and economics of these electronic processes that dictate the success of solar energy utilization. As the reader will find, it is very efficient to convert solar energy to low quality energy, however, higher quality is much less efficient. The definition of low and high quality energy is defined here as, high quality energy can be used to generate low quality while the opposite is not true (at least with any reasonable technology that would attract economic investment).

Low quality energy is generation of hot water for residential or industrial use at a temperature of approximately 50°C, which occurs at around 70% efficiency! The efficiency is defined as every 100 Watts that strike a device are converted into 70 Watts of useful power suggesting that creation of hot water is very efficient.

Higher quality energy, as generated with photovoltaics, have devices that are less efficient at approximately 10%. This efficiency is dictated in part by the fact that the physics of photovoltaic devices demands that the material should begin absorbing solar energy above a certain energy level to allow a potential (voltage) to be developed. Without this potential no power can result since power is the product of voltage and current. Conversely, current is maximized by absorbing all the solar energy since this will excite the most electrons.

So, a tradeoff between voltage and current results. A larger voltage means less solar energy is absorbed to produce current since part of the solar spectrum is not absorbed. Conversely a lower voltage means a higher current, and it is this tradeoff that sets the upper limit on the power conversion efficiency of a contemporary solar cell at approximately 31%. The reader will fully appreciate this after reading later chapters. There are also inefficiencies that occur after the solar radiation is absorbed and the best laboratory devices can operate at around 20% efficiency, while devices in the field are 12% efficient (this is for crystalline Silicon, by far the dominant product sold today).

In reality the solar cell is about 39% efficient in terms of what the maximum possible efficiency would be (*i.e.* 39% ≈ 12%/31%) and compares well to a coal-fired power plant which can extract 35–40% of coal's inherent energy content to produce electricity. Thus, they are comparable at this level albeit one that changes the playing field. Discussion in later chapters will consider the physics, chemistry and engineering of solar energy and the devices used to harvest its energy; in this chapter a brief introduction is given to solar energy and its value to humankind.

1.1 Should solar energy be harvested?

Is harvesting solar energy required? The answer to this question depends to whom you ask it. If the responder is someone who tracks energy resources such as coal, natural gas or petroleum then the answer is, 'No.' There are reserves that can last for a reasonably long time at the present usage rate with coal being the longest lasting, perhaps for over 1000 years. Thus, society will not run out of energy resources.

Will population increase cause a strain on existing energy resources requiring more? According to the United Nations, the population of the Earth will continue to grow until about the year 2100 where there will be approximately 9 billion people on Earth. This is the result of the *medium* projection with the high and low projections giving different numbers. There are many figures given for the amount of people the Earth can support, however, this is most likely the correct magnitude. There will be about 30% more people on Earth in the near future and assuming the energy use per person remains the same this represents a large energy use increase, however, there are still plenty of resources. Yet, developing countries will use more energy per person as their economies move forward generating an even higher energy demand. Even if the world energy demand increases by a factor of two in the near future, accounting for all factors, there are still plenty of energy resources available, why should solar energy be considered?

The true challenge is greenhouse gas emissions generated by burning fossil fuels and solar energy is one technology that can mitigate their generation. As a person on Earth, it is difficult to fathom that we can affect the planet and in particular the atmosphere. The Earth is huge! The ecosystem is full of checks and balances and what we do can surely be small and forgiven by the planet. The difficulty is that the average person does not realize the *magnitude* of energy use and the associated atmospheric emissions.

The magnitude of the emission rate will be addressed in Example 1.1; let's begin and ascertain if the Earth is in fact a constant commodity or is a living, evolving organism. As can be seen in Fig. 1.1, the so-called Keeling Curve, the atmospheric concentration of Carbon Dioxide (CO_2) is an increasing function of time as measured at the Mauna Loa volcano in the state of Hawaii. Indeed, just recently, the atmospheric concentration of CO_2 reached 400 ppm (by mole fraction) which is most likely the highest concentration since the Pliocene epoch approximately three million years ago. Furthermore, air bubbles trapped in the Antarctic ice show the concentration of CO_2 has oscillated between 180 and 280 ppm over the last 800,000 years and so 400 ppm is a wild deviation from this tight band. Furthermore, it appears as if the concentration has not been above 400 ppm for at least 20 million years. So, the Earth has evolved to a state over the last 100 years or so that has not been seen for millions of years.

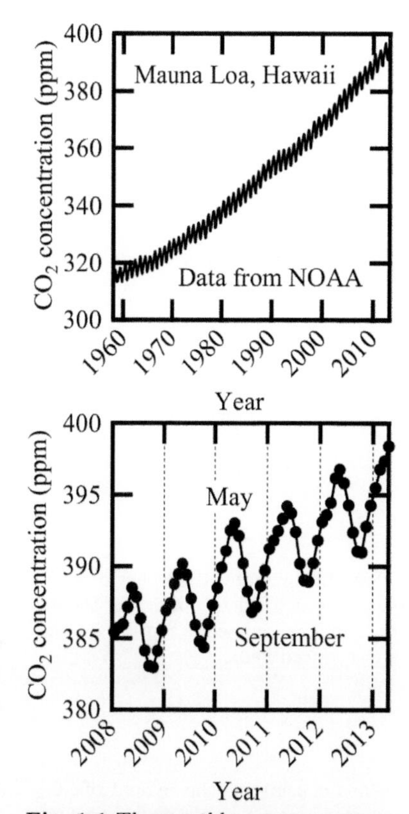

Fig. 1.1 The monthly average concentration (in ppm on a molar basis) of Carbon Dioxide as a function of date. Data for the entire data range (upper) and over a smaller time scale to show the seasonal oscillation (lower). Source: U.S. Department of Commerce, National Oceanic and Atmospheric Administration (NOAA).

Yet, the state of the Earth is not the same as it was even three million years ago since, at that time, the polar ice caps were much smaller and the sea level much higher. Presently though, the arctic ice is melting and not recovering, according to the Polar Science Center at the University of Washington, with the Arctic losing about 280 km^3 of ice per year. Assuming the ice is 2.5 m thick (the average thickness of winter ice in the Arctic) this represents an area of about 335×335 km^2 which is about the area of the state of Virginia or the country of Ecuador.

The energy to melt this ice is immense. The heat of fusion for water is 334 J/g and this represents an energy of almost 10^{20} J and is comparable to the United States of America's (USA's) annual (primary) energy consumption of 97.2 Quads (quadrillion or 10^{15} BTUs) in 2011 or again 10^{20} J! However, the seasonal ice loss from April to September in the Arctic is 16,400 km^3 which is almost 60-times larger than the permanent ice loss demonstrating exactly how much energy the polar ice cap requires every year to melt. Yet, the fact remains there is permanent ice loss and an amount of energy equivalent to that used by the USA is supplied to permanently melt it every year. At the present rate the arctic ice will be lost in around 50 years, since the mean volume is about 13,000 km^3, that is if the loss rate does not accelerate. So, the Earth may be driven to the same state as it was three million years ago by having smaller polar ice caps, higher seas generated by the ice melt and more atmospheric CO_2.

It seems clear that the Earth is evolving to another state. It is not static and is living, as seen in Fig. 1.1 through the annual fluctuation of CO_2 concentration from May to September and back again. This is due to plants growing in the spring and summer as photosynthesis uses it to generate biomass. As they die in the fall, CO_2 is generated through decay, yet, just as there is an annual permanent ice loss in the Arctic, there is an overall upward trend of 1.75 ± 0.02 ppm per year increase. Is this a significant increase and could it be generated by the burning of fossil fuels? The information developed in the following example can be used to answer this question.

Example 1.1

Determine the World Carbon Dioxide yearly generation rate and compare it to the increase seen at the Mauna Loa site.

Two pieces of information are needed; the total amount of CO_2 generated by burning fossil fuels and the number of moles of molecules making up the atmosphere, the later quantity is considered first.[2] The atmosphere can be considered a spherical shell, yet, the challenge is to define its thickness. For the present, assume it is 16 km thick, where the atmospheric density is approximately 10% that at sea level, and has an average density of 0.7 kg/m^3. The density of air at standard temperature and pressure is about 1.3 kg/m^3 while in space it is zero, so, 0.7 kg/m^3 is an average density justified *a posteriori*.[3] The molecular weight of air is approximately 29 g/mol and the radius of the Earth (R_E), 6380 km, to yield 2.0×10^{20} mol-air.

A more accurate estimate is made by realizing the atmospheric pressure at the Earth's surface P_a is about 10^5 Pa and by a force balance one has

$$m_a g = 4\pi R_E^2 P_a$$

where m_a is the mass of the atmosphere and g, the acceleration due to gravity. The effect of curvature was neglected since the atmospheric thickness is much less than the radius of the Earth. Solving for m_a yields 5.2×10^{18} kg-air or 1.8×10^{20} mol-air. This later figure will be used for the number of moles of air in the atmosphere, and for future reference the atmospheric thickness will be assumed equal to 16 km since using this thickness arrives at a very similar mass/mole value.

The amount of CO_2 generated per year can be estimated by knowing how much primary energy is consumed per year and multiplying by the amount of CO_2 produced by this consumption. Primary energy means the energy used to make the energy we use. In other words, consider a coal-fired power plant that burns coal at an efficiency η of 35% to generate electricity. If the power plant generates 100 MW of electricity then the amount of primary energy consumed is 100 MW/0.35 or 286 MW.[4]

The amount of primary energy used for the entire World was 511 Quads in 2010 or 5.42×10^{20} J. Now 0.32 kg-CO_2 are generated for each kW-h of coal burned and about three-quarters of this are generated when oil is burned, natural gas produces approximately 0.19 kg-CO_2/kw-h (see Table 1.1). An average value of 0.24 kg-CO_2/kW-h is assumed yielding 3.6×10^{13} kg-CO_2; the Energy Information Administration estimates 3.3×10^{13} kg-CO_2, validating our crude estimate.[5] This is 7.5×10^{14} mol-CO_2 released to the atmosphere every year.

[2]The abbreviation for moles is mol and mol-air means *moles of air*.

[3]The approximate atmospheric thickness is needed when considering the *air mass index* discussed in Section 2.4 and is introduced here.

[4]When considering the amount of CO_2 generated from a power plant one must calculate the primary energy which is the energy required to power the plant accounting for its efficiency. One must divide the generated power by the efficiency and a good first estimate is to use $\eta = 0.35$.

Table 1.1 Carbon Dioxide emissions from burning various substances.

Substance	CO_2 amount (kg-CO_2/kW-h)
Wood	0.39
Coal	0.32
Oil	0.24
Natural Gas	0.19

[5]This value is different to that given by the Carbon Dioxide Information Analysis Center of Oak Ridge National Laboratory which is 3–4 times less, $\approx 9 \times 10^{12}$ kg-CO_2. It is unclear why the two values are different, however, our estimate from the burning of fossil fuels agrees fairly well with the EIA's estimate, so, this value will be used.

Taking the ratio of the molar release rate of CO_2 per year and the total moles of the atmosphere finds 4.2 ppm! This is larger than the yearly rise in CO_2 monitored at Mauna Loa volcano of 1.75 ppm. The NOAA estimates that about 30–50% of this CO_2 is absorbed by the oceans and, to the order of this analysis, accounts for the CO_2 generation and where it eventually resides. However, the result is that the oceans are becoming more and more acidic as a result of this absorption producing adverse effects which could produce long-term damage, if not irrevocable devastation of the oceanic ecosystem.[6]

It seems undeniable that the burning of fossil fuels is responsible for the increase in CO_2 in the atmosphere which has increased the total amount to a level not seen for millions of years. Furthermore, the increase has correlated exactly with the industrial revolution lending further credence to the fact that CO_2 emissions are human generated.

[6] The NOAA actually calculates the CO_2 concentration on a dry air basis which may increase our estimate of 4.2 ppm. However, our estimate is reasonably sound since the United State Geological Survey states there are 12,900 km^3 of water in the atmosphere which will not significantly change the above estimate.

Carbon Dioxide has a long lifetime in the atmosphere since it is a very stable chemical compound. It certainly will not go away in a human lifetime and so that generated today will be here for hundreds of years. The accumulation is certain unless mitigation of burning fossil fuels occurs by using energy technologies that do not release CO_2. Solar energy technologies approach this criterion and could be used to reduce CO_2 emissions.[7]

One could argue CO_2 is a natural chemical compound and there is little concern for its release. Are you willing to determine what state the Earth will achieve without knowing the outcome? This is where we are now with the highest CO_2 concentration seen in over a million years and the effect is cumulative since CO_2 will remain with us for a very long time. This concentration in turn could dictate the global climate.

[7] According to the National Renewable Energy Laboratory (NREL) solar energy technologies generate 0.045 kg-CO_2 per kW-h of electricity generated, which is substantially less than that generated with a coal-fired power plant.

Calculation of how much CO_2 will warm the atmosphere is involved and complicated. One avenue to consider is that CO_2 absorbs radiation emitted by the Earth and traps it since it has an absorption band very near the maximum of the radiation spectrum emitted by the Earth. The mean surface temperature of the Earth[8] is T_E = 288 K and according to the Wien displacement law given in Chapter 2

$$\lambda T_E|_{\max} \approx 2.9 \times 10^6 \text{ nm-K} \tag{2.7}$$

one can predict the wavelength λ where the maximum amount of radiation will be emitted by the Earth, assuming the Earth is a black body radiator. Performing this calculation finds a wavelength of 10.1 μm which is near a strong absorption band for CO_2 at 13–18 μm. It can be argued that the CO_2 concentration is already large enough to absorb all this radiation, which is true, and any increase will not affect the Earth's temperature. The effect is subtle though and having a higher CO_2 concentration means the radiation *emitted* from the Earth will be absorbed *nearer* the Earth's surface. When the radiation is re-emitted by the atmosphere then it is less likely to escape to space and will be reabsorbed by the Earth to ultimately produce global warming.

[8] If there were no greenhouse gas warming then the effective surface temperature of the Earth would be 255 K instead of the average present temperature of 288 K. We need some greenhouse gases to keep us warm. The manuscript by Lacis et al. ('Atmospheric CO_2: Principal control knob governing Earth's temperature,' Science **330** (2010) 356) elucidates the effect of CO_2 on the Earth's temperature quite well.

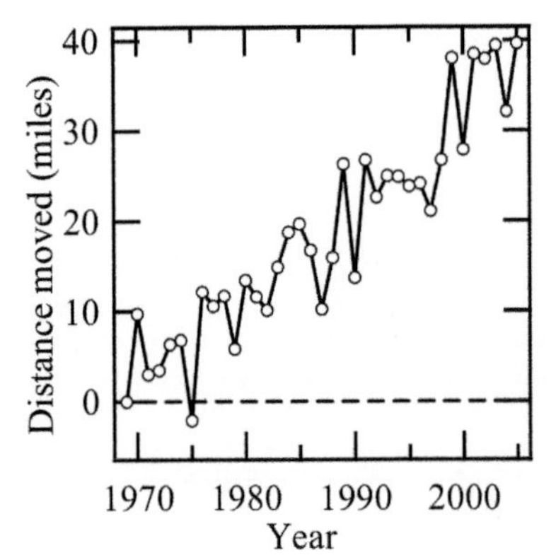

Fig. 1.2 Average northerly distance moved at the *Christmas Bird Count* for 305 widespread bird species in North America. Source: National Audubon Society in a special report entitled 'Birds and Climate Change, Ecological Disruption in Motion (2009).'

Also, weaker CO_2 absorption bands near this one may have a greater influence at a higher concentration and close the *window* where compounds like water do not absorb much radiation. This could also increase the amount of energy absorbed by the atmosphere prohibiting its release to space.

An indicator of how the Earth will change over time is given in a report by the National Audubon Society who monitor the number of birds seen in the *Christmas Bird Count.* The count is done on a voluntary basis and at the same time all over the USA where the number and type of birds are determined (recently, the 113th count was performed between December 14, 2012 through January 5, 2013). Data were analyzed over a period of more than 35 years and are provided in Fig. 1.2, and over time birds have moved more to the north in terms of where they spend their (northern) winters. Anecdotal evidence has suggested that birds, insects and other fauna have changed their residence during the various seasons. This graph demonstrates via a large amount of data that birds are spending their winters much further north. In fact, more than 60 bird species have moved over 100 miles further north in the winter to indicate that the continental USA is warming. This could certainly be due to the large CO_2 concentration and its recent, on a geological time scale, rapid increase.

1.2 The magnitude of energy consumption

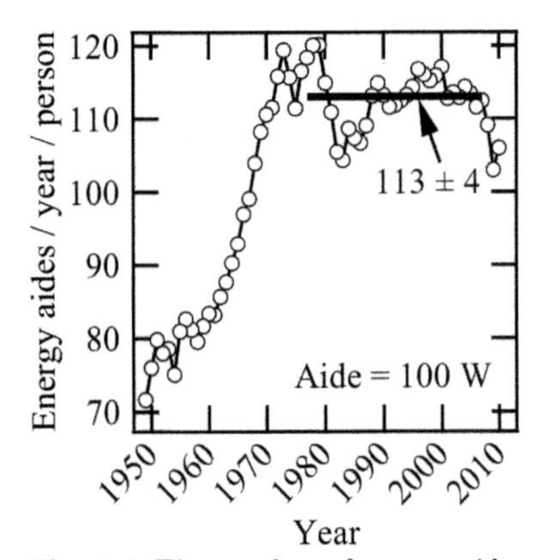

Fig. 1.3 The number of energy aides needed to service each citizen of the USA. Between 1977 and 2007 this number has remained fairly constant at 113 ± 4. Note how downturns in the economy influence energy consumption. Source: U.S. Energy Information Administration.

Citizens of the USA require approximately 100 quads of energy each year. A quad is a quadrillion BTUs of energy or 10^{15} BTU. Although one can argue the archaic nature of the BTU as an energy unit, one cannot deny that any number having fifteen zeros after it is probably fairly large. How large is it?

To answer this question we must break it down to a manageable number. To understand the magnitude of energy consumption or power, as well as what it represents, consider how much power a human being requires to live. Power is energy use per unit time and a human requires about 2000 kcal of nutritional energy per day, which translates to approximately 100 W as shown in Example 1.2. The total energy usage for the USA is plotted as a function of time in Fig. 1.3 and is normalized by the population and the power required for human life.

This graph shows how many people would surround each person, say as energy aides, providing power for them to use every day. Most recently, within the last 30 years, it has remained constant and there would be 113 energy aides around each and every citizen of the USA 24 hours a day, seven days a week, 52 weeks a year imagine how hot a room full of people would get if the aides were with them.

Example 1.2

How much power does a human exert?

A human eats (or consumes) about 2000 kcal of energy content per day. This is the amount of energy required for life. Power is energy per unit time and to convert this to a power we do some unit conversions as shown below

$$\frac{2000\ \text{kcal}}{\text{day}} \times \frac{1000\ \text{cal}}{\text{kcal}} \times \frac{4.184\ \text{J}}{\text{cal}} \times \frac{\text{day}}{8.64 \times 10^4\ \text{s}} = 96.9\ \text{W} \approx 100\ \text{W}$$

So, a human being is almost exactly equivalent to a 100 W light bulb! Note that elite athletes can generate much more power so they eat more. A cyclist in the Tour de France can generate 500 W of power over sustained distances.

Now that a very large number can be related to a manageable number it is clear that an effective strategy to solving this energy challenge and atmospheric CO_2 emission is *conservation.* If each person in the USA were to conserve the equivalent of one 100 W light bulb's worth of energy per day then that would represent approximately 1% of the energy required by that society. This is a very useful strategy and one worth pursuing. This number is also a harbinger of a more chilling scenario.

Not all societies use as many energy aides as the USA does, or in other words the per capita energy consumption of all other humans is less. Consider the population of the World at approximately 7 billion people. If all societies were to consume as much energy as the average citizen in the USA ($113 \times 100\ \text{W} \times 3.16 \times 10^7\ \text{s/year} = 3.57 \times 10^{11}\ \text{J/year}$) then this would require 2370 Quads of energy per year. The World energy consumption is presently only 511 Quads. This is more than a quadrupling of the primary energy consumption suggesting that, eventually, as all societies in the World achieve energy equality, fossil fuel reserves will be more rapidly depleted and CO_2 emissions significantly increased. There are many global societal issues this energy *entitlement* presents. Can all societies demand the lifestyle this energy consumption represents? Will this occur in the future based on an energy portfolio that only includes fossil fuels?

As can be seen in Fig. 1.4 the USA government does not consider much of a change in its energy portfolio to occur in the next 30 years. There is a slight increase in renewable energy, however, it is only minimal and certainly will not address the CO_2 emissions in the near future.

1.3 The future for solar energy

No economic market is a free market for any country in the World and this includes the USA. Various instruments are used to aid and regulate industries and the energy industry is no exception. Most individuals have heard of energy regulatory issues and regulations that aid in the reduction of CO_2, as well as other chemical emissions to the atmosphere. This is certainly a valuable effort and should be continued.

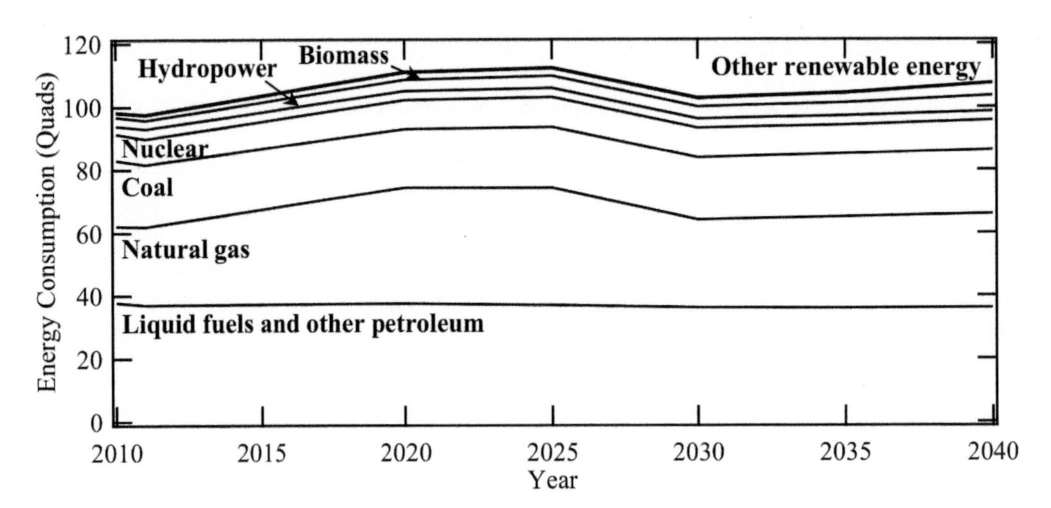

Fig. 1.4 Energy consumption projection for the USA from the U.S. Energy Information Administration. Other renewable energy includes solar and wind technologies. A Quad (10^{15} BTU) is approximately 10^{18} J or 1 EJ (see Appendix A for various conversion factors).

Yet, energy technologies have also been aided as they develop their inherent technology to become more efficient as well as to develop the infrastructure required for widespread use. The graph in Fig. 1.5 shows the energy subsidy for various technologies in the USA as they grew during their first 30 years of receiving a subsidy. *Oil and Gas* as well as *Nuclear* energy all received substantial subsidies, and continue to do so. A caveat on the subsidy the Oil and Gas industry still receives is that part of the subsidy includes maintaining the strategic oil reserve and to give cheap (if not free) energy to those who cannot afford it. The industry does obtain a subsidy over that required to perform these admirable functions to aid its continued existence and growth and so does not feel the full effect of market forces.

The purpose of this discussion is not to criticize subsidies. It is merely to point them out and if any given technology is to succeed in a market that has robust technologies already in place, with an infrastructure which is already developed, then some help will be required. It is clear from the graph that *Renewables* have not received as many subsidies as the others have, particularly *Biofuels* which have recently received a huge influx of funds.

Are Biofuels a panacea for energy use? Firstly, their introduction into gasoline at 10% reduces the fuel mileage of automobiles by 3–4% (this is for E10 gasoline, source: Energy Information Administration). Thus, the consumer is paying for ethanol at both ends, primarily for the subsidy and secondarily through poorer fuel mileage! Can biofuels provide all the liquid fuel requirements? The next example provides information to help answer this question.

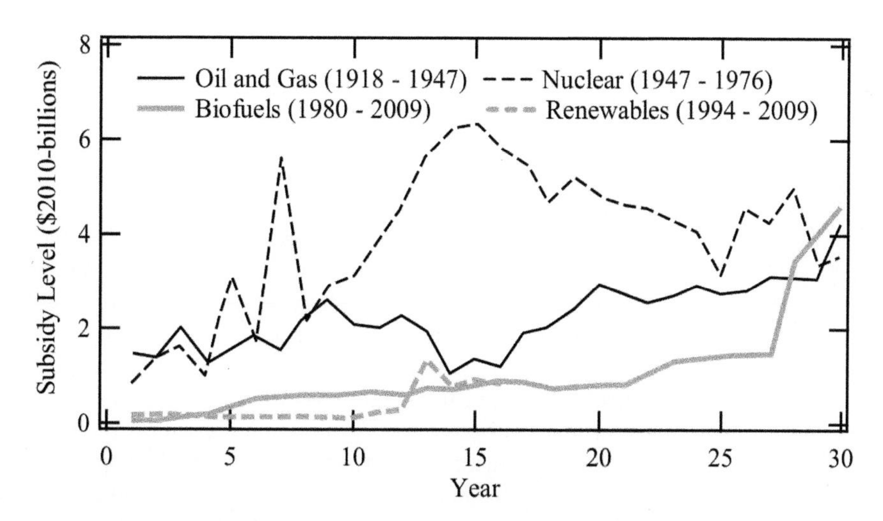

Fig. 1.5 Energy subsidies for the first 30 years of a given energy technology after subsidies were appropriated to that technology in the USA. Source: N. Pfund and B. Healy, 'What would Jefferson do? The historical role of federal subsidies in shaping America's energy future,' **Sept.** (2011) Double Bottom Line Investments.

Example 1.3

What is the power density, in terms of energy produced per unit area, of various energy technologies?

This will be done firstly for biofuels, in particular bioethanol, and the fermentation of corn into ethanol is considered. So, the energy density will be in terms of how much energy can be derived from each acre of crop. This will be compared to the areal power density supplied by the Sun.

Corn is one of the more efficient crops to produce ethanol and so represents a best case scenario. The power content that can be gained from corn per unit area will be determined and compared to the power density of solar radiation as well as other technologies. Power density is defined as the number of watts generated per unit area which will make perfect sense as a figure of merit for energy production from biofuels and the Sun. It is an artificial one for the other technologies and is introduced for energy sources like petroleum or coal to demonstrate, as a country, how much energy the USA uses per unit area for comparison.

Corn can be produced to yield 150 bushels per acre and 2.5 gallons of ethanol can be obtained from a bushel of corn. So, one can obtain 375 gallons of ethanol per acre. The energy content of ethanol is 89 MJ/gallon and if we assume there is one crop per year then the energy density is only 0.25 W/m^2, quite low.

The Sun can produce much more power per unit area, taking a conservative estimate of 500 W/m^2, and dividing by two to account for day–night cycles, one has 250 W/m^2. Assuming the solar device is only 10% efficient then there is 25 W/m^2 available and this power is 100 times more dense than for bioethanol.

Table 1.2 Areal power density of various energy technologies.

Technology	Power density (W/m^2)
Solar	25.0
Bioethanol	0.25
Petroleum	0.15
Natural Gas	0.10
Coal	0.081
Bioethanol*	0.056

*Assuming only 1.8×10^{12} m^2 cropland area is available (see text). So, the density was renormalized by the factor 1.8×10^{12} $m^2/8.08 \times 10^{12}$ m^2

A number of energy technologies are compared in Table 1.2 and, for petroleum, natural gas and coal, the contemporary energy consumption in a year for the USA was used to find a power and was then divided by the area of the continental USA (8.08×10^{12} m^2) to determine the power density.

This is an interesting comparison of power densities as none approach that of the Sun. If you are an engineer, and you were to *start* an energy infrastructure, which technology would you choose? Clearly the answer is solar energy since it is of order one-hundred times more dense (at today's usage level) than the others. Even if we increased the use of petroleum, natural gas and coal by a factor of ten (a chilling thought in terms of CO_2 emissions) their *density* is still a fraction of solar energy's.

This calculation shows that even at 10% efficiency the Sun can produce a lot of energy. The results of the analysis can also be used to determine how much area would be required to replace all the petroleum used by motor vehicles in the USA per year. This is 3.19×10^9 barrels per year, the energy content of a barrel of gasoline is 5.9×10^9 J, so is equivalent to a power of 6×10^{11} W. Assuming the power density of ethanol is 0.25 W/m^2 then an area of 2.4×10^{12} m^2 is required or about one-quarter of the area of the continental USA. According to the United States Department of Agriculture, 1.8×10^{12} m^2 is presently used for cropland and would all have to be farmed to produce 75% of the ethanol required for transport.

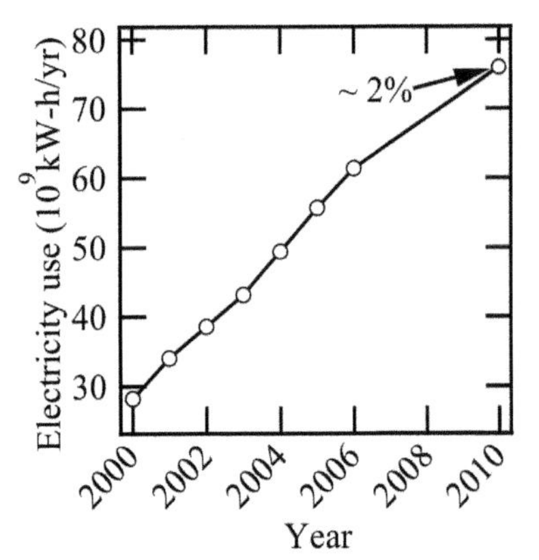

Fig. 1.6 Total electricity use of data servers in the USA as a function of year. This includes electricity to operate the servers and all ancillary equipment plus cooling which accounts for approximately 50% of the electricity use. Source: U.S. Environmental Protection Agency, Report to Congress on Server and Data Center Energy Efficiency, Public Law 109–431 (2007) and James Glanz, 'Power, Pollution and the Internet,' New York Times, Sept. 22, 2012.

Solar photovoltaic (PV) electricity has an advantage over other energy technologies since it inherently produces direct current (DC). Coal, nuclear energy and natural gas are used to produce heat that boils water to generate high pressure steam to turn a turbine (a wind turbine uses wind instead of steam). The turbine is connected to a generator that has a shaft with permanent magnets on it that rotates within wound wires. The rotating magnetic field produces alternating current (AC) within the wires and this is how most electric power is generated. Digital devices (cameras, phones, computers, ...) ultimately use low voltage DC (LVDC) power and not AC, so, in a contemporary house or building AC power is passed through a transformer–rectifier combination to convert AC to DC power.

Thus, PV generated electricity can be used to directly power digital devices at a reduced cost. Part of the future for solar PV could be to power these devices and data servers throughout the World. The data given in Fig. 1.6 shows that the electricity consumption of the data servers used to store information for the World Wide Web is significant and accounted for 1.5% of USA consumption in 2006 and almost 2% in 2010! If the linear trend continued until the present day, at a 5.5×10^9 kW-h/yr increase, then it will account for approximately 2.5% of electricity consumption, which is considerable. The World Wide Web is not a green technology! It is better for you to walk to a store and buy a product than to order it online on many levels.

Assuming the same power density for solar PV energy as in Example 1.3, 25 W/m^2, then an area of about 20×20 km^2 would have to be used to supply power to the data centers. This is not an unreasonable area to cover and could isolate data centers from the electrical grid. Furthermore, usage of solar PV electricity to directly power digital devices at the home or workplace with specially designed power systems could kickstart use of solar PV. This would require LVDC power to be available to the user from solar cells mounted on the building. It is believed that LVDC power is an alternate power system that will grow in the future and, importantly, this will be a distributed power system, that is locally generated and used.

The natural question arises, when considering the future of solar energy: What research and development is being performed to ensure its success? Over the last century much research has been done to optimize the production and use of fossil fuels and as can be seen in Fig. 1.7 the research enterprise for solar energy has not been consistent. The topic *Solar Energy* was used as a search topic in the Web of Science® for each year from the late 1940s to the present day for publications from the World and the USA.

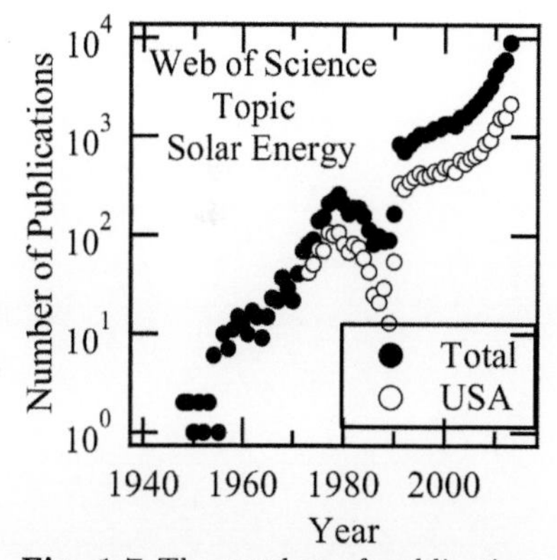

Fig. 1.7 The number of publications from the Web of Science for the entire World (total) and for the United States of America (USA) as a function of year for the topic *Solar Energy*. Specific data for the USA is not available before 1970.

Research was severely curtailed in 1980 as the result of a policy decision in the USA as measured by this technique. In fact, by 1990 in the USA there were only 12 publications with this topic. There was a sudden increase in publications around that time and a cursory review of some of them shows that the researchers were apparently concerned with extraterrestrial solar energy, quite possibly for weapon systems, although this is not a statistical analysis of the data. Regardless, terrestrial solar energy became the main topic after the year 2000 and it has continued to grow. It is interesting that a policy decision effectively set research on terrestrial solar energy back by 20 years, according to this technique of assessing research activity.

It is certain that solar energy requires more research and development to allow its success and integration into the energy industry. Scientific discoveries are needed, yet, good engineering as well as manufacturing expertise needs to be pursued. It is hoped that the scientist and engineer entering this field can benefit from this monograph to ensure that the future of solar energy is bright.

1.4 Conclusion

In this chapter an argument was made that reduction of CO_2 emissions is required or the Earth will be driven to a state not seen for millions of years. The use of solar energy could mitigate these emissions. Although one may dispute that CO_2 in the atmosphere contributes to global climate change, despite the clear argument given above, it is obvious that the increased concentration is human-made and what state the Earth ultimately achieves when the CO_2 concentration is very high is not known.

Table 1.3 Lead time for construction of various energy technologies to make electricity. Source: U.S. Energy Information Administration, AEO 2012 Electricity Market Module Assumptions Document.

Technology	Lead time (years)
Nuclear	6
Coal	4
Wind (offshore)	4
Geothermal	4
Natural gas	3
Wind	3
Solar (thermal)	3
Solar (PV)	2

Now, for the sake of argument, assume the World yearly electricity consumption doubles in 50 years which amounts to an increase of 18.5×10^{12} kW-h. If 500 MW coal-fired power plants were to be used to make this electricity then this equates to 4220 plants and one must be erected every 4.3 days! The challenge for future energy technology may not be the cost of construction or $/kW-h or even reduction in CO_2 emissions, it may be how fast can they be built. It takes about four years to obtain the permits for a coal-fired power plant and even longer for a nuclear power plant, as shown in Table 1.3. If it was decided to start building these plants today then four years worth of plants, equivalent to 340 plants, would have to start having their approval right now. Solar thermal and photovoltaic power plants require less time and photovoltaic plants have half the permitting time of a coal-fired plant.

The photovoltaic power plant has another advantage, it does not require cooling water to generate electricity. Water will be a valued commodity in the future and to waste it for a power plant may not be something allowed in the future.

There is also a less measurable parameter important to energy technologies and that is whether the public would want that technology next to their neighborhood. Coal-fired power plants need to be next to a railroad track for coal delivery (see Example 9.1 to find that a 500 MW power plant needs an 825 m long coal train to deliver coal every day) as well as a water source like a river or lake. This places serious limits as to the *neighborhood* where the plant can be located and the neighbors it inconveniences. Honestly answer this question, 'Would you rather have a coal-fired power plant next to your house or a solar PV array?'

A drawback to solar energy utilization is that the Sun does not shine for 24 h every day. This makes it an intermittent power source requiring storage that can be either chemical or thermal in nature. Energy storage will not be covered in this book, however, it is important and is a research field in its own right. Generation of Hydrogen from photovoltaic devices is one way to chemically store energy for later use and molten salts can be used to store energy for a heat power cycle. Both of these, as well as batteries, can be used in the future requiring a solar energy infrastructure to be developed.

As an example of the present energy infrastructure consider when your car requires gasoline. You don't drive to a refinery to fill up the tank. The refinery may be temporarily closed for maintenance, not making your grade of gasoline at that moment or any of numerous reasons that would prohibit you obtaining fuel and you would have to wait: equivalent to the Sun not shining. Rather there is a robust storage and distribution system available to you so you can drive to a gasoline station and obtain your fuel. Fossil fuels have had over 100 years for their infrastructure to be put in place and it may well take that long for the same to occur for solar energy.

The first step in adopting any technology is the hardest and adopting fossil fuels as an energy source was mitigated by individuals becoming wealthy. How can a similar economical spark occur to start the alternative energy revolution? A vibrant education, research and entrepreneurial landscape must be fostered to make small steps until a critical mass is formed. Only then will alternative energy sources, in particular solar energy, become a reality, as economic realities drive progress.

Widespread adoption of alternative energy sources could occur with the very poorest on the planet first since any energy source at all will be better than none. By the very nature of where they live, and perhaps for political and economic reasons, the energy must be distributed rather than available from a central source, like a coal-fired power plant. However, the research and development focus must center on cost rather than efficiency to supply energy to the billions of people living in abject poverty on the planet. Furthermore, this must be done in a sustainable manner. Although one may read this as ecologically sustainable, it must be economically sustainable and not rely on monetary donations. In other words an alternative energy economy must be in place for it to succeed either in the poorest or wealthiest countries. Although this monograph is not an economic tool, it is hoped the information within it will drive informed decision-making, and even fabrication of new and novel designs, to boost this economy.

1.5 General references

N.S. Lewis and G.G. Nocera, 'Powering the planet: Chemical challenges in solar energy utilization,' Proc. Nat. Acad. Sci. **103** (2006) 15729.

G. Lüthi *et al.*, 'High-resolution carbon dioxide concentration record 650,000–800,000 years before present,' Nature **453** (2008) 379.

G. McDonald *et al.*, 'The long term impact of atmospheric carbon dioxide on climate,' Report to the Department of Energy, Technical Report JSR-78-07 (1979).

C.K. Prahalad, 'Fortune at the Bottom of the Pyramid: Eradicating Poverty Through Profits,' Dorling Kindersley Pty. Ltd. (2006). This is a useful book to read to understand the economics of the poorest on Earth.

V. Quaschning, 'Renewable energy and climate change' J. Wiley & Sons (2010).

See also the Intergovernmental Panel on Climate Change, Fifth Assessment Report (AR5), Climate Change 2013: The Physical Science Basis.

Exercises

(1.1) Determine the time that CO_2 will last in the atmosphere. In other words, how long does it take CO_2 to degrade?

(1.2) Are parts per million (ppm) by number of moles and volume the same for a gas? The mixing ratio is sometimes used in atmospheric science and it is defined as the number of moles of a substance to the number of moles of everything else (not including the given substance). Sometimes the mixing ratio is similarly defined by mass rather than moles so some care should be taken to determine the definition. At what concentration limit will ppm and mixing ratio be equivalent?

(1.3) Estimate how much plant and tree material decays every fall in the northern hemisphere.

(1.4) How many trees would have to be planted each year to eliminate the increase in CO_2 seen at the Mauna Loa volcano assuming this is the global concentration of CO_2? What percentage of the World land area is this?

(1.5) Estimate how much the World ocean height will change if all the ice in the arctic melts.

(1.6) If the average temperature of ocean water increases by 1 and 10 °C how much will the ocean height change?

(1.7) Since 1920 the mean sea level has risen by 3.20 ± 0.25 mm/year in Lewes, Delaware, USA, estimate what is the temperature rise over a 100 year period assuming this is due to climate warming only.

(1.8) Compare the combustion energy density of ethanol and gasoline on a per mass and volume basis; which is a more energy dense fuel?

(1.9) Estimate how much E10 gasoline costs the average person in the USA each year in terms of reduced fuel mileage when driving their automobile. Multiply this result by the number of people in the USA and estimate how much money is taken out of the economy each year. Assume E10 and pure gasoline cost the same amount.

(1.10) Determine how much CO_2 a person generates per year through exhalation, how much all the people on Earth generate in a year and finally how many parts per million in the atmosphere this is.

(1.11) If all the electricity used in the World were generated by burning wood or natural gas, how much CO_2 would be released into the atmosphere in ppm? Compare this to the amount being released now.

(1.12) The National Audubon Society has a fact sheet entitled, 'Global Warming Campaign: Top Ten Things You Can Do.' Write down the ten items with a brief explanation of what they are and how much each item will reduce CO_2 emissions if followed. If everyone in the average household in your country followed these guidelines, how much CO_2 would be reduced in terms of atmospheric (molar) ppm?

(1.13) If all the coal and liquid petroleum products used as an energy resource in the USA in one year were placed in an area the size of the continental USA, how thick would each layer be?

(1.14) Determine how much cement manufacture and gas flares contribute to the emission of CO_2. Describe why cement manufacture is a source of CO_2 emission and what a gas flare is and why it is performed. Hint: The data from Oak Ridge National Laboratory are quite useful.

(1.15) If all the coal-fired power plants were replaced by solar cells to produce electricity in the USA, how much would it cost and how much would the CO_2 emissions be reduced in terms of ppm in the atmosphere? Compare the cost to the USA's defense budget.

(1.16) In Section 1.1 it was mentioned that 280 km^3 of ice was being melted each year in the Arctic. Assume when all the ice is melted that the energy required to melt the ice now can be used to heat the atmosphere, how many °C will the atmosphere increase in temperature each year? Note the heat capacity of air is remarkably independent of pressure until extremely low pressure and can be assumed constant for the purposes of this calculation.

(1.17) As mentioned on Section 1.1 the seasonal ice loss in the Arctic is 16,400 km^3, if all the Arctic ice is lost then this energy buffer no longer exists. More precisely, the energy used in melting the ice must go somewhere else and heat it up. Assume it goes to heat up the atmosphere, estimate how much it will increase in temperature. Hint: See Exercise 1.16.

(1.18) When all the Arctic ice melts and does not return in the northern winter, estimate the reduction in shipping times from China to Europe.

(1.19) Professor Quaschning has introduced the term "anytime energy" in his book (see *General references* above) as a way to emphasize how much energy we use from the Sun. Calculate how much energy the Sun contributes to your house by its warming effect in the winter and compare it to the average electricity use for a residence. Is it a significant amount?

(1.20) Compare the amount of water used to produce electricity from various technologies such as biodiesel, coal-fired power plant, etc. Much information can be obtained from this article: W.D. Jones, 'How much water does it take to make electricity?' IEEE Spectrum, April 1, 2008.

Solar radiation

2

In this chapter we discuss how the Sun makes energy, or equivalently radiation and light, through a series of reactions that eventually reaches the Earth's surface and is utilized by a device. The amount of radiation generated is suitably modeled as a black body spectrum, according to Planck's Law, assuming the Sun operates at 5793 K. As this radiation propagates outwards from the Sun only part of it reaches the Earth, which is determined analytically by considering the Earth–Sun geometric relation. Ultimately the radiation must penetrate the atmosphere where it is scattered and absorbed and the spectrum is modified to represent the actual energy density that reaches the Earth's surface. The radiation intensity is still quite large although the effective black body temperature has been reduced to 5359 K. Finally expressions are developed to determine the amount of radiation that can be used by a given device, assuming it absorbs over a given energy or wavelength range, where it is translated into useful power. So, in this chapter, we consider the formation of energy within the Sun and how much can be used by a solar powered device.

2.1 Fusion in the Sun

The Sun makes radiation by converting mass into energy through the reaction of smaller elements to make heavier, or in other words *fusion* occurs. When the lighter Hydrogen nuclei (or protons, ^{1}H) are fused together to make a Helium nucleus (sometimes called an alpha particle, ^{4}He) a slight amount of mass is lost which is converted into energy. The mass loss is incredibly small for each reaction, of order 5×10^{-28} kg, however, it produces a huge amount of energy, 6.5×10^{14} J/kg. Compare this to burning gasoline which produces a mere 5×10^7 J/kg! How can nuclear fusion produce such a large amount of energy (or light)? The answer lies with Einstein and his famous equation $E = mc^2$, energy E equals mass m times the speed of light squared c^2.

The Sun is approximately 90% Hydrogen that reacts in the overall reaction of four Hydrogen nuclei to make the next lightest element, Helium's, nucleus. This occurs in a series of reactions though since the likelihood of four particles colliding to form one is extremely small. It should be recognized, most of the reactions occur in the core which has a radius about one-quarter of the Sun's external radius yet contains 40% of its mass. The core provides 90% of the energy making it the hottest part, due to this extreme condition, with a temperature estimated to be over 10^7 K!

The high temperature dissociates hydrogen into its individual components, ^{1}H, and an electron, e^-, accounting for why it is nuclei reacting rather than atoms. Furthermore, it provides enough kinetic energy for two protons to get close enough ($\approx 10^{-15}$ m) allowing strong interactions to hold them together on occasion. Subsequently, when they successfully stick together, one of the protons changes into a neutron. These two steps, protons sticking together and changing, are rare enough to account for why the Sun is 4.6 billion years old, otherwise it would have burned-up a long time ago.

This reaction is the first in a chain of reactions as shown in the sequence below. The Fusion-1 reaction generates a positively charged electron, or positron e^+ and a massless neutrino ν. The reaction takes about 10^9 years for success while the Annihilation reaction occurs quite quickly, as does Fusion-2 which requires approximately a second to occur due to the large amount of ^{1}H nuclei present and that change to a neutron is not required. These two reactions produce energy in the form of gamma photons (γ) and a light isotope of Helium, ^{3}He. The subsequent Fusion-3 reaction requires a relatively short 10^6 years, compared to Fusion-1, which rapidly leads to the Disintegration reaction of the unstable Beryllium nucleus ^{6}Be yielding Helium, ^{4}He, and two protons.

FUSION REACTIONS IN THE SUN: THE PROTON–PROTON CYCLE

^{1}H	+	^{1}H	$\longrightarrow$	^{2}H	+	e$^+$	+	ν		Fusion-1
e$^-$	+	e$^+$	$\longrightarrow$	2γ						Annihilation
^{2}H	+	^{1}H	$\longrightarrow$	^{3}He	+	γ				Fusion-2
^{3}He	+	^{3}He	$\longrightarrow$	^{6}Be						Fusion-3
		^{6}Be	$\longrightarrow$	^{4}He	+	^{1}H	+	^{1}H		Disintegration

This amazing reaction sequence is called the proton–proton cycle; how does it produce so much energy? To understand this, write down the overall fusion reaction and the mass of each component

$$\begin{array}{ccccc} 4\ ^1\mathrm{H} & + & 2\ \mathrm{e}^- & \longrightarrow & ^4\mathrm{He} \quad + 2\,\nu + 6\,\gamma \\ 6.693\times10^{-27}\ \mathrm{kg} & + & 1.8\times10^{-30}\ \mathrm{kg} & \longrightarrow & 6.645\times10^{-27}\ \mathrm{kg} + 0\ \mathrm{kg} + 0\ \mathrm{kg} \end{array}$$

Summing the mass over the complete reaction one finds a very small mass loss of -4.8×10^{-29} kg. Remember $E = mc^2$ which mathematically states mass and energy are different forms of the same thing being related by the speed of light squared. The mass is converted into energy which is eventually directed out of the Sun. So, even though the mass loss is small, the speed of light is not, since it is 2.998×10^8 m/s in magnitude. Thus, each reaction produces 4.5×10^{-12} J of energy which on a per mass of reactants basis is approximately 6.5×10^{14} J/kg.[1]

The Sun, and all stars, are quite complicated objects and the above is a cursory description of the reactions that occur within it. Furthermore, the Sun has an internal control system that regulates the fusion

[1] One should realize that if a reaction on Earth is exothermic then a small amount of mass is lost, while if it is endothermic an immeasurable amount of mass is gained within the system under consideration! Strictly speaking, the mass balance is not always followed.

reactions. Suppose the reaction rate increases due to a hot spot developing. This results in thermal pressure that pushes the core outward with the expansion tending to cool the core. Since nuclear fusion is quite sensitive to temperature the reaction rate slows and gravity pulls the material inward and the process repeats. This internal control system is quite important and without it the Sun would effectively not operate. The Sun also has quite an interesting structure and three zones surround the core: the radiative zone, the convection zone and the corona, that each play integral parts in the overall production of radiation that reaches the Earth.

2.2 Extraterrestrial solar radiation

The Sun can be considered to be a black body radiator, meaning it will absorb all energy which comes to it, however, it radiates energy characteristic of its own internal processes alone.[2] As such, it radiates a spectrum of energies that depend only on a characteristic temperature which for the Sun is 5793 K. Using *Planck's law* one finds, n, the number of photons # emitted by a black body per unit time per unit sr (steradian which are the units of a solid angle, there are 4π sr in a sphere, see Fig. 2.1) per unit energy E per unit area A

$$n = \frac{2}{h^3c^2}\frac{E^2}{\exp(E/k_BT_S)-1}\,[=]\,\frac{\#}{\text{s-sr-J-m}^2} \tag{2.1}$$

where h is Planck's constant, c, the speed of light, k_B, Boltzmann's constant, T_S, the Sun's temperature and [=] means 'has dimensions of.'

Any who have read the chemistry/physics/optics literature, or even the solar energy literature, can find the nomenclature confusing and what to call n is not apparent. Here we will denote it as *photon number rate*. If one considers a differential solid angle, $d\Omega$, energy, dE, and area, dA, then $nd\Omega dEdA$ is the number of photons emitted within these small bounds per unit time. So, n is akin to the velocity of water in a pipe with area dA (and solid angle $d\Omega$ and energy range dE).

The solid angle and area are independent variables and related by the distance r from the radiation source to the receiving body (of projected area A) as can be seen in Fig. 2.1. So, as a body is moved further from the source a smaller solid angle is required to cover it according to, $\Omega \propto A/r^2$. The 'proportional to' symbol is required since one must consider the projection of the body's surface area to that of a spherical surface which depends on their relative geometries.

A more useful relation is to integrate n over the solid angle ($d\Omega = \sin(\theta)d\theta d\phi$ in spherical coordinates). However, the integral is performed to include only those photons that reach the Earth from the Sun with the angular range adjusted accordingly. Noting the photon number rate must be multiplied by $\cos(\theta)$, to ensure the unit normal locating the differential area at the Earth's surface is aligned with the perpendicular (See Fig. 2.1), one finds the integrated photon number rate,[3] N_p

[2] A cavity with an aperture in it can be used to simulate black body radiation. If the cavity and aperture are designed such that any radiation entering the aperture is not reflected back then it is *absorbing* all the radiation and so is a black body, at least along the axis of the aperture as shown in the figure below. Thus, radiation emitted from it follows a black body spectrum given by eqn (2.1) which will be a function only of the cavity, aperture, materials of construction and device temperature. United States Patent 4,317,042 gives a nice description of black body simulators. Since a cavity can be used to simulate a near perfect black body this type of radiation is sometimes called cavity radiation.

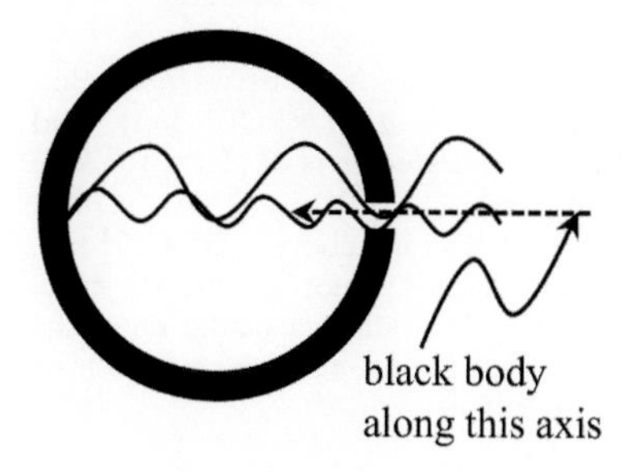

[3] If N_p were subsequently integrated over all energy levels one would obtain the photon flux J which is the number of photons impacting a surface per unit time per unit area (#/s-m^2).

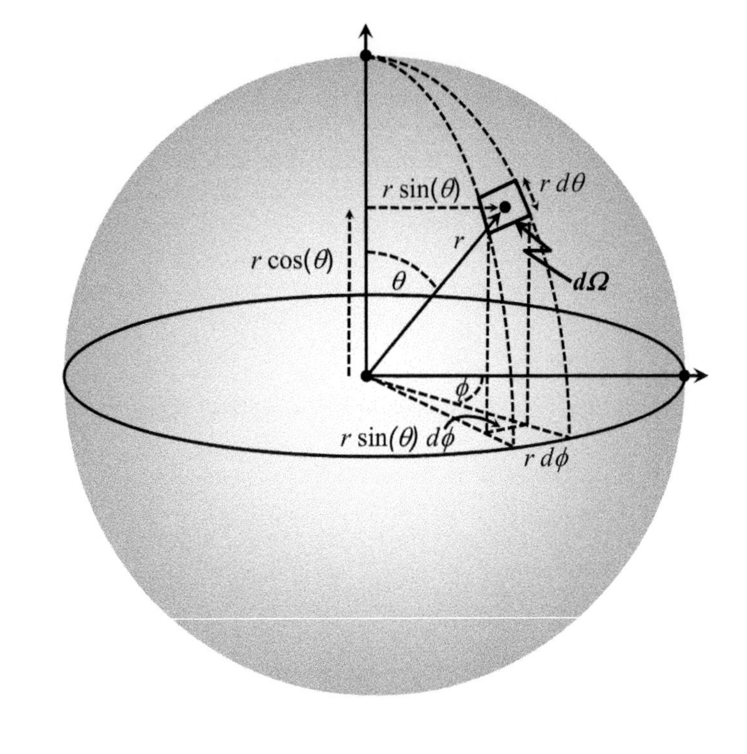

Fig. 2.1 Diagram showing the radius r and angles θ and ϕ used to describe a sphere in three dimensions. The solid angle $d\Omega$ is an infinitely small, non-dimensional 'area' on the surface of a sphere. The dimensional area of the infinitesimal rectangle dA is $r^2 \sin(\theta) d\theta d\phi$, so, the solid angle is defined as $d\Omega = dA/r^2 = \sin(\theta) d\theta d\phi$ on the sphere surface.

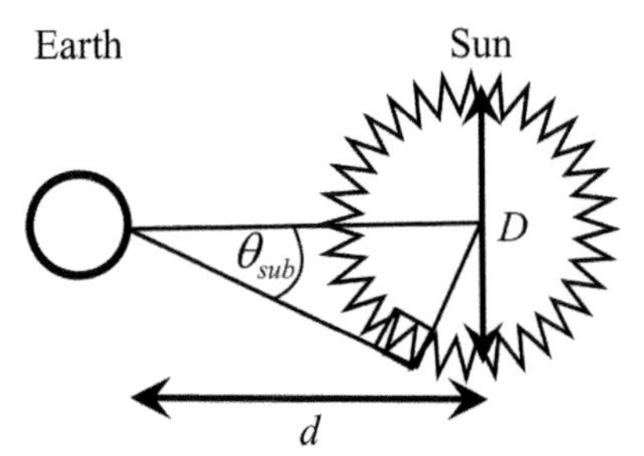

Fig. 2.2 The distance between the Sun and Earth is d (1.5×10^8 km), and the Sun's diameter is D (1.4×10^6 km) then $\tan(\theta_{sub}) \approx \theta_{sub} = D/2d = 0.265°$. The Earth's diameter is 1.28×10^4 km, so the figure is not drawn to scale! Knowing θ_{sub} allows calculation of the geometric factor $g_S \equiv \pi \sin(\theta_{sub})^2 = 6.720 \times 10^{-5}$.

$$N_p(E) = \frac{2\pi \sin(\theta_{sub})^2}{h^3 c^2} \frac{E^2}{\exp(E/k_B T_S) - 1}$$
$$\equiv \frac{2g_s}{h^3 c^2} \frac{E^2}{\exp(E/k_B T_S) - 1} \; [=] \; \frac{\#}{\text{s-J-m}^2} \tag{2.2}$$

where θ_{sub} is the angle subtended between the Sun and Earth ($\theta_{sub} = 0.265°$), see Fig. 2.2, and g_S is a geometric factor equal to $\pi \sin(\theta_{sub})^2$. Were we situated right at the Sun's surface, meaning $d = 0$ in the figure and $\theta_{sub} = 90°$, we would be subjected to all the Sun's radiation. Fortunately we are not at that spot and receive only about 20 millionths of the Sun's radiative power.

Example 2.1

Find the number of photons striking the Earth's atmosphere per second in a square meter with an energy in the middle of the visible spectrum. Assume an energy range of ±1%.

We must find $N_p(E)dE \approx N_p(E)\Delta E$ at an energy corresponding to a wavelength of *ca.* 500 nm. First note that

$$E = \frac{hc}{\lambda} = \frac{6.634 \times 10^{-34}\,\text{J-s} \times 2.998 \times 10^8\,\text{m/s}}{500\,\text{nm}} \times \frac{10^9\,\text{nm}}{\text{m}} = 3.978 \times 10^{-19}\,\text{J}$$

then use eqn (2.2) with $g_S = \pi \sin(\theta_{sub})^2 = 6.720 \times 10^{-5}$ to find

$$N_p(E) = \frac{2 \times 6.720 \times 10^{-5}}{[6.634 \times 10^{-34}\,\text{J-s}]^3\,[2.998 \times 10^8\,\text{m/s}]^2} \times \frac{[3.978 \times 10^{-19}\,\text{J}]^2}{\exp\left(\dfrac{3.978 \times 10^{-19}\,\text{J}}{1.381 \times 10^{-23}\,\text{J/K} \times 5793\,\text{K}}\right) - 1}$$

$$N_p(E) = 5.653 \times 10^{39} \frac{\#}{\text{s-J-m}^2}$$

The energy range is $\Delta E = 0.02 \times 3.978 \times 10^{-19}$ J $= 7.956 \times 10^{-21}$ J, making

$$N_p(E)\Delta E \approx 4.5 \times 10^{19} \frac{\#}{\text{s-m}^2}$$

This is a very large number of photons striking the Earth's atmosphere in a second for a very small energy range, accounting for why the Sun can supply the World's yearly energy needs in one hour (see page 1).

Even more useful is a quantity we will call the *energetic irradiance* or I, defined as $E \times N_p(E)$,

$$I(E) = E \times N_p(E) = \frac{2g_s}{h^3c^2} \frac{E^3}{\exp(E/k_BT_S) - 1} \;[=]\; \frac{\#}{\text{s-m}^2} \tag{2.3}$$

and is the photon number rate at a given energy weighted by that energy. Now integrating I over all energy levels one arrives at the extraterrestrial power per unit area or *irradiance*, given the symbol P_{ext}

$$P_{ext} = \int_0^\infty I(E)dE = \frac{2\pi^4}{15} g_s \frac{k_B^4}{h^3c^2} T_S^4 \equiv \sin(\theta_{sub})^2 \sigma_S T_S^4 \;[=]\; \frac{\text{W}}{\text{m}^2} \tag{2.4}$$

where σ_S is Stefan's constant equal to 5.670×10^{-8} W/m^2-K^4. This equation is sometimes called the *Stefan-Boltzmann law.* The Sun radiates an incredible power of 63 MW/m^2 at its surface ($\theta_{sub} = 90°$), quite a large number, and equivalent to a small power plant we use on the Earth to generate electricity within a square meter. Yet, at the edge of the Earth's atmosphere the power is much less and equal to 1366 W/m^2 ($\theta_{sub} = 0.265°$). This is the *solar constant* and represents the irradiance, or energy per unit time per unit area, reaching the Earth.[4] One should remember this is the extraterrestrial power and a significant amount of radiation is absorbed or scattered by the atmosphere lowering the irradiance even more before reaching us on the surface and will be discussed in the next section.

The irradiance was determined by integrating the energetic irradiance over all energy levels. We find using energy units as the independent variable useful since, as will become apparent later, operation of photovoltaic devices depends strongly on the energy absorbed and this is the natural variable.[5] However, the irradiance is frequently given in terms of

[4] Due to imperfections in the Earth's orbit around the Sun, P_{ext} depends on the time of year and can be approximately described by

$$P_{ext} = 1366 \frac{\text{W}}{\text{m}^2}\left[1 + 0.033\cos\left(\frac{360°n}{365}\right)\right]$$

where n is the day number with January 1 having $n = 1$.

[5] As shown in Example 2.1, the relation between energy E and wavelength λ is $E(\text{J}) = hc/\lambda = 1.99 \times 10^{-16}/\lambda(\text{nm})$, which is Planck's law.

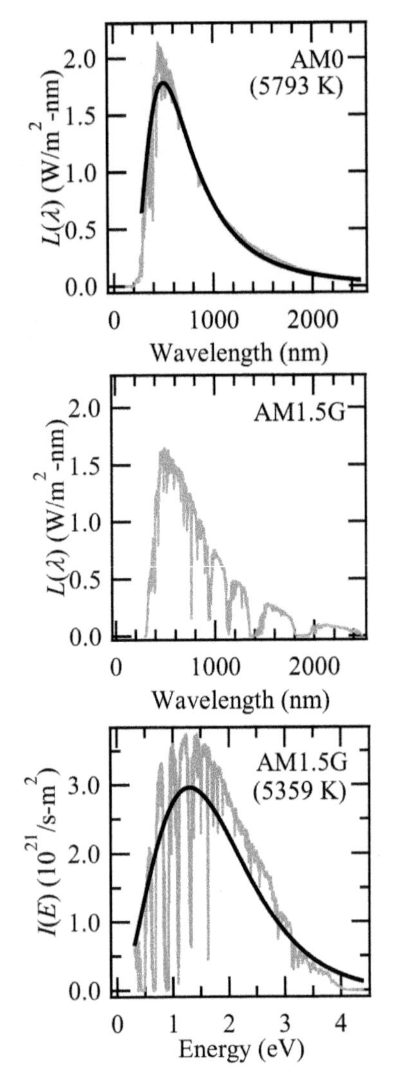

Fig. 2.3 (upper) The extraterrestrial solar spectrum AM0 compared to Planck's law given in eqn (2.6) by the line assuming a black body temperature of 5793 K. In this graph $L(\lambda)$ is plotted as a function of λ. (middle) The spectral irradiance standard AM1.5G for the solar spectrum at the Earth's surface. (lower) The standard AM1.5G terrestrial solar spectrum compared to eqn (2.2) at a black body temperature of 5359 K. The energy unit of the electronvolt (eV) was used (see Section 2.5).

the radiation wavelength (λ), rather than energy (E), in what is called the spectral irradiance ($L(\lambda)$). Both the energetic irradiance and spectral irradiance are defined such that if integrated over all energy levels or wavelengths, respectively, one arrives at the irradiance in W/m^2 or

$$P_{ext} = \int_0^\infty I(E)dE = \int_0^\infty L(\lambda)d\lambda \, [=] \, \frac{\text{W}}{\text{m}^2} \tag{2.5}$$

The relation between the two is $L(\lambda) = [E/\lambda] \times I(E)$. Since $E = hc/\lambda$ one can find

$$L(\lambda) = \frac{2g_S hc^2}{\lambda^5} \frac{1}{\exp(hc/\lambda k_B T_S) - 1} [=] \frac{\text{J-s-m}^2/\text{s}^2}{\text{m}^5} [=] \frac{\text{W}}{\text{m}^2\text{-nm}} \tag{2.6}$$

This may be the more familiar expression for Planck's law, rather than eqn (2.2), to the student. The later dimensions are used so when $L(\lambda)$ is integrated with respect to the wavelength, in nm, one arrives at the irradiance in W/m^2.

Comparison to the real solar spectrum is relatively straightforward and in this case the extraterrestrial spectrum is denoted as AM0 meaning *Air Mass Zero*. As mentioned above the actual amount of radiation reaching the ground is less than the extraterrestrial given by eqns (2.2) or (2.6) and the air mass index is used to demonstrate the amount of atmosphere between the point of observation and the Sun. Clearly, 'zero' means there is no atmosphere present and this will be discussed in more detail in Section 2.4.

The AM0 spectrum is shown in Fig. 2.3 and comparison to eqn (2.6) reveals reasonable agreement, especially at the longer wavelengths. If the true AM0 spectrum were graphically integrated with wavelength (eqn (2.5)) then the irradiance is 1366 W/m^2 in agreement with integrating eqn (2.6) over all wavelengths. In fact, the Sun's temperature was set to 5793 K so the appropriate irradiance was determined when the Stefan-Boltzmann relation (eqn (2.4)) is used.

The spectrum maximum can be found by taking the derivative of eqn (2.6) with respect to λ to find

$$\lambda T_S|_{\max} \approx \frac{hc}{5k_B} = 2.9 \times 10^6 \text{ nm-K} \tag{2.7}$$

assuming $\exp(-hc/\lambda k_B T_S)$ is small (a good assumption for our Sun). This is called the *Wien displacement law* and for a 5793 K black body the maximum occurs at $\lambda = 500$ nm, which agrees well with the true AM0 spectrum's maximum location (see Fig. 2.3). Lower black body temperatures shift the maximum to longer wavelengths which is what effectively happens at the Earth's surface to the terrestrial spectrum.

2.3 Earth and Sun geometry

To understand how much extraterrestrial radiation reaches us on the surface of the Earth we must consider the Earth's geometrical relation

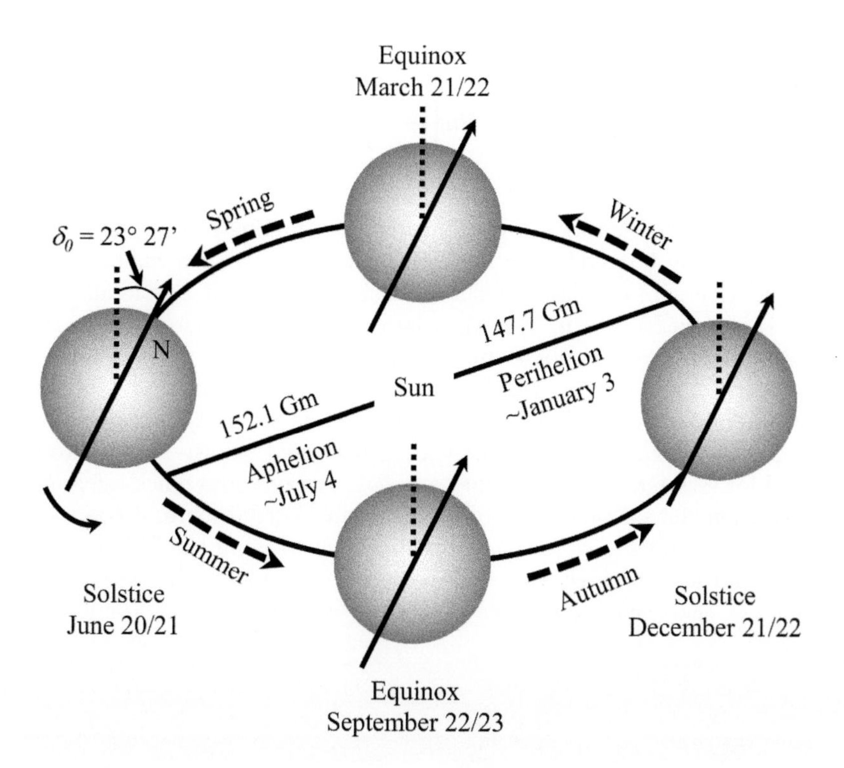

Fig. 2.4 The Earth–Sun geometry showing the angular declination of $23°27' = 23.45°$ for the Earth relative to the ecliptic plane formed by the path of the Earth's orbit around the Sun. Also shown are the points of closest and farthest distance, the perihelion and aphelion, respectively, as well as the equinoxes and solstices. The decimal equivalent of $23°27'00''$ or 23 degrees, 27 minutes and 00 seconds is: $23° + 27/60° + 00/3600° = 23.45°$.

with the Sun. The Earth revolves around the Sun producing the various seasons we experience because it has an angular declination (δ_0) relative to the normal defining the ecliptic plane, that is the plane defined by the revolution. Reference to Fig. 2.4 reveals that the north pole points towards the Sun in the summer and away in the winter making the length of day and average temperature change accordingly.

The declination is actually quite large at $23°27'$ making the Sun change position in the sky by twice this amount over a year. For example, the angular altitude of the Sun at solar noon, or the time when the Sun is highest in the sky above the horizon for the northern hemisphere $\alpha_{S,N}$, is easily calculated to be in the range

$$\alpha_{S,N} = 90° - |L| \pm \delta_0 \; [=] \; \text{deg.} \tag{2.8}$$

where L is the latitude. The latitude L is taken as negative in the southern hemisphere and in general, with reference to Fig. 2.1, one finds $L = 90° - \theta$ and θ has values from 0° to 180°. Obviously the Sun is highest above the horizon during the summer and δ_0 is taken as positive during this season. Remember this date range is six months apart between the northern and southern hemispheres (see below). At the two equinoxes $\alpha_{S,N}$ is merely $90° - |L|$, if we were on the equator the Sun would be directly overhead at 90° on these days for solar noon.

An approximate relation to determine the effective declination angle between the south pole to north pole axis and the Sun δ is

$$\delta \approx \operatorname{sgn}(L)\,\delta_0 \sin\left(\frac{[284+n]}{365} \times 360^\circ\right) \tag{2.9}$$

with n being the number of days from the beginning of the year (January 1 is $n = 1$ making $\delta = -23.1^\circ$ in the northern hemisphere) and the function $\operatorname{sgn}(\bullet)$ takes the sign of L, by definition if L is identically zero then $\operatorname{sgn}(L)$ returns 1. This relation traces the angle between the Sun and Earth reasonably accurately with an angle of zero near the two equinoxes. The graph in Fig. 2.5 shows the solar declination as a function of time in a way known as the *analemma* which is centuries old, the other axis *Equation of Time* will be discussed below. This figure is useful for estimates of the declination angle and provides quick reference as to when the solstices and equinoxes occur. Furthermore, if one were to focus a camera at a certain place in the sky, say on the Sun at solar noon, and take a picture the first day of each month, the Sun would trace out the analemic pattern during the year on the film (or digital frame).

Example 2.2

Find the angle the Sun makes with the horizon when it is highest in the sky for Hoboken, N.J., U.S.A. and Brisbane, QLD, Australia on July 21 and February 14.

The analemma and eqn (2.9) can be used to find the angle the Sun makes with the horizon at solar noon (*i.e.* when it is at its highest point in the sky). In general, the angle is called the solar altitude for any time and is given the symbol α_S (see Fig. 2.9 below), in this example it is calculated at solar noon, when the Sun is highest in the sky, with eqn (2.8)

$$\alpha_{S,N} = 90^\circ - |L| + \delta \tag{2.8}$$

which has been generalized to any day, not just at the solstices, so the sign of the declination angle is important. The latitudes for Brisbane and Hoboken are $-27^\circ28'16''$ and $40^\circ44'38''$, respectively, and with reference to Fig. 2.5 or eqn (2.9) one can determine

Date	Brisbane	($L = -27.47^\circ$)	Hoboken	($L = 40.74^\circ$)
July 21:	$\delta = -20.44^\circ$	$\alpha_{S,N} = 42.09^\circ$	$\delta = +20.44^\circ$	$\alpha_{S,N} = 69.70^\circ$
Feb 14:	$\delta = -13.62^\circ$	$\alpha_{S,N} = 76.15^\circ$	$\delta = -13.62^\circ$	$\alpha_{S,N} = 35.63^\circ$

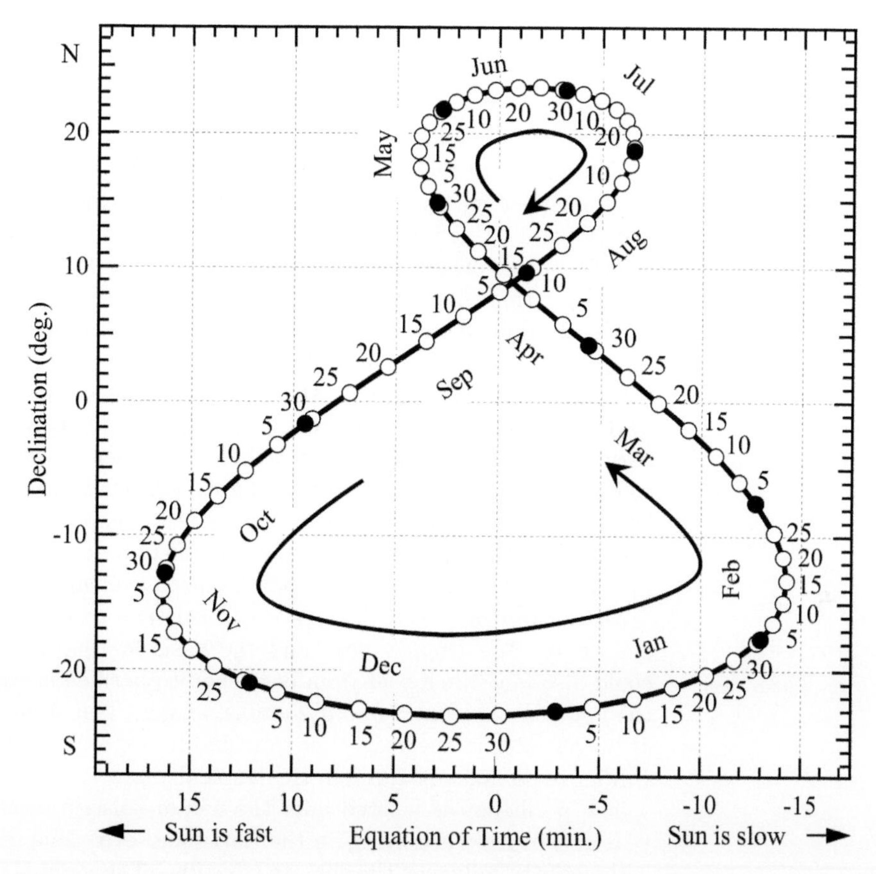

Fig. 2.5 The analemma showing the solar declination and equation of time as a function of day in the year at solar noon. Filled circles represent the first day of each month. When the Sun is slow the value for the equation of time must be taken as positive in eqn (2.24) and when fast should be taken as negative. The declination angle in the southern hemisphere is obtained by multiplying the angle by −1. The autumnal equinox is inaccurate in this graph even though a more accurate equation than eqn (2.9) was used, this is a challenge to all equations when considering the Earth–Sun geometry.

It is clear that the solar altitude depends strongly on the time of year as well as your position on the Earth. In three months it can change by 30° which is quite a large value. During other times of the year the change is much less, for example, in the winter and summer. Finally, comparing the values given in the table to one of the numerous calculators available, one will find there is some deviation. However, it is relatively small and within the tolerance acceptable to the accuracy required here.

Of course, the change in angle affects the number of hours between sunrise and sunset (N_h) which is found from geometrical relations

$$N_h \approx 2 \times \frac{24\ \text{h}}{360°} \times \cos^{-1}(-\tan(|L|) \times \tan(\delta))\ [=]\ \text{h} \tag{2.10}$$

where L is again the latitudinal angle above the equator. The factor of '2' is included in the equation since the cosine function is between 0° and 90° while the Sun will move up to 180° through the sky. Note this equation must produce an angle in degrees rather than radians for it to give the correct number of hours. The equinoxial hours of sunlight in a day is 12, while, if one were located 40° above the equator, the day ranges from 9 to 15 hours for the winter and summer solstice, respectively.

Certainly the variation of the effective angle between the Sun and Earth produces the largest effect on the local climate and the ability to harvest solar radiation. However, there are other factors and reference to Fig. 2.4 shows the perihelion and aphelion, which are the closest and farthest distances to the Sun, respectively, we experience on the Earth since its orbit around the Sun is not a perfect circle. These dates change by a reasonable amount over time measured over millennia.

Measuring the year from solstice to solstice, or equinox to equinox, one defines the *tropical year* and the basis for many calendars. One could also measure a year from perihelion to perihelion thereby defining the *anomalistic year* to find it 25 mins. longer. The perihelion regresses, or moves later in the year, over time and in approximately 21,000 years will move completely through the tropical year.

The perihelion is located near the winter solstice merely due to the year at which we are living on the Earth and over time migrates, which is thought to influence the climate (this line of thought is controversial). There are other factors that affect the climate such as the declination angle changing, called precession, caused by the shape of the Earth (it's not a perfect sphere) and gravitational pull from the Sun and Moon. The north star Polaris was not the true north star as little as 5000 years ago due to precession. For our purposes though, these factors will be ignored and the declination angle will be assumed constant at $23°27'$ ($\approx 23.45°$) and the Earth will be assumed to be 150 Gm (1.5×10^8 km) from the Sun.

2.4 Terrestrial solar radiation

Extraterrestrial radiation does not suffer from atmospheric adsorption and scattering while terrestrial does, and fortunately so, or the Earth would be a much hotter place. The terrestrial spectrum is shown in Fig. 2.3 and one finds many dips caused by water and other chemicals absorbing the radiation. Also, the overall magnitude of the spectrum is decreased because of this, which is readily seen in Fig. 2.3 by comparison of the upper and middle graphs that are on the same scale. A standard terrestrial spectrum exists, required for simulation and testing purposes, and is denoted as AM1.5G as shown in the figure.

In this case the number 1.5 refers to an air mass 1.5-times that directly above you or in other words the sunlight must travel a distance of 1.5 times the atmospheric thickness. Figure 2.6 is used to demonstrate that the distance x relative to the distance h represents the air mass index ($AM \equiv x/h$). So, if the Sun is directly overhead one expects AM1.0, or an air mass index of one, and it can be easily calculated for other angles as

$$AM \approx \frac{1}{\cos(\theta_Z)} \tag{2.11}$$

although this is inaccurate when the zenith angle θ_Z approaches 90°. Instead one can use the *law of cosines* on the triangle $R_E \to x \to h + R_E$ to find

$$AM = -\frac{R_E}{h}\cos(\theta_Z) + \left\{\left[\frac{R_E}{h}\cos(\theta_Z)\right]^2 + 2\frac{R_E}{h} + 1\right\}^{1/2} \tag{2.12a}$$

and

$$\cos(\theta_Z) = \frac{2\frac{R_E}{h} + 1 - AM^2}{2\frac{R_E}{h}AM} \approx \frac{1}{AM} \tag{2.12b}$$

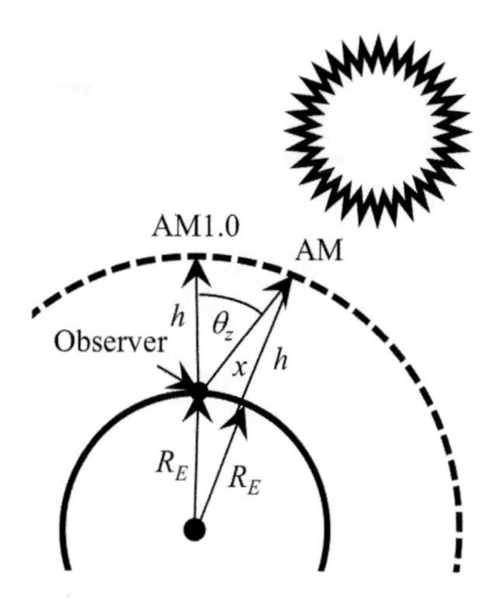

Fig. 2.6 Geometry of the Earth and its atmosphere demonstrating the air mass index. Here R_E is the Earth's radius, h, the atmospheric thickness and x, the atmospheric thickness for various angles of the Sun from the vertical or zenith, θ_Z.

where R_E is the Earth's radius. Both eqns (2.11) and (2.12a) give similar values when using the radius of the Earth (6380 km) and atmospheric thickness (≈ 16 km) for most angles from the vertical. Only when θ_Z approaches 90° do they significantly deviate. At this condition one obtains AM28.2 which is the largest possible value (a more accurate value will be obtained if the curvature of the Earth is taken into place as well as elevation above sea level).[6]

[6]The actual atmospheric thickness is difficult to quantify, at 16 km above sea level the atmospheric density is about 10% of that at sea level while at 22 km it is 5%. See Example 1.1 where an effective atmospheric thickness of 16 km was suggested.

The AM1.5 standard has an angle of 48.2° from the vertical or zenith, yet, other variables are considered in this standard atmosphere. Included is that the receiving surface is at an angle of 37° from the horizontal and also positioned southward, facing directly towards the equator. The influence of receiver angle and location on light absorption will be discussed below.

The AM1.5 spectrum derives from other, older spectra developed because approximately half the solar production resources occurred above AM1.5 and half below. Standard water and ozone concentrations are also assumed as well as other atmospheric conditions. Given this, the total terrestrial irradiance (eqn (2.5), P_{ter} in this case though) is determined to be 967 W/m^2 by graphically integrating the spectrum

$$P_{ter} = \int_0^\infty I(E)dE = \int_0^\infty L(\lambda)d\lambda, \text{ e.g. for AM1.5G} \tag{2.13}$$

similarly to what was done in eqn (2.5). The standard condition is then finally corrected to give a total irradiance of 1000 W/m^2, by multiplying the energetic irradiance or spectral irradiance with a common scaling factor of 1000/967 = 1.034. Although this may seem somewhat

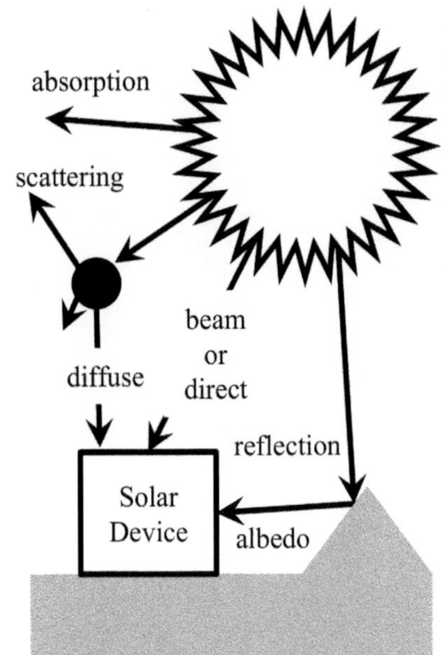

Fig. 2.7 Solar radiation will reach a solar device in three different ways; either *directly* from the Sun, reflection from the surrounding terrain to produce an *albedo* component and scattered from particles or molecules for a *diffuse* component. Some radiation is *absorbed* by the atmosphere and doesn't reach the device at all.

arbitrary, the AM1.5G spectrum is an extremely useful tool for standardizing simulations and characterization devices, as well as giving a rational approximation to the available terrestrial solar power available. The reader may notice that the letter G has now appeared next to AM1.5 and it stands for *Global.* This is related to the type of solar radiation reaching the device; is it directly from the Sun, scattered by dust or reflected from the ground? The spectrum can be developed to include one or all of these components.

The amount of radiation that actually reaches the Earth's surface depends on several factors, as shown in Fig. 2.7, and defines the global spectrum. Several components contribute with the first being the *direct* or *beam* component, which is the radiation that reaches a device directly from the Sun and is given the symbol B_{ter}. Some of the solar radiation will be absorbed by molecules and objects, which accounts for the extraterrestrial irradiance P_{ext} being reduced to B_{ter}. Photons scattered by atmospheric components contribute to the *diffuse* component.[7] Finally, the last component is reflected by the surroundings to reach the device and is called the *albedo*.[8]

The AM1.5G is the standard spectrum where G stands for global meaning that the beam (or direct) and diffuse components are included with both contributing to P_{ter}. Solar cells and solar hot water heaters can both use this type of radiation, so, the spectrum would be used to simulate their performance. Another standard spectrum exists called the AM1.5D with D implying that only the direct beam component is included. This spectrum is used for solar devices that include optics since the diffuse component cannot be focussed and reflected to the point of interest on the solar device.

The radiation intensity that comes from the Sun to the observer directly is a function of many variables: cloud cover, pollution level, ozone concentration, water concentration, etc. and there exist many models to describe it. Here the beam component is written in the simplest possible manner

$$B_{ter} = \tau \times P_{ext} \tag{2.14}$$

where τ is an overall transmission coefficient for radiation propagating through the atmosphere. Note the difference between B_{ter} and P_{ter}. While B_{ter} is the amount of radiation propagating directly from the Sun to a point on the Earth, P_{ter} is the total radiation at that point which may include the diffuse component too.

In general τ will be the product of that for absorption τ_{abs}, scattering from molecules, dust and water droplets (such as in clouds),[9] τ_{scat} (≈ 0.95), and reflection from the atmosphere, τ_{refl} (≈ 0.8) and is mathematically expressed by

$$\tau = \tau_{abs} \times \tau_{scat} \times \tau_{refl} \tag{2.15}$$

If B_{ter} is 700 W/m^2 then $\tau \approx 0.5$, since P_{ext} is 1366 W/m^2, making $\tau_{abs} \approx$

[7] The scattered component is that which makes the sky blue since shorter wavelength light is more readily scattered according to Lord Rayleigh's prediction. Scattering can also be used to predict the weather; find out the origin of the expression; 'Red sky at night, sailors' delight, red sky in the morning sailors' warning.'

[8] What is presented here is a very simplified calculation, sophisticated models and software exist to find the various components.

[9] Clouds will scatter radiation to reduce transmission. Even if the cloud cover is very large, radiation will still power a photovoltaic device unless it is extremely dark. Solar powered devices that require focussing the Sun's direct beam will not function in this case. Cloud cover, and its lack thereof, is a major consideration in the placement of solar powered devices.

0.8. The Beer–Lambert–Bouguer law for absorption can be written (see Chapter 5)

$$\tau_{abs} = \exp(-\alpha x) \equiv \exp(-\langle \alpha h \rangle AM) \tag{2.16}$$

where α is the atmospheric absorption coefficient and $\langle \alpha h \rangle$ represents the average absorption coefficient for the atmosphere over all wavelengths for its thickness. One then finds $\langle \alpha h \rangle$ is approximately 0.1 for AM1.5G conditions meaning the absorption coefficient of the atmosphere is quite small.

B_{ter} will be the amount of radiation reaching the observer at a point, or very small area as if you were looking directly at the Sun (not advised). Since most solar powered devices are quite large and rarely lie flat on the ground, some geometry must be considered because of this and the fact that the Sun almost always makes an angle to the vertical. Basically, a device that makes an angle with a radiation source will present less area to that source and the challenge is to find that angle for the time and date of interest, which is what is considered now.

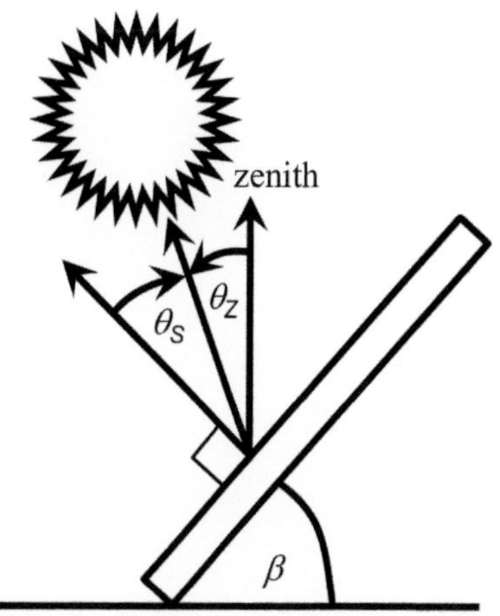

Fig. 2.8 A solar device is placed at an angle β to the horizontal while the Sun makes an angle θ_Z to the zenith and θ_S to the device's surface normal. If the Sun is at solar noon such that $\omega = 0$, one finds $\beta = \theta_S + \theta_Z$. The angle ω is the hour angle defined below, see Fig. 2.9.

Consider a flat solar device that makes an angle β with the horizontal as shown in Fig. 2.8. Now consider the three radiation components that could strike the device: the direct or beam component $B(\beta)$, the diffuse component $D(\beta)$ and the albedo component $A(\beta)$, where we can write

$$P_D(\beta) = B(\beta) + D(\beta) + A(\beta) \tag{2.17}$$

$P_D(\beta)$ is the amount of irradiance that the device can use. To simplify the discussion in the beginning let the device lie flat on the ground ($\beta = 0$) and in this case θ_S will equal the magnitude of the zenith angle θ_Z and $A(0)$ will be identically zero since reflected radiation from the surrounding terrain can not reach the device in most cases. So, it is found that

$$P_D(0) = B(0) + D(0) \tag{2.18}$$

The beam component is easily found from

$$B(0) = B_{ter} \times \cos(\theta_Z) \tag{2.19}$$

to account for the device area presented to the Sun. Of course, if θ_Z is 90° then the largest direct beam component is found.

The challenge now is to find θ_Z as a function of time, which concerns geometrical relations and a correction called *the equation of time* which has a long history. First, one can write

$$\cos(\theta_Z) = \sin(\delta)\sin(|L|) + \cos(\delta)\cos(|L|)\cos(\omega) \tag{2.20}$$

where, as before, δ, is the angular declination, L, the latitude and ω, the hour angle given by (see Fig. 2.9)

$$\omega = \frac{360°}{24\,\text{h}} \times [t_{LST} - 12\,\text{h}] \tag{2.21}$$

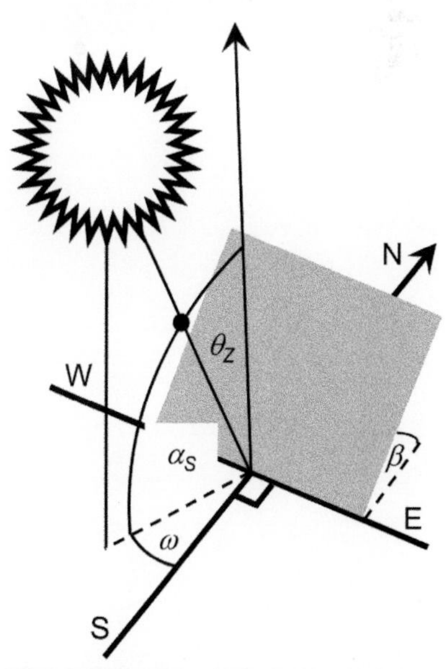

Fig. 2.9 Angles used to describe the position of the Sun for a southward facing solar device inclined at an angle β.

with t_{LST} being the local solar time in hours. The hour angle converts the local solar time to the Sun position in degrees. The device in the figure is pointing directly to the south and although it is relatively easy to correct for deviations from this direction, it merely adds more complicated geometrical relations that need not be considered here. There are many treatises that deal with this and other complicating factors that the interested reader may consult. Since many solar devices are frequently pointed as close to a southward direction as possible, or to the north for the southern hemisphere, this complication is not considered here.

Solar time is rooted in the fact that each day has a point when the Sun is at its highest point, called the meridian, which occurs at solar noon, $t_{LST} = 12$ h and $\omega = 0$. A solar day is the time between subsequent meridians, which can change depending on the time of year and can deviate by a reasonable amount from the standard clock-based day of 24 h. Since solar 'high' noon, or the meridian, will occur when the Sun is at its highest position then by definition $\omega = 0$ at solar noon and the solar morning has $\omega < 0$ and the solar afternoon has $\omega > 0$. From eqn (2.20) it is possible to define the sunset angle ω_S by letting $\theta_Z = 90°$

$$\cos(\omega_S) = -\tan(|L|) \times \tan(\delta) \tag{2.22}$$

An equivalent sunrise angle is equal to $-\omega_S$ by definition. These angles can be used to derive eqn (2.10) to give the number of possible hours of sunlight.

One can find t_{LST} from

$$t_{LST} = t_L + t_{cor} \tag{2.23}$$

where t_L is the local (clock) time and t_{cor}, the correction for *longitudinal* position, eccentricity of the Earth's orbit and its axial tilt as well as daylight savings time. One can now see the challenge that must be faced. The Earth's orbit is not perfect and the local time must be made such that the Sun's position can be accurately known so we write t_{cor} as

$$t_{cor} = \frac{24\,\text{h}}{360°}[\Lambda_{STM} - \Lambda] + t_{EOT} - t_{day} \tag{2.24}$$

with Λ representing the longitudinal position, Λ_{STM} is the local standard time meridian, which is referenced to the prime meridian (at Greenwich, UK), t_{EOT} is the equation of time value and t_{day} is the correction for daylight savings time (if it is one hour later than standard time then t_{day} is +1 h). One finds Λ_{STM} from

$$\Lambda_{STM} = \frac{360°}{24\,\text{h}}[t_L - t_{GMT}] \tag{2.25}$$

where t_{GMT} is Greenwich mean time. This equation will merely tells you the longitudinal segment of the sphere which you occupy. Your longitudinal position will indicate the part of the segment that you occupy and this is the first term in eqn (2.24). The daylight savings time correction, t_{day}, is easily understood and now we require the equation of time to correct for imperfections in the Earth's orbit. This equation has been developed over hundreds of years and there are many 'equations of time' and we present a simple one here

$$t_{EOT} = 0.165\sin(2\Delta) - 0.126\cos(\Delta) - 0.025\sin(\Delta)\ [=]\ h \qquad (2.26a)$$

where

$$\Delta = \frac{360°}{365}[n - 81] \qquad (2.26b)$$

with n being the day from the beginning of the year (January 1 is $n = 1$). The equation yields a negative value when the Sun is slow and to use this in eqn (2.24) one should take the value as positive, when t_{EOT} is positive then this means the Sun is fast and this value should be taken as negative.

Now we have all the information to calculate ω. This will allow the direct or beam component to be calculated $B(0)$ from eqns (2.19) and (2.20).

The diffuse radiation is determined by accurate measurement of $B(0)$ and $D(0)$ which lead to the development of a correlation with the *clearness index*, K_T defined by

$$K_T = \frac{P_D(0)}{P_{ext}\cos(\theta_Z)} \qquad (2.27)$$

where $\cos(\theta_Z)$ is added to the denominator to make it equivalent to the extraterrestrial radiation falling on a horizontal plane. In some cases a correction factor for the eccentricity of the Earth's orbit around the Sun is placed in the denominator that influences the solar constant P_{ext}, see note 4. In general, the clearness index varies from 0.3 for overcast climates to 0.7 - 0.8 for very sunny climates at high elevation.

There are many correlations relating the clearness index to the amount of diffuse radiation, one such is that of Orgill and Hollands, see Fig. 2.10

$$\frac{D(0)}{P_D(0)} = \begin{cases} 1.0 - 0.249K_T & \text{if } 0 \le K_T \le 0.35 \\ 1.557 - 1.84K_T & \text{if } 0.35 \le K_T \le 0.75 \\ 0.177 & \text{if } 0.75 \le K_T \end{cases} \qquad (2.28)$$

A simple expression that produces 15 - 20% diffuse radiation on a perfectly clear day

$$\frac{D(0)}{P_D(0)} = \frac{0.85}{1 + 50 \times K_T^6} + 0.15 \qquad (2.29)$$

which has the advantage of being a continuous function rather than piecewise continuous and adequately represents the data. Of course, as the clearness index approaches zero one obtains complete diffuse radiation as expected while as it approaches 1 there is 15–20% diffuse radiation.

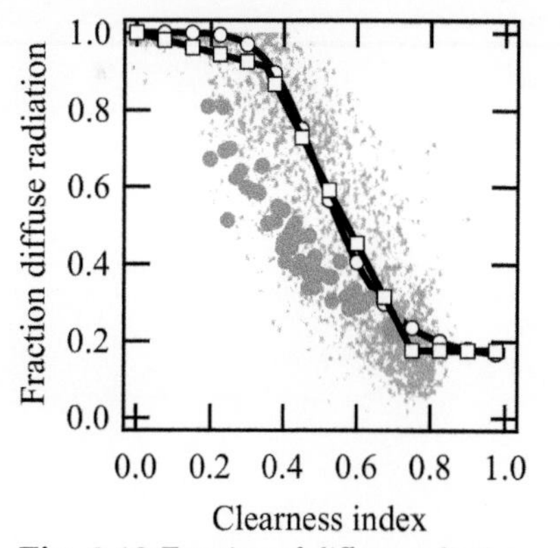

Fig. 2.10 Fraction of diffuse radiation ($D(0)/P_D(0)$) as a function of clearness index (K_T). The present correlation is eqn (2.29) and is compared to that of Orgill and Hollands (Solar Energy **19** (1977) 357). The data are from J.A. Duffie and W.A. Beckman, 'Solar Engineering of Thermal Processes,' 3rd Edition, John Wiley & Sons (2006) for Cape Canaveral, FL, U.S.A. (small gray symbols) and B.Y.H. Liu and R.C. Jordan, Solar Energy **4** (1960) 1 (large gray symbols).

Example 2.3

Determine the total, beam (direct) and diffuse radiation reaching a horizontally placed solar device on April 30 at 10 am in Newark, DE, U.S.A. assuming the overall direct beam transmission through the atmosphere is 0.6.

The first quantity that must be found is B_{ter} from eqn (2.14)

$$B_{ter} = \tau P_{ext} \tag{2.14}$$

which will be substituted into eqn (2.19)

$$B(0) = B_{ter} \times \cos(\theta_Z) \tag{2.19}$$

The zenith angle is given by

$$\cos(\theta_Z) = \sin(\delta)\sin(L) + \cos(\delta)\cos(\text{Ł})\cos(\omega) \tag{2.20}$$

To find this we note that April 30 is the 120th day of the year and the declination angle from eqn (2.9) is 14.58° or it can be estimated from Fig. 2.5 to similar effect. The latitude for Newark, DE is 39°41′1″ N which is 39.68°. The hour angle now must be found

$$\omega = \frac{360°}{24\,\text{h}} \times [t_{LST} - 12\text{ h}] \tag{2.21}$$

which also requires these equations

$$t_{LST} = t_L + t_{cor} \tag{2.23}$$

$$t_{cor} = \frac{24\,\text{h}}{360°}[\Lambda_{STM} - \Lambda] + t_{EOT} - t_{day} \tag{2.24}$$

$$\Lambda_{STM} = \frac{360°}{24\text{ h}}[t_L - t_{GMT}] \tag{2.25}$$

$$t_{EOT} = 0.165\sin(2\Delta) - 0.126\cos(\Delta) - 0.025\sin(\Delta)[=]h \tag{2.26a}$$

where

$$\Delta = \frac{360°}{365}[n - 81] \tag{2.26b}$$

The standard time meridian Λ_{STM} is −75° since Newark is 5 h behind Greenwich mean time, while t_{EOT} is +0.04649 h indicating the Sun is slightly fast compared to local time. The longitude for Newark is $\Lambda = 75°45'0'' = 75.75°$ and by combining these one finds, noting daylight savings time is in effect

$$t_{corr} = \frac{24\,\text{h}}{360°}[-75° - \{-75.75°\}] - 0.04649\,\text{h} - 1\,\text{h} = -0.9964\,\text{h}$$

Now t_{LST} = 10.00 h − 0.9965 h = 9.004 h allowing the hour angle to be calculated

$$\omega = \frac{360°}{24\,\text{h}} \times [9.004 - 12]\,\text{h} = -44.95°$$

Notice it is negative in value indicating it is in the morning. The zenith angle is

$$\cos(\theta_Z) = \sin(14.59°)\sin(39.68°) + \cos(14.59°)\cos(39.68°)\cos(44.95°)$$
$$\cos(\theta_Z) = 0.6879 \Rightarrow \theta_Z = 46.53°$$

Now we can find $B(0)$

$$B(0) = 819.6\frac{\text{W}}{\text{m}^2} \times 0.6879 = 563.8\frac{\text{W}}{\text{m}^2}$$

To find $D(0)$ and ultimately $P_D(0)$ requires use of a model and we use eqn (2.28) which is combined with eqns (2.14), (2.18), (2.19) and (2.27) to arrive at

$$P_D(0) = \left[0.152 + \{0.0230 + 0.543\tau\}^{1/2}\right] P_{ext}\cos(\theta_Z) \tag{2.31}$$

assuming $0.35 \leq K_T \leq 0.75$. Now it is easily found that

$$P_D(0) = 0.511 \times P_{ext} = 698.3\frac{\text{W}}{\text{m}^2}$$

which is valid for the clearness index region assumed ($K_T = 0.74$). The diffuse component is found from eqn (2.18)

$$D(0) = P_D(0) - B(0) = 698.3 - 563.8, \frac{\text{W}}{\text{m}^2} = 134.5\frac{\text{W}}{\text{m}^2}$$

A final comment, the clearness index was quite close to the validity of the correlation used. If the high clearness index limit of $D(0)/P_D(0) = 0.177$ in eqn (2.28) is used one obtains

$$P_D(0) = 1.22\tau P_{ext}\cos(\theta_Z) = 685\frac{\text{W}}{\text{m}^2}$$

which is close to the value obtained with the other correlation.

Very rarely are solar devices located horizontally on a surface and they are typically tilted to maximize the amount of radiation falling on them. A rule of thumb is to tilt them at an angle given by $\beta = |L| + 10°$ and point them towards the equator. It is shown in Fig. 2.11 how much radiation can be obtained by tilting a device at an angle to the horizontal, while facing due south. This can increase $P_D(\beta)$ by order 50%, which is quite substantial.

A tilted device will receive the direct beam and diffuse components as well as the albedo. The beam component can be written

$$B(\beta) = B_{ter}\cos(\theta_S) \tag{2.32}$$

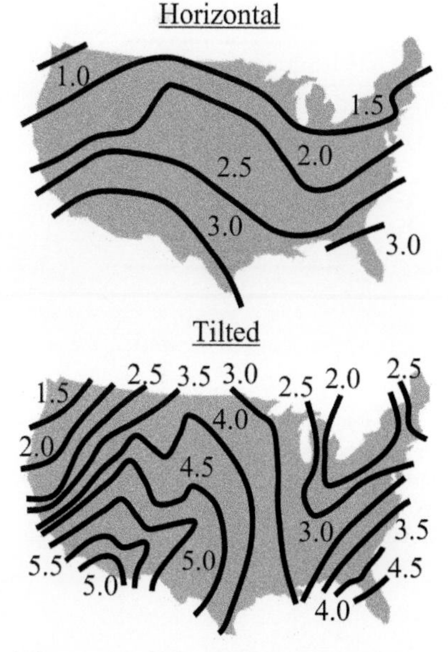

Fig. 2.11 Solar irradiation for the United States on a horizontal and tilted surface, demonstrating the difference between the two, during the month of December. The numbers next to the lines are the irradiation in kW-h/m^2-day. Note these are only approximate values and presented to show the large difference between using a tilted and horizontally placed solar device. Adapted from D.K. McDaniels, 'The Sun: Our future energy source,' John Wiley & Sons (1984).

where θ_S is the angle that the unit normal for the device makes with the direct beam radiation from the Sun as shown in Fig. 2.8. The geometric manipulation to find θ_S is not difficult, however, it is cumbersome and so the final result is given here

$$\cos(\theta_S) = \cos(\theta_Z)\cos(\beta) + \sin(\theta_Z)\sin(\beta)\cos(\omega) \tag{2.33}$$

Obviously, when $\beta = 0$ one finds $\theta_S = \theta_Z$. Reference to Fig. 2.8 allows the following relation to be found

$$\beta = \theta_S + \theta_Z, \qquad \omega = 0 \tag{2.34}$$

at solar noon when $\omega = 0$ which is readily verified with eqn (2.33). The relations above allow one to determine $B(\beta)$, now the other components for the solar radiation incident on a tilted solar device will be considered.

Diffuse radiation is not typically the same in all directions. For example, right next to the direct beam from the Sun, molecules and dust scatter light more or less in a forward direction. This causes a diffuse component right near the direct beam. This is called circumsolar diffuse radiation and will be more intense than diffuse radiation elsewhere in the sky. In spite of this, we use an approximation that the diffuse component, as well as the albedo, are isotropically (evenly) distributed throughout the sky, see Fig. 2.8, allowing two simple relations to be written

Table 2.1 Albedo factors for various surfaces.

Surface	ρ_A
asphalt	0.15
lawn	0.20
concrete	0.30
sand	0.2–0.6
snow	0.5–0.9
water	0.08–1.0

$$D(\beta) = \frac{1+\cos(\beta)}{2} D(0) \tag{2.35}$$

and

$$A(\beta) = \rho_A \frac{1-\cos(\beta)}{2} [B(0) + D(0)] \tag{2.36}$$

where ρ_A is the albedo factor describing the reflective ability of various surfaces, see Table 2.1.[10] If β tends to zero the diffuse component is maximized since the device can gather all the isotropic diffuse radiation while the view factor for the albedo suffers and this component tends to zero.

[10] Do you know why the albedo factor changes so much for water? Hint: Consider the effect of wind and the seasons. Would you want to design a solar device near water that depends on reflected radiation?

Example 2.4

Determine the total, beam (direct), diffuse and albedo radiation reaching a tilted solar device for the same conditions in Example 2.3. Assume the device is tilted at an angle of 50° from the horizontal and faces directly south and that the albedo factor is 0.25.

The angle the device makes with the Sun is found from eqn (2.33)

$$\begin{aligned}\cos(\theta_S) &= \cos(\theta_Z)\cos(\beta) + \sin(\theta_Z)\sin(\beta)\cos(\omega)\\ &= \cos(46.53^\circ)\cos(50^\circ) + \sin(46.53^\circ)\sin(50^\circ)\cos(-44.95^\circ)\\ \cos(\theta_S) &= 0.8357 \Rightarrow \theta_S = 33.31^\circ\end{aligned}$$

One can now calculate the beam, diffuse and albedo components from eqns (2.32), (2.35) and (2.36) to be 684.9, 110.5 and 31.2 W/m^2, respectively. The total radiation presented to the device is obtained from eqn (2.17)

$$P_D(\beta) = 684.9 + 110.5 + 31.2, \frac{\mathrm{W}}{\mathrm{m}^2} = 826.6\frac{\mathrm{W}}{\mathrm{m}^2}$$

If the device were merely placed horizontally on the ground there would only be 698.3 W/m^2 available to it. However, placing it at an angle to the horizontal increases its potential power density by almost 20%. This is a substantial increase. One should also note how much the albedo component contributes, as well as the increase in the beam component due to tilting the device.

The above discussion demonstrates that the angle β tremendously affects the power available to a solar device, is there an optimum? To determine this we first assume the Sun is at its highest point ($\omega = 0$) to write eqn (2.17), while inserting the various terms and noting $\theta_S = \beta - \theta_Z$ from eqn (2.34)

$$P_D(\beta) = B_{ter}\cos(\beta - \theta_Z) + \frac{1+\cos(\beta)}{2}D(0) + \rho_A\frac{1-\cos(\beta)}{2}[B(0) + D(0)] \tag{2.37}$$

Now take the derivative of $P_D(\beta)$ with respect to β and set it equal to zero to arrive at an optimum angle β_{opt}

$$\tan(\beta_{opt}) = \frac{B_{ter}\tan(\theta_Z)}{B_{ter} + \frac{1}{2}D(0) - \frac{1}{2}\rho_a[B(0) + D(0)]} \tag{2.38}$$

after some trigonometric substitution. The optimum angle for Example 2.4 turns out to be 47.6° and so having a β of 50° is not too far from the optimum angle, at least at solar noon.

If the diffuse component and the albedo reflectance are negligible then one finds $\tan(\beta_{opt}) = \tan(\theta_Z)$ or $\beta_{opt} = \theta_Z$ so the device should be directly perpendicular to the beam component to generate the most power. This condition should hold for most conditions on Earth since the beam component is the largest and sole source of power. Only when the diffuse and albedo components become significant is the optimum angle different to θ_Z.

If the sky is very cloudy and no beam component is present then the optimum angle is 0°. This is because the diffuse radiation component (as well as the albedo) was assumed to be isotropic and with the device lying flat on the ground allows it to harvest diffuse radiation from the entire sky above. If you live in a place that is very cloudy for most of the year then β_{opt} may very well be zero! Intermediate cases occur depending on the magnitude of the various terms and β_{opt} will take values between zero and θ_Z. Of course, the hour angle was assumed to be zero and if this assumption is relaxed then eqn (2.38) becomes more complicated.

2.5 Useful irradiance

The irradiance is determined by considering the contribution of all the energy levels (eqns (2.5) and (2.13)). However, a given solar powered device may only absorb certain energy levels of the Sun's radiation and the power one could extract from it may be much less than all that is present. So, we consider a power density that is generated from the absorbed radiation with energy levels between E_1 and E_2 to find the useful irradiance P_Δ

$$P_\Delta = \int_{E_1}^{E_2} I(E)dE \tag{2.39}$$

By calculating P_Δ in this manner one assumes all the radiation is perfectly absorbed within the energy limits, which may not be the case, and represents an efficiency of 100% for the energy levels considered. A non-ideality factor will be considered in subsequent chapters when real materials are considered.

The integral can be evaluated after integration by parts and then expanding $\ln(1-e^{-x})$ in a Maclaurin series to arrive at

$$\frac{P_\Delta}{K} = \left[x^3 \ln(1-e^{-x}) - 3\sum_{n=1}^{\infty} \frac{n^2x^2 + 2nx + 2}{n^4} e^{-nx} \right]_{x=x_1}^{x_2} \tag{2.40}$$

Here $K = 15\sin(\theta_{sub})^2 \sigma_S T_S^4/\pi^4$ and $x_i = E_i/k_B T_S$ with E_i representing either of the two energy limits in the integral. If the limits are 0 and ∞ then the bracketed term on the right-hand side of the equation yields $\pi^4/15$, the sum is the Riemann zeta function ($\zeta(4)$), and eqn (2.4) results.

Consider the limits in eqn (2.40) to be near the visible spectrum of radiation with wavelengths between 300 and 800 nm, then one can determine the energy, say E_1, to be of order 5×10^{-19} J, which is an extremely small number and unwieldy to use. Instead define a new energy unit, the electronvolt or eV,[11] that is the amount of kinetic energy gained by an electron when accelerated through a potential of one volt. In this case the energy is of order 2.5 eV in the visible spectrum and $k_B T_S$ is 0.5 eV making x of order 5.

[11]The electron volt is defined as

$$1 \text{ eV} = 1.602 \times 10^{-19} \text{ J}$$

and Planck's law can be written

$$E(\text{eV}) = \frac{1240}{\lambda(\text{nm})}$$

The magnitude of the numbers leads to convenient approximations in eqn (2.40) since $x \approx 5$. One can expand $\ln(1-e^{-x})$ in a Maclaurin series and use only the first term in the expansion, as well as that in the summation, to find, with a small error

$$\frac{P_\Delta}{K} = \left[x_1^3 + 3x_1^2 + x_1 + 6\right]e^{-x_1} - \left[x_2^3 + 3x_2^2 + x_2 + 6\right]e^{-x_2} \tag{2.41}$$

If the second limit is $x_2 \to \infty$, then the second term on the right-hand side tends to zero and one can find the useful irradiance if all the radiation above E_1 is absorbed. Letting E_1 and E_2 tend to zero and infinity, respectively, does not lead to the Stefan-Boltzmann law (eqn (2.4)) due to the approximations used, yet it is not too inaccurate since it is only different by a factor of 1.08 from the true value.

Example 2.5

A solar cell can only absorb radiation that has an energy greater than 1.1 eV, determine the absolute maximum amount of power that it can generate.

This will be done by using eqn (2.41) and assuming AM1.5G radiation is present (T_S = 5369 K). First the value of x_1 must be determined which is equal to $1.1\,\text{eV}/[8.617\times10^{-5}\,\text{eV/K}\times 5369\,\text{K}] = 2.377$, of course $x_2 \to \infty$. The value of K must also be found

$$\begin{aligned} K &\equiv \frac{15}{\pi^4}\sin(\theta_{sub})^2 \sigma_S T_S^4 \\ &= \frac{15}{\pi^4}\sin(0.265°)^2 \times 5.670\times10^{-8}\,\text{W/m}^2\text{-K}^4 \times [5369\,\text{K}]^4 \\ K &= 155.2\,\text{W/m}^2 \end{aligned}$$

One can now find the useful irradiance with eqn (2.41)

$$\begin{aligned} P_\Delta &= K\left[x_1^3 + 3x_1^2 + x_1 + 6\right]e^{-x_1} \\ &= 155.2\,\text{W/m}^2 \times \left[2.377^3 + 3\times2.377^2 + 2.377 + 6\right] \times \exp(-2.377) \\ &= 155.2\,\text{W/m}^2 \times [13.43 + 16.95 + 2.377 + 6] \times 9.283\times10^{-2} \\ &= 558.4\,\text{W/m}^2 \end{aligned}$$

All the numbers in the square brackets were kept to show that they are all of the same order and none can be ignored. This is a substantial power suggesting the device is over 55% efficient! Unfortunately, this maximum power is not possible with a solar cell as will be found out in Section 3.4 and the maximum efficiency for a solar cell that absorbs all radiation above 1.1 eV is only 31% due to an effect called *thermalization.*

Equation (2.41) is also very useful to easily find the power available over certain energy ranges, such as occur in photosynthesis, for example. As one can see in Table 2.2, the maximum useful irradiance occurs near a wavelength of 500 nm for the limited wavelength range considered at each point, in agreement with the Wien displacement law. Plants absorb near wavelengths 400 and 600 nm, right on either side of the maximum and could produce at most of order 150 W/m^2 in power. Since the

Table 2.2 Energy available over limited wavelength ranges for the AM1.5G solar spectrum.

Wavelength range (nm)	Mid-range energy (eV)	P_Δ (W/m^2)
300 ± 25	4.1	42
400 ± 25	3.1	78
500 ± 25	2.5	90
600 ± 25	2.1	83
700 ± 25	1.8	71
800 ± 25	1.6	58

available power is ≈ 1000 W/m^2 one can immediately ascertain that plants can only be ≈ 15% efficient with the true efficiency closer to order 1% due to an overall inefficient process, although light absorption itself may be fairly efficient. Also, bioethanol production has a power density of approximately 0.25 W/m^2, as determined in Example 1.3, so, it is only approximately 0.025% efficient!

2.6 Time averages

The irradiance that can fall on a device $P_D(\beta)$, as well as its beam, diffuse and albedo components, were all considered as instantaneous values. In other words, they have units of energy per time per area. While important, the general user is interested in how much energy per area is available over a given time period. So, much effort in the solar industry is invested in finding the average energy per unit area available for given locations and specific time frames. This may be done over hourly, daily or monthly time frames.

Indeed this is a difficult subject and requires analysis of statistical data of observations over years. To make a model of this, one must first account for imperfections in the Earth's orbit since the extraterrestrial radiation is not constant over a year. Consider the simplest possible quantity by determining P_{ter} over a given time period from t_1 to t_2 which we write as

$$\Sigma P_{ter} = \int_{t_1}^{t_2} \left[1 + 0.033 \cos\left(\frac{360^\circ n}{365} \right) \right] \tau(E,t) I(E,t) dt \, [=] \, \frac{\mathrm{J}}{\mathrm{m}^2} \tag{2.42}$$

using the relation given in note 4. The symbol ΣP_{ter} is used to highlight the difference in dimensions and the fact that the power density is being integrated over time. To do this correctly the fact that the transmittance $\tau(E,t)$ is a function of energy and time must be taken into account as well as $I(E,t)$. In addition, other factors must be considered, such as the time of year, local weather patterns and the like that influence P_{ter} which make this a difficult topic.

These factors are not ignored here. Their consideration is necessary and can be considered by the student if need be through consultation of other texts, so, they will not be considered in depth here.

2.7 Conclusion

Solar radiation or insolation is produced by a series of reactions at the very core of the Sun. Through Einstein's famous equation $E = mc^2$ we can calculate that an incredibly small mass loss from this fusion reaction produces a very large energy that finds its way to the Earth's surface. Planck's law can be used to adequately model the energy distribution of the radiation and it was found that via the Earth–Sun geometry and attenuation of the irradiance by the atmosphere the magnitude of the radiation is significantly reduced in magnitude.

Solar powered devices can use this radiation, however, further geometric relations are needed to find the projected area of that device with the beam component of the radiation. Since the Sun moves through the sky during the day, and each day has unique conditions, a reasonably accurate method was presented to determine the angle the device makes with the beam at any time for any day. Tilting the device from the horizontal generally increases the power that can be generated and this can be significant as proved through analytical calculation.

Finally, a solar powered device, in general, cannot absorb radiation over all energy levels and an analytical expression was presented to account for this. Photosynthesis was considered to determine how much energy plants would absorb over discrete energy levels. The relation for useful irradiance becomes very important when semiconductors are considered in later chapters as they can only absorb radiation above certain energies.

2.8 General references

J.A. Duffie and W.A. Beckman, 'Solar energy of thermal processes,' John Wiley and Sons, 3rd Edition (2006)

M. Iqbal, 'An introduction to solar radiation,' Academic Press, New York (1983)

D.K. McDaniels, 'The Sun: Our future energy source,' John Wiley and Sons, 2nd Edition (1984)

A.B. Meinel and M.P. Meinel, 'Applied solar energy, An introduction,' Addison-Wesley Publishing Company, Reading, Mass. (1976)

F.W. Taylor, 'Elementary climate physics,' Oxford University Press (2005)

Exercises

(2.1) Determine the mass density in the Sun's core.

(2.2) What is the Earth's rotational velocity on its surface as it revolves through a day in km/h?

(2.3) What is the Earth's angular velocity around the Sun as it orbits in km/h?

(2.4) Take the individual reactions in the proton–proton cycle and balance them to arrive at the overall reaction given in the text.

(2.5) Assuming the Sun will continue to radiate 63 MW/m^2 during its remaining life, how long will it last?

(2.6) How much of the Sun's mass is lost by providing energy to the Earth in a day?

(2.7) Find the photon flux described in note 3. How many photons hit the Earth in one day?

(2.8) Derive the equation for the spectral irradiance given in eqn (2.6).

(2.9) Derive the Wien displacement law given in eqn

(2.7). Assume the law is applied for radiation with wavelengths near the visible part of the spectrum allowing you to make an approximation.

(2.10) If you treated the Earth and yourself as a black body radiator what wavelength would correspond to the spectrum maximum and what would be its irradiance?

(2.11) Write Stefan's constant in terms of variables and determine its value to four significant figures.

(2.12) The photons released by the Sun can actually induce a pressure, p. In fact, in a derivation of eqn (2.1) one assumes an ideal gas of photons exists, which has all the properties of a gas made of atoms or molecules. One can write

$$p = \frac{8\pi^5}{45} \frac{k_B^4}{h^3 c^3} T_S^4 \equiv \sigma_p T_S^4$$

Calculate σ_p and find the pressure induced by solar radiation. Is this significant? What would the temperature have to be to create one atmosphere pressure?

(2.13) Integrate the AM1.5G spectrum available from the National Renewable Energy Laboratory's web site to find P_{ter} by using the trapezoidal rule. See eqn (2.13) for the appropriate integral. Give your result in W/m^2.

(2.14) Integrate the AM1.5D spectrum available from the National Renewable Energy Laboratory's web site to find the total irradiance for direct (beam) and circumsolar radiation. The AM1.5G spectrum has an irradiance of 1000 W/m^2, determine how much of this spectrum is diffuse radiation by subtracting the two values. See eqn (2.13) for the appropriate integral.

(2.15) The AM1.5G spectrum is available from the National Renewable Energy Laboratory's web site, however, it is given as the spectral irradiance. Convert the spectral irradiance to the energetic irradiance, create a graph of energetic irradiance versus energy (in eV), and indicate on the graph several absorption bands for water, oxygen, ozone and carbon dioxide. This graph should look like the middle graph in Fig. 2.3.

(2.16) The Earth System Research Laboratory of the National Oceanic and Atmospheric Administration gives an approximate correction due to refraction of the Sun's rays for the solar altitude α_S. This correction is the result of the Sun's rays traveling through the vacuum of space and being bent by the Earth's atmosphere due to the difference of refractive index (Snell's law). The effect explains why the Sun's rays can be seen even when the Sun is below the horizon. The correction is given by

$$\frac{1^\circ}{3600''} \left[\frac{58.1''}{\tan(\alpha_S)} - \frac{0.07''}{\tan(\alpha_S)^3} + \frac{0.000086''}{\tan(\alpha_S)^5} \right]$$

for $5^\circ \le \alpha_S \le 85^\circ$. Determine the magnitude of this correction.

(2.17) Explain why these are special latitudes on the Earth: Tropic of Cancer, Arctic Circle, Tropic of Capricorn and Antarctic Circle.

(2.18) Find a date and location on the Earth where there will be no shadows at solar noon.

(2.19) Lamm (Solar Energy **26** (1981) 465) gives an accurate Equation of Time that accounts for leap years which can be written

$$t_{EOT} = \sum_{k=0}^{5} \left[A_k \cos\left(\frac{2\pi k n}{365.25}\right) + B_k \sin\left(\frac{2\pi k n}{365.25}\right) \right]$$

with the coefficients given in the table below and n is the day number with $n = 1$ being January 1 of the leap year and $n = 1461$ being December 31 of the fourth year. Compare this equation to the simpler one given in eqn (2.26) by finding the absolute average deviation (AAD) between the two over a period of four years. The AAD is the absolute value of the difference between two values for a given day averaged over the entire four year period.

k	A_k (h)	B_k (h)
0	2.0870×10^{-4}	0
1	9.2869×10^{-3}	-1.2229×10^{-1}
2	-5.2258×10^{-2}	-1.5698×10^{-1}
3	-1.3077×10^{-3}	-5.1602×10^{-3}
4	-2.1867×10^{-3}	-2.9823×10^{-3}
5	-1.5100×10^{-4}	-2.3463×10^{-4}

(2.20) Assume the date is July 24 and you are at a latitude of 40° N, plot the total, beam (direct), diffuse and albedo radiation (in W/m^2) reaching the solar device as a function of incline angle β at solar noon. Assume the overall transmission coefficient τ is 0.5 and ρ_A is 0.3. What is the optimum angle and how much power is lost or gained compared to using the rule-of-thumb value of $L + 10^\circ = 50^\circ$.

(2.21) Calculate the total, beam (direct), diffuse and albedo radiation falling on a device over a year's time in W/m^2 for the 15th of each month at a latitude of 40° and at solar noon. Assume the device can be at any angle from 0° to 90° in 10° increments and compare the maximum to the rule-of-thumb value of $|L|+10^\circ = 50^\circ$. Take the average over the year to determine the average total irradiance available for the comparison. Assume the overall transmission coefficient τ is 0.5 and ρ_A is 0.3.

(2.22) Much of Europe and the United States are located at a latitude of 42.5° N while Asia is located at 30° N, which region has more power for a device mounted horizontally at their solar noon at the winter and summer solstices and the equinoxes. Assume the overall transmission coefficient is 0.7 in the summer and changes linearly with time to 0.4 in the winter.

(2.23) A material absorbs radiation from the Sun for energy levels above 1.1 eV, what is the maximum power (in W) one could extract from this material if it covered an automobile? Assume the AM1.5G spectrum is available to the material. Could it provide all the power for this automobile to function? (*n.b.* This material represents Silicon if it operated as an ideal solar cell.)

(2.24) Diffuse radiation, which is irradiance scattered by particles, water droplets and molecules, has a spectrum of irradiance versus photon energy which is slightly *blue shifted* meaning that there is slightly more irradiance at higher energies than expected based on the beam component. Explain why.

(2.25) Equation (2.28) was used in the derivation of eqn (2.31). Generalize eqn (2.28) to a form of $D(0)/P_D(0) = a + bK_T$, where a and b are constants, and find the equivalent to eqn (2.31) containing a and b rather than numerical coefficients. Insert values for a and b from eqn (2.28) which are valid for the three clearness index regions. Now make a graph of K_T versus τ; does this graph reveal anything unexpected?

(2.26) Use eqn (2.29) and plot the overall transmission coefficient τ versus clearness index K_T.

(2.27) Derive eqn (2.38).

Basic principles

In this chapter the methods for harvesting solar energy are reviewed by dividing them into three categories for discussion purposes: light absorption, photovoltaic devices and solar thermal systems; later chapters will build on the discussion presented here. Before considering this, a brief treatise of thermodynamics is presented in the first two sections, specifically the First and Second Laws of Thermodynamics for both closed and open systems are discussed. This may or may not be of interest, depending on the reader's background, yet, even those with a rudimentary knowledge of thermodynamics should glance at this part of the chapter to become familiar with the nomenclature.

The most useful result in the first two sections is eqn (3.12) that is particularly helpful for analyzing solar thermal applications rather than solar photovoltaic. Basically one uses this equation, for the applications in this monograph, to determine how much a fluid will heat up as it flows into and out of a process when heat and work is applied to it. One should not dismiss these sections completely if the reader's interest is solely photovoltaics since the principles presented here, and later chapters, can be used to determine the temperature of a photovoltaic device placed in the field. The efficiency typically decreases with an increase in temperature, so, this may be a valuable calculation when designing a device.

Later sections are used to give a fundamental understanding of harvesting solar energy. Interaction of radiation (light) with matter is considered in Section 3.3 which will be built upon in Chapter 5. Similarly, photovoltaics are introduced in Section 3.4 where again a deeper account is given in Chapter 6, and solar thermal applications in Section 3.5 with more detail in Chapters 7 to 9. This type of presentation will give the reader some time to digest information before tackling the subject in greater detail. Firstly, though, an introduction to thermodynamics is presented.

3.1 Thermodynamics–Closed systems

Consider systems that do not allow mass flow across their boundary first, which is applicable to many solar processes including a photovoltaic device or solar cell. A Carnot cycle is such a system and has heat Q_H supplied from a reservoir kept at a constant high temperature T_H to an engine which produces work W and rejects heat Q_L to a constant low temperature T_L reservoir. As shown in Fig. 3.1, the *First Law of*

Solar Energy, An Introduction, First Edition, Michael E. Mackay

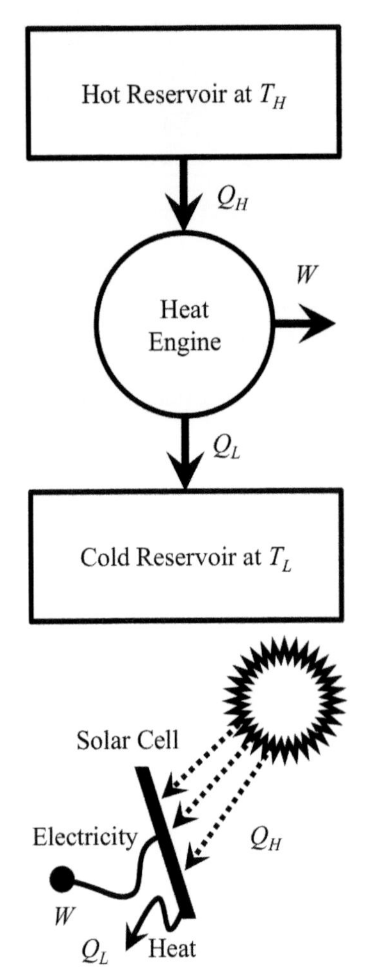

Fig. 3.1 The Carnot cycle or heat engine compared to a photovoltaic device. Heat from the *Hot Reservoir* (Sun) supplies heat (Solar Radiation) to the *Heat Engine* (Photovoltaic Cell) at a constant temperature to make *Work* (Electricity) while heat is rejected to the *Cold Reservoir* at constant temperature (heat transferred to the surrounding atmosphere).

Thermodynamics or FLOT is required to understand how a Carnot cycle works and can be written

$$\Delta U = Q + W [=] \text{ J} \qquad \text{(FLOT)} \tag{3.1}$$

where the symbol Δ means 'change of,' U is the internal energy and Q and W are heat and work, respectively. The symbol [=] means 'has dimensions of' where we have used Joules.

In order to understand the FLOT one must first understand exactly what internal energy is. Internal energy is merely the bonding and association of atoms and molecules, in a system like the heat engine, and represents the state of their interactions. For example, the internal energy for water vapor (steam) is very different to liquid water since the distance between the molecules is quite dissimilar in these two states of matter. Water at 1 bar pressure and 99.6°C has an internal energy of 417.5 kJ/kg and 2506 kJ/kg in its saturated liquid and vapor states, respectively. Their difference is the internal energy of vaporization. Another view of internal energy can be gained *via* molecular dynamics simulations. Particles (that represent atoms or molecules) are put in a figurative box and given thermal and potential energy. The thermal energy yields kinetic behavior and the potential results from attractive or repulsive forces. van der Waals and hydrogen bonding forces represent forms of an attractive potential. Newton's laws of motion are assumed to operate and the sum-average of all the kinetic and potential energies is the internal energy for that system. Internal energy is the fundamental energy measure of atoms and molecules while other thermodynamic quantities such as the enthalpy, introduced below, are a combination of parameters.

So the reader can get a better, more firm, grasp of internal energy and what it means, this final discussion on how it is measured is given. The above two examples demonstrate why the internal energy would change for various states of matter and the microscopic forces that are integral to its magnitude, how it is measured will reinforce exactly what it is. This is accomplished by using the FLOT with no work term, thus, $\Delta U \equiv Q$ and heat introduced into or taken from the system is equivalent to the change of internal energy. The reference state is chosen so it is unequivocally defined and for water it is typically taken as its triple point and the internal energy is defined as zero here.[1] Thus, to find the internal energy of liquid water at room temperature and pressure, heat is carefully added to a well insulated, constant volume container containing water at its triple point. When room temperature and pressure are obtained the total amount of heat added **is** its internal energy, this is about 100 kJ/kg for liquid water.

Amazingly the internal energy is derived from the FLOT! Then the internal energy and FLOT are used to model processes. This seems unfair. The truth is one cannot prove the FLOT is true, it is merely shown to work since the internal energy derived in the above manner is used in a predictive manner for dissimilar processes and found to be accurate. So, it is accepted as a truth.

[1] Gibbs' phase rule is

$$f = c - p + 2$$

where f is the number of degrees of freedom, c, the number of components and p, the number of phases. Water at its triple point, where vapor, liquid and solid are in equilibrium, has $f = 0$ and so is uniquely defined. The triple point occurs at 0.01°C and 611.73 Pa.

Energy transfer across a system's boundary, due to a temperature difference is defined as heat, while work can be described as something whose sole effect on the surroundings (everything outside of a system) could be moving a weight. Electricity generated by a photovoltaic device is considered work since, if it crosses the system's boundary, then it could be connected to a motor that can raise a weight (see Fig. 3.1).

Heat and work have signs associated with them and heat supplied to a system is considered positive while if the surroundings do work on the system it is considered positive, a photovoltaic device would therefore have a negative work term (see Fig. 3.1).

Consider heat being supplied to a system such that it rises in temperature, and does no work, then the FLOT can be written, $\Delta U = Q$, as mentioned above. Since the heat flow is positive into the system the internal energy increases because the atomic and molecular associations have changed. Physically one can imagine the atoms and molecules will vibrate more enthusiastically due to the increase of thermal energy given to them which increases the internal energy.

Likewise, take an incandescent light bulb as the system, and supply electricity to it, it would heat up due to the work being applied to the system. Here the FLOT reduces to $\Delta U = W$, if we ignore heat loss (or gain). Since the sign of work is considered positive then ΔU is again positive as the atoms in the system increase in temperature. Another example of positive work supplied to a system is a stirrer used to mix components in a tank. If the tank is taken as the system then work is being supplied to the system and it is positive in value. The example below reinforces the sign of heat and work and the concept of a system boundary.

Example 3.1

Can you cool a room by opening the door on a refrigerator and leaving it on?

Consider the refrigerator placed in a well insulated room shown in Fig. 3.2. The first system needed to solve this problem is the interior of the refrigerator labeled *System 1* in the figure. Even though the door is open one can expect that the interior cavity will be colder than the surroundings merely due to the way a refrigerator works, so, it is possible that the refrigerator will cool the surrounding room.

Now take *System 2* into account, this is the entire room, we can write the FLOT as

$$\Delta U = W$$

The heat term is neglected since we assumed the room was *well insulated*, which means adiabatic, so, there are no heat losses, as an approximation. A contemporary refrigerator uses 500 kW-h per year to operate, which will be 1.4 kW-h per day. There are 3600 kJ/kW-h and so approximately 5000 kJ are required in one day.

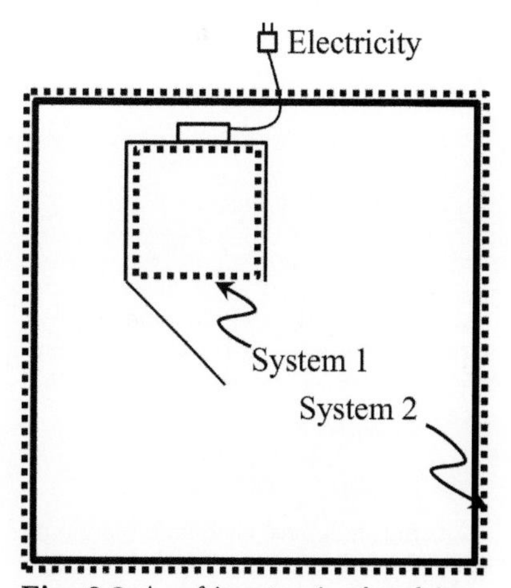

Fig. 3.2 A refrigerator is placed in a room and the door opened, will this cool the room? There are two systems that should be considered, the first is the internal cavity of the refrigerator and the second the entire room. Analysis is given in Example 3.1.

This is the work term and it is positive in value since work is being supplied to the room by electric current flow to the refrigerator. Assume no air leaks out of the room (this is a closed system since it is isolated and the process is adiabatic) and that the room is 3 m wide by 7 m long by 2 m high, making a total air volume of 42 m^3. The density of air is approximately 1.2 kg/m^3 and so there is a mass m of about 50 kg of air in the room (did you know there is this much air in a room?). Since the increase in internal energy equals the work supplied, 5000 kJ, one can find the temperature rise in one day ΔT. This is done by noting the change in internal energy ΔU equals the system mass m times the constant volume heat capacity C_v and the temperature rise or[2]

$$\Delta U = mC_v\ \Delta T \tag{3.2}$$

The inherent assumption in arriving at eqn (3.2) is that C_v is constant over the temperature range of interest which is typically a good assumption. Now one can write the FLOT as

$$\Delta U \equiv mC_v\Delta T, \Rightarrow \Delta T = \frac{W}{mC_v} \approx \frac{5000\,\text{kJ}}{50\,\text{kg} \times 1.0\,\text{kJ/kg-K}} = 100\text{ K!}$$

noting $C_v \approx 1$ kJ/kg-K for air. Imagine if this were done in the 1970s when refrigerators required over 2000 kW-h per year to operate! There are several assumptions made in the above example. The energy required to operate the refrigerator would be much greater if the door were left open since it would continuously operate and so the temperature rise would be much larger. In addition, the room is not adiabatic nor closed and heat and air would leak out of it to temper this temperature rise. This does demonstrate how much energy is supplied to a house by appliances though.

[2] Heat capacity is the ability of a system to store energy, as eqn (3.2) implies. ΔU is the energy stored for a change in temperature ΔT; if C_v is high then more energy is stored for a given ΔT. Generally, the more disorder a system has the more *states* the molecules can assume to store more energy. Imagine atoms having the ability to rotate and vibrate to store energy when they are less densely packed in a liquid compared to their crystalline counterpart. This is why liquid water has a higher heat capacity than solid water or ice. Steam has a low heat capacity since the density of atoms is quite low compared to either water or ice regardless of all the states the atoms can have. An interesting manuscript on heat capacity and disorder (really density) is, A. Chumakov *et al.*, 'Role of disorder in the thermodynamics and atomic densities of glasses,' Phys. Rev. Lett. **112** (2014) 025502-1-6.

The FLOT is useful and allows one to know what is possible from a thermodynamic view, whereas the *Second Law of Thermodynamics* or SLOT demonstrates what is not possible and is written as

$$\Delta S \geq 0\ [=]\ \frac{\text{J}}{\text{K}} \qquad \text{(SLOT)} \tag{3.3}$$

with S representing the entropy which has unusual units that become clear below when it is defined. Note $\Delta S \geq 0$ applies to the system under consideration as well as the surroundings and the entropy change is that for the entire universe when a process occurs. If the process is reversible then one can write $\Delta S = 0$, while irreversibility implies an entropy increase.

Entropy is usually associated with an increase in disorder from a statistical mechanics view popularized by the great scientist Boltzmann. Although this is true, one should view entropy as associated with energy distribution.[3] Thus, a process will occur if its entropy or the system's ability to distribute energy increases.[4] Indeed the differential form

[3] An untidy study area does not necessarily have greater entropy than a neat one, unless its energy distribution is increased.

[4] Nothing inflames academic passions more than discussing thermodynamics and entropy. We are aware the above definition will certainly draw criticism. This will be taken as its definition here with the *caveat* that when we begin to understand thermodynamics someone points out a counter-example that thoroughly undoes our resolve.

of entropy is written $dS = \delta Q/T$, which has heat or energy in its very definition. The symbol δ means a path dependent derivative, this is not discussed further here and reference to any thermodynamics, physical chemistry or physics textbook will give an adequate discussion of entropy and path dependent derivatives.

Consider the entropy loss for the high temperature reservoir in Fig. 3.1, that is given by $\Delta S_H = -|Q_H|/T_H$ assuming it is a reversible, isothermal process. Reversible means, in this case, that if the same amount of heat were added to the high temperature reservoir then it would revert to be in exactly the same state as before the heat was removed. We write the signs for heat and work explicitly here and ΔS_H is negative since Q_H is negative (the absolute value signs around Q_H are added to explicitly show the sign for entropy; the same convention will be used for W, however, the absolute value signs will be eliminated from here on). Similarly, let heat be reversibly added to the low temperature reservoir maintained at T_L, so, its entropy change would be $\Delta S_L = Q_L/T_L$.

Assuming the heat engine operates reversibly one can write the First and Second Laws of Thermodynamics as[5]

$$\text{FLOT: } \Delta U = Q_H - Q_L - W, \tag{3.4a}$$

$$\text{SLOT: } \Delta S = Q_H/T_H - Q_L/T_L \tag{3.4b}$$

[5]The signs for Q and W in the First and Second Laws of Thermodynamics will be written before the magnitude. So, $-|Q_H|$ is equivalent to $-Q_H$.

The assumption, in this ideal situation, is that the reversible heat engine remains unchanged in this process so ΔU and ΔS are both zero. In this case one finds, $W = Q_H - Q_L$ and $Q_L/T_L = Q_H/T_H$ and ultimately that

$$\eta \equiv \frac{W}{Q_H} = 1 - \frac{T_L}{T_H} \tag{3.5}$$

where η is the heat engine's efficiency. Efficiency is written here as the amount of work one obtains for the amount of heat (energy) one puts in. Temperature is measured in the absolute scale and it is clear that any process will not operate at 100% efficiency except under the most unusual circumstance.

It is frequently asked, "Why does one need the low temperature reservoir, won't the efficiency increase?" Assuming this, one finds the FLOT appears fine: $W = Q_H$ which is larger than above. However, the SLOT reduces to: $Q_H = 0$ for a reversible process making the work zero. The process simply does not do anything.

Now consider a photovoltaic device as shown in Fig. 3.1. Assume the Sun provides heat (energy crossing the system boundary of the photovoltaic device) at 5359 K (a reasonable black body radiation temperature and that which represents the AM1.5G spectrum for terrestrial radiation, see Section 2.4) while the photovoltaic device exchanges heat to the surroundings at the ambient temperature, 300 K. Its efficiency will be $1 - 300\,\text{K}/5359\,\text{K} = 94.4\%$. This is the absolute maximum efficiency at which a photovoltaic device can operate. This is certainly impressive, however, we will discover later the efficiency is much less than this due to a variety of non-idealities that occur, including the photovoltaic device increasing in temperature which reduces its efficiency.

Example 3.2

What is the temperature rise of a photovoltaic device during the first minute of operation?

First the photovoltaic device is assumed to operate adiabatically so no heat escapes its boundary making $Q_L = 0$ in Fig. 3.1. We are not assuming the cell works reversibly and the Second Law reduces to $\Delta S > 0$ which is not very useful. The FLOT can be written

$$\Delta U = Q_H - W$$

where Q_H is the energy crossing the system boundaries (considered as heat above) and W is the work obtained, which is the electricity generated. We will estimate Q_H and W to find the change in internal energy, then from the definition of the heat capacity, determine the temperature rise ΔT ($\Delta T = \Delta U/[m \times C_v]$). The mass of the system is assumed to be m.

To do this example we need some information. Assume the photovoltaic module or device operates at 15% efficiency and that the Sun supplies 1000 W/m^2 of radiation (this will be related to Q_H). A typical module is 1.6 m by 0.8 m in size, so, the power input is 1000 W/m^2 × 1.6 m × 0.8 m which equals 1280 W. Over one minute this is approximately 7.7×10^4 J and is equal to Q_H.

The work one obtains in a minute is $0.15 \times 7.7 \times 10^4$ J = 1.2×10^4 J and is in the form of electricity as mentioned above. This yields

$$\Delta U = 7.7 \times 10^4 \text{ J} - 1.2 \times 10^4 \text{ J} = 6.5 \times 10^4 \text{ J}$$

The module typically weighs approximately 15 kg and the heat capacity of Silicon is approximately 700 J/kg-K, so, the temperature rise is

$$\Delta T = \frac{\Delta U}{mC_v} = \frac{6.5 \times 10^4 \text{ J}}{15 \text{ kg} \times 700 \text{ J/kg-K}} \approx 6 \text{ K}$$

The solar panel will increase in temperature about 6 K in only one minute! The efficiency of a solar panel typically decreases by approximately 0.5%/K making a considerable decrease in efficiency in a very short time.

Of course, the module will lose heat equivalent to Q_L in the heat engine either by thermal radiation or by simple heat transfer to the surroundings, so, the above estimate provides the upper limit in temperature increase. Photovoltaic device installers are now placing water-filled pipes on the back of the panels to absorb some of the heat, which makes sense to increase the module efficiency and to produce hot water for domestic use. How much hot water one can obtain can be estimated with principles discussed below and in Example 3.3.

3.2 Thermodynamics–Open systems

Many solar powered devices flow a heat transfer fluid through concentrated sunlight as the heat or energy source for a power plant or even a hot water heater. These can be continuous flow systems and an analysis technique is required for their design. We will consider only steady-state operation of devices here, which significantly simplifies their consideration.[6] Yet, this is a reasonable approximation since solar devices are designed to work during the day and we are not interested in the start-up of their operation or control principles.

[6] The unsteady-state FLOT is used in Section 8.4.

Reference to Fig. 3.3 shows an open system where a mass flow rate of $\dot{m}_1$ enters and $\dot{m}_2$ exits. Also, the mass enters through a conduit (pipe) of area A_1, at height h_1, Temperature T_1, pressure P_1, specific volume $\hat{V}_1$, specific enthalpy $\hat{H}_1$ (discussed below) and specific entropy $\hat{S}_1$, similar variables are given for the exit conditions. By specific we mean *per unit mass*, or per kg, making the dimensions of specific volume m^3/kg which is the inverse of the mass density. During the process, energy or heat is added at a rate $\dot{Q}$ (it can be extracted too) and work is extracted at a rate $\dot{W}$ (it can also be added to the system) and both have dimensions of J/s or W.

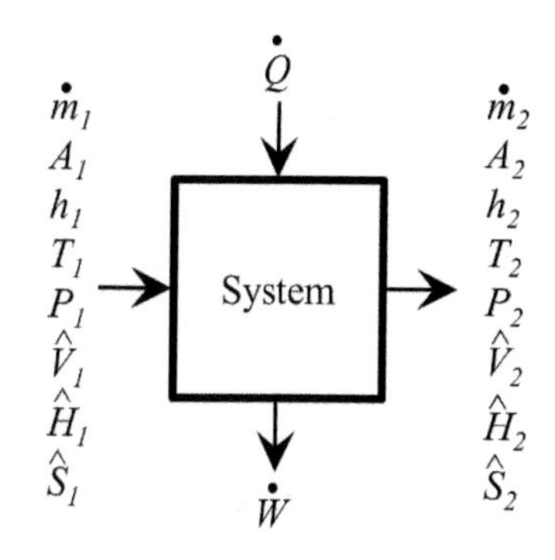

Fig. 3.3 An open system where material enters under condition 1 and exists under 2, during the process heat is added and work extracted.

The first consideration is the mass balance. For the system in Fig. 3.3 it is quite simple since at steady state whatever goes in, must come out (ignoring any infinitesimal mass loss or gain through chemical reactions)

$$\dot{m}_1 = \dot{m}_2 \equiv \dot{m} \; [=] \; \frac{\text{kg}}{s} \tag{3.6}$$

Multiple inlets and outlets are possible and the equation can be suitably modified by summing over all the inlets on one side and outlets on the other.

The FLOT is a little more complicated. With a flow system the material can enter and exit at any given velocity and so the kinetic energy (KE) must be considered since it can change significantly. Imagine water flowing through a nozzle at the end of a hose, the velocity in the hose prior to entering the nozzle is much lower than after exiting it. This would represent a change in KE for a flowing system. The potential energy (PE) can also change when the mass enters and exits at different heights. Potential energy is particularly important when pumping water to the top of a hill, for example. The KE and PE are fundamental energies too, just like the internal energy, and these three are added together to make up the *total energy* of the material entering or exiting the system.

One usually writes the KE and PE as $\frac{1}{2}mv^2$ and mgh, respectively, where v is the velocity of the system with mass m and g is the gravitational acceleration. The system could be a ball that is thrown upwards, for example. Since we are concerned with a flow system these are written as $\frac{1}{2}\dot{m}v^2$ and $\dot{m}gh$ making the total rate of energy $\dot{E}$ equal to

$$\dot{E} = \dot{m}\hat{U} + \frac{1}{2}\dot{m}v^2 + \dot{m}gh = \dot{m}\left[\hat{U} + \frac{1}{2}v^2 + gh\right] \;[=]\; \mathrm{W} \tag{3.7}$$

which will be applied to the entrance and exit of the system. In this case the system would have mass flowing into and out of it at a mass flow rate $\dot{m}$ and velocity v at height h. The velocity is given by

$$v = \frac{\dot{m}\hat{V}}{A} \tag{3.8}$$

requiring knowledge of the material's specific volume $\hat{V}$.

The mass may experience a pressure change and this could compress it, for example. This is a work term and we denote it as pressure-volume work or $\dot{W}_{PV}$ and it can be written as $P(dV/dt)$ where (dV/dt) is the time t rate of volume change. By our definition of work, doing work on a system is positive, so we must write

$$\dot{W}_{PV} = -P\left(\frac{dV}{dt}\right) \;[=]\; \frac{\mathrm{Pa\text{-}m^3}}{\mathrm{s}} \equiv \mathrm{W} \tag{3.9}$$

since compression will decrease the volume and this should be considered positive work. Although compression is not important for solids and liquids, gases are quite compressible and $\dot{W}_{PV}$ can be a significant contribution. So, this term can be written as, using the variables in Fig. 3.3,

$$\dot{W}_{PV} = -[\dot{m}_2 P_2 \hat{V}_2 - \dot{m}_1 P_1 \hat{V}_1] \;[=]\; \frac{\mathrm{kg}}{\mathrm{s}} \times \mathrm{Pa} \times \frac{\mathrm{m^3}}{\mathrm{kg}} \equiv \mathrm{W} \tag{3.10}$$

It is hoped the reader accepts the imprecision of this derivation and accepts the above equation for $\dot{W}_{PV}$, as well as subsequent equations. Rather than performing a true derivation, detail has been removed for conciseness. Any undergraduate thermodynamics textbook will give a more acceptable derivation.

The FLOT can now be written as

$$\begin{aligned}\dot{m} \times \left[\hat{U}_2 + \frac{1}{2}v_2^2 + gh_2\right] - \dot{m} \times \left[\hat{U}_1 + \frac{1}{2}v_1^2 + gh_1\right] = \\ \dot{Q} + \dot{W}_m - [\dot{m}_2 P_2 \hat{V}_2 - \dot{m}_1 P_1 \hat{V}_1]\end{aligned} \tag{3.11}$$

The work term has been written as the sum of the pressure-volume work and any other work which is *mechanical* in nature $\dot{W}_m$. This is work such as that done by a pump or a stirrer or electricity generated by a photovoltaic device. A pump or stirrer would do work on the system and are positive in value while electricity from a photovoltaic device is negative work.

Rather than having to calculate $\dot{m}P\hat{V}$ each time we do a calculation we can define a new quantity, specific enthalpy, denoted as $\hat{H}$, that equals $\hat{U} + P\hat{V}$, allowing us to write the above equation as

$$\dot{m} \times \left[\hat{H}_2 + \frac{1}{2}v_2^2 + gh_2\right] - \dot{m} \times \left[\hat{H}_1 + \frac{1}{2}v_1^2 + gh_1\right] = \dot{Q} + \dot{W}_m \qquad (3.12)$$

Introducing enthalpy merely makes the calculations less cumbersome and is tabulated along with the internal energy. This is the working equation and one that will be used to analyze and design the operation of solar powered devices. The relation can be called the *Energy Balance.*

The SLOT for an open system can be derived in a similar manner, however, we merely write the final form here

$$\dot{m}\hat{S}_2 - \dot{m}\hat{S}_1 \geq \frac{\dot{Q}}{T} \qquad (3.13)$$

As mentioned above, the SLOT indicates what is not possible and is not as *useful* as the FLOT, however, it does have its utility. For example if the process is considered reversible and adiabatic then we would obtain $\hat{S}_2 = \hat{S}_1$ that would constrain the exit state. The example below shows the utility of both the FLOT and SLOT for open systems.

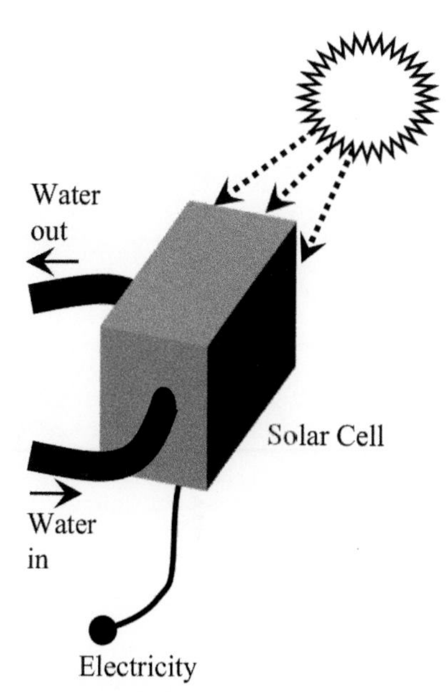

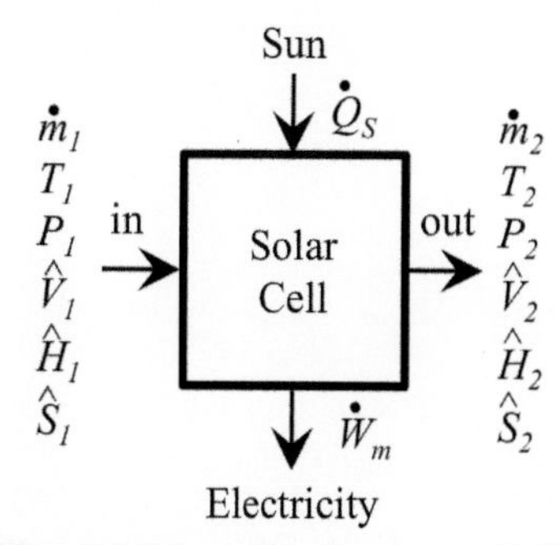

Fig. 3.4 Schematic of a solar cell that heats water and the open system diagram used to describe it.

Example 3.3

In Example 3.2 we calculated the temperature rise of a photovoltaic device in one minute. Now consider the same photovoltaic device and let water flow underneath it to capture the waste heat and determine if it can significantly increase its temperature if it enters the system at 6×10^{-5} *m*^{3}*/s (approximately 1 gallon per minute) as shown in Fig. 3.4*

Since the specific volume of water is approximately 10^{-3} m^3/kg one finds the mass flow rate equal to 6×10^{-2} kg/s. Now the energy balance or FLOT can be simplified to

$$\dot{m}\left[\hat{H}_2 - \hat{H}_1\right] = \dot{Q}_S - \dot{W}_m$$

where the subscripts 1 and 2 represent the inlet and outlet, respectively. We have neglected the KE and PE and have assumed the Sun supplies heat at a rate $\dot{Q}_S$ and electricity is extracted at a rate $\dot{W}_m$. In addition, it is assumed that all the waste heat is directed to heat the water.

In the previous example the photovoltaic device was assumed to be 1.28 m^2 in area, and assuming a standard AM1.5G spectrum (see Section 2.4) which supplies 1000 W/m^2 of power, one has $\dot{Q}_S$ = 1280 W. The photovoltaic device operates at 15% efficiency so $\dot{W}_m$ = 192 W. The water's change in specific enthalpy is approximated by $\hat{H}_2 - \hat{H}_1 = C_p\Delta T$, where C_p is the constant pressure heat capacity for water (4.187 kJ/kg-K), allowing the energy balance to be written as

$$\Delta T = \frac{\dot{Q}_S - \dot{W}_m}{\dot{m}C_p} = \frac{1280 - 192, \text{W}}{6 \times 10^{-2} \text{ kg/s} \times 4187 \text{ J/kg-K}} = 4.3 \text{ K}$$

One can conclude from this equation that as more work is performed (more electricity extracted) there is less heat available to heat the water!

The temperature rise is not that large and one typically desires hot water at 50 °C, so, if it enters the system at 20 °C then it would only be 24.4 °C at the exit. To use this strategy to produce hot water on demand the area would have to be increased by a factor of 7 which is not out of the question. One would have many solar panels on a roof and this could surely work. However, to obtain hot water during the night and at times when the insolation is not that large, one would have to heat the water during periods of high insolation and store it in a well-insulated vessel. This is what is done in a solar hot water heater, the water flows through the device gaining energy from the Sun and circulates water from a well insulated reservoir whose contents gradually increase in temperature. So, this type of solar hot water heater would work with the *caveat* that if the photovoltaic device becomes more efficient to produce electricity then there would be less heat available to generate hot water!

Finally, we did not use the SLOT as it does not add much to the discussion. The way we designed the system above had already constrained the variables enough that we did not need another equation. Furthermore, the heat transfer most likely will not occur reversibly and isothermally and the SLOT is written as

$$\dot{m}\hat{S}_2 - \dot{m}\hat{S}_1 \geq \frac{\dot{Q}_S}{T} \tag{3.13}$$

which is not particularly helpful.

3.3 Light absorption

Harvesting solar energy is a complicated process and dictated by two of the greatest discoveries of quantum physics from the beginning of the 20th century: light consists of small packets of energy called photons and the behavior of electrons is described by energy levels and wave functions. These two phenomena are intimately coupled when a material is exposed to solar radiation since the photon energy must match the energy levels available to the electrons for any effect to occur.

Electrons are the part of a material that first interacts with light or radiation in general. After all what else is there in a material? It is made of atoms that may be bonded with other atoms and the atomic interactions are dictated by the electron cloud. Deep inside the atom is the nucleus that is shielded by the electron cloud. The neutrons and protons are in a highly constrained state and are bonded quite tightly together; this is fortunate or we could have atomic fission which is certainly undesirable. So, the electron controls what will eventually occur: it may become an excited free electron to enable a photovoltaic device or it may pass the energy it absorbs to the crystal lattice[7] subsequently falling back to its original energy level. In this case the material heats

[7]The term crystal lattice is used here to mean the atoms that are bonded together and is a general term to denote a solid material.

up since the lattice vibrations are increased by absorbing this energy and this is a process particularly useful for solar thermal applications.

The energy of a photon E is related to its frequency ν (with units cycles/s) through $E = h\nu$, where h is Planck's constant. Most of the solar energy harvested by materials and devices is through processes that occur one photon at a time. That is, we consider light to be made of a very large number of photons that act independently and whose energy is harvested through an electronic excitation between two energy levels, E_i and E_f, within the material

$$E = h\nu = E_f - E_i \tag{3.14}$$

where E_i is the electron's energy in its initial state and E_f is the energy of the final state. To understand how solar energy is harvested from the Sun, one should consider the likelihood that photons are absorbed. In other words, solar energy with frequency ν must be absorbed by the material to raise an electron to a higher energy level[8] of E_f; what materials can do this?

[8]Careful with the terminology *higher energy level.* Electron energies are taken as negative, meaning that an electron at a lower energy level has a more negative value and is most likely physically closer to the atomic nucleus. A higher energy level is less negative and the electron is most likely farther from the nucleus and so encounters less nuclear pull. Of course, a free electron, whose motion is not guided by any nucleus, has zero (potential) energy.

The answer requires an understanding of energy levels, energy bands and the probability that a photon causes an electronic transition between two energy levels. The reader has probably encountered energy levels associated with a single atom consisting of a proton and electron such as in a Hydrogen atom. This system can be easily described through quantum mechanics to yield wave functions, that allow only discrete energy levels E_n to give the *Bohr model* of an atom

$$E_n = -\frac{m_e q^4}{8\epsilon_0^2 h^2 n^2} = -\frac{2.179 \times 10^{-18}\,\mathrm{J}}{n^2} = -\frac{13.60\,\mathrm{eV}}{n^2} \tag{3.15}$$

where m_e is the mass of an electron, q, the charge of an electron, ϵ_0, the permittivity of free space and n the principal quantum number describing the energy level of the orbiting electron ($n = 1, 2, 3, \ldots$). The characteristic energy is quite small when expressed in Joules, so, a new energy unit is frequently used, the *electron-volt* or eV. It is equal to the amount of kinetic energy an electron gains when accelerated through a potential of one volt. In terms of a conversion factor, 1 eV $= 1.602 \times 10^{-19}$ J. In our convention the electron's energy is taken as negative when attracted to a nucleus accounting for the sign in the equation. This fits in with the discussion about higher energy above, if n is larger then the electron has a less negative value and so is raised to a higher energy.

Equation 3.15 correctly predicts the absorption spectrum for the Hydrogen atom. There are only certain energy levels available to an atom of Hydrogen and so only photons of a given energy, or frequency, ν_{ab}, can cause an electronic transition from an energy state with quantum number, n_i, to n_f after absorption. This can be written

$$\tilde{\nu}_{ab} = \frac{\nu_{ab}}{c} = \frac{\mu q^4}{8\epsilon_0^2 h^3 c} \times \left[\frac{1}{n_i^2} - \frac{1}{n_f^2}\right] \tag{3.16}$$

by using eqns (3.14) and (3.15). Here we have recognized frequency is given by c/λ and rearranged the equation to yield the wave number that is absorbed, $\tilde{\nu}_{ab}$, which has units of inverse length, a convention frequently used by chemists. The group of constants in front of the square brackets in eqn (3.16) is called the Rydberg constant, $R_\infty = q^4/8\epsilon_0^2 h^3 c$, and is numerically equal to 109,737 cm^{-1}. The symbol μ represents the reduced mass of the electron which is equal to $1.05\,m_e$ for a Hydrogen atom and is essentially a correction factor to this simple theory. A transition from $n_i = 1$ to any other principal quantum number, $n_f = 2, 3, 4, \ldots$, will yield certain absorption transitions and is called the *Lyman* transitions while transitions from other initial quantum numbers are named after those who discovered them, as shown in Fig. 3.5.[9] If atoms operated independently then absorption of light would only occur at very distinct frequencies or energy levels.

[9] Actually the spectra were determined by emission as the excited electron fell to a smaller principal quantum number.

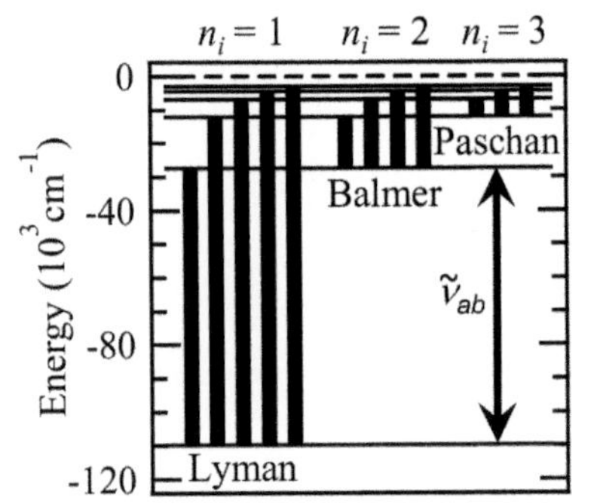

Fig. 3.5 Absorption of radiation by a Hydrogen atom occurs for discrete energy levels depending on the initial and final principal quantum numbers. Various spectra are labeled according to the scientist who discovered them. The arrow labeled $\tilde{\nu}_{ab}$ is calculated using eqn (3.16) with $n_i = 1$ and $n_f = 2$ and represents the energy (wave number) of the radiation necessary to excite an electron from this lowest energy state to the first unoccupied state.

When atoms are forced together to make molecules and/or the atoms and molecules are condensed and/or reacted to make liquids or solids the atoms can no longer act individually from an electronic point of view. The *Pauli exclusion principle* comes into play and no two electrons within this system of atoms can have the same set of quantum numbers. Of course two electrons can have almost all the same quantum numbers with spin being the one that makes their overall state different. However, there are only two spin states, up or down, which means they spin clockwise or counter-clockwise. So when many atoms are put in close proximity, the electrons change their energy levels very slightly making the transition from a lower to higher energy level blurred, and the unique transitions shown in Fig. 3.5 are not present. This is discussed in greater detail in the next chapter.

One now considers a transition from the Highest Occupied Molecular Orbital or HOMO to the Lowest Unoccupied Molecular Orbital or LUMO (as a chemist may state) or from the Valence Band to the Conduction Band (as a physicist may state). The energy that must be absorbed for this transition is denoted as the band gap energy, E_g, as shown in Fig. 3.6. The band gap energy is completely analogous to $\tilde{\nu}_{ab}$ above.

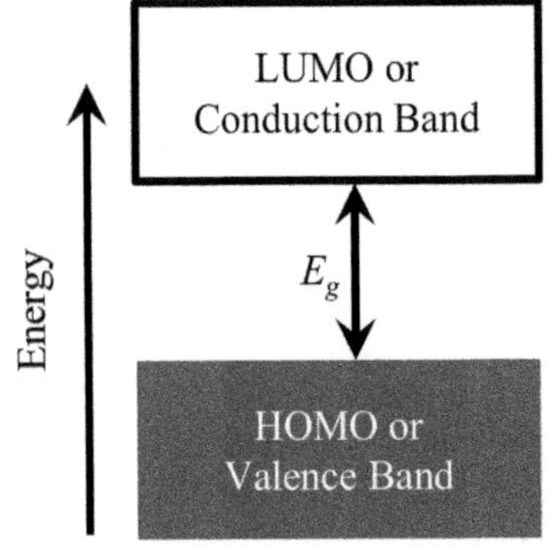

Fig. 3.6 In a molecule, liquid or solid the lower energy levels around a single quantum level spread to account for the Pauli exclusion principle and the electrons form an energy band called the Highest Occupied Molecular Orbital (HOMO) or Valence Band. The lowest energy states available to any excited electron similarly spread and are called the Lowest Unoccupied Molecular Orbital (LUMO) or Conduction Band.

Consider Silicon, in the third row of the periodic table, which absorbs solar radiation at wavelengths shorter than approximately 1130 nm corresponding to a band gap energy of about 1.1 eV. This statement should be clarified. The band gap energy means that photons with energy less than its value cannot be absorbed since there are no states available to the electrons. This is what quantum mechanics tells us, as does Fig. 3.5. Photons above this energy of 1.1 eV will be absorbed to various degrees dependent on details of the atomic arrangement.

Example 3.4

Using the information in this section estimate the energy levels that Group IV elements will absorb.

The answer to this is very involved and complicated quantum mechanical calculations would have to be carried out since electron orbitals hybridize (change) as the closely packed atoms are forced near to each other and covalent bonds form. For now let's persist and upon rearranging eqns (3.15) and (3.16) into energy units we find

$$E_g \approx 13.60\,\text{eV} \times \left[\frac{1}{n_i^2} - \frac{1}{n_f^2}\right] \tag{3.17}$$

The question now is, what values do we use for n_i and n_f? Some readers familiar with these sort of calculations may realize that we are applying eqn (3.17) to elements that have an atomic number greater than 1, this will be addressed by merely assuming that the excess nuclear charge is shielded by lower lying electrons. We are interested only in the electron which is most easily excited.

We can guess that the tightly bound electrons near the nucleus, that have smaller principal quantum numbers, will be difficult to excite to a higher energy level and expect that the more loosely bound electrons in the outer shells of the electron cloud are the most likely to be excited by radiation absorption. So, we will grade each element by the principal quantum number of its outermost electrons and assume they are excited to a level with a principal quantum number one higher in magnitude. Again this is a very rough calculation, yet, it is important to recognize the connection between what one may have learned in elementary chemistry or physics to such a sophisticated calculation.

Consider the element Carbon first. It is in the second row of the periodic table and the principal quantum number for its outermost electrons is 2. Assume that the initial and final quantum numbers for the electron after absorbing a photon are 2 and 3, respectively, then we obtain $E_g \approx 2$ eV. The actual band gap energy for Carbon, in the form of diamond, is 5.5 eV which is much higher than our calculated result. In spite of this difference the order of magnitude is correct and we press on with the discussion.

Silicon, which is a row three element, with $n_i = 3$ for the outermost electrons, and right below Carbon, one would find a much lower band gap of about 0.7 eV. The actual value is 1.1 eV, as mentioned above, which is fairly close to the calculated value. Furthermore, it is lower than that of diamond, supporting the concept eqn (3.17) presents, an element with a larger principal quantum number for its outermost shell will (most likely) have a lower band gap energy. Right below Silicon is Germanium which has a band gap energy of 0.67 eV and with $n_i = 4$ we calculate a band gap energy of 0.3 eV. Finally, Tin and Lead don't have a band gap energy since they are metals and from the equation we find 0.2 eV and 0.1 eV, respectively.

Actually eqn (3.17) is not correct, which was alluded to above, and is strictly for a Hydrogen atom which has an atomic number Z of 1. Since all the above elements have a larger number of protons than one there will be more pull of the nucleus on the electrons, however, it is not as large as one would expect. The equation should be written

$$E_g \approx 13.60\,\text{eV} \times Z_{eff}^2 \left[\frac{1}{n_i^2} - \frac{1}{n_f^2} \right] \tag{3.18}$$

where Z_{eff} is the effective charge number an electron would feel. Electrons in the first orbital will have a Z_{eff} close to the true atomic number. Progressing to the outermost orbitals, the charge is shielded by the inner shell electrons and the effective charge number is reduced. By using eqn (3.17) it was implicitly assumed that the inner electrons shielded all the nuclear charge except a charge of 1. The effective charge number and band gap energy calculated with eqn (3.18) are given in Table 3.1.

This doesn't work well demonstrating that more detailed calculations are warranted, indeed it's amazing that eqn (3.17) seems to work better. This may be due to the way the effective charge is calculated, the ionization energy is measured experimentally and E_g in eqn (3.18) is taken as this energy and n_f is set to infinity. Then Z_{eff} can be determined which will clearly overestimate E_g.

Table 3.1 Effective charge and band gap energy calculations for the Group IV elements. The atomic number and true band gap energy are given parenthetically after the element symbol with the next two columns showing the effective atomic number for the outermost electron and the band gap energy calculated with eqn (3.17) (or in the parentheses is a calculation with eqn (3.18)), respectively. The band gap energy is given in eV.

Element (Z, E_g)	Z_{eff}	E_g-calc
C (6, 5.5)	1.82	1.9 (6.3)
Si (14, 1.11)	2.32	0.7 (3.6)
Ge (32, 0.64)	2.09	0.3 (1.3)
Sn (50, 0)	3.67	0.2 (2.2)
Pb (82, 0)	4.43	0.1 (2.0)

Regardless, if we accept that eqns (3.17) and (3.18) give a reasonable estimate, and view Fig. 2.3 to look at which band gap energy would absorb the most radiation, it would appear that we would want to use an element with the lowest possible band gap energy, like Germanium or even Lead, rather than Silicon. Remember, radiation can only be absorbed for energies greater than the band gap energy and all the energy below E_g will not be absorbed. As will become clear in the next section this will not create a very good photovoltaic device and it turns out that Silicon of all the group IV elements is the most suitable.

Before considering photovoltaics we briefly mention the absorption efficiency which is an important measure of material suitability. The photon is not always absorbed by an atom and the internal quantum efficiency of a material $IQE(\nu)$ is the efficiency with which light is absorbed at frequency ν and is a number between zero and one. A perfect absorber, or black body, absorbs light with $IQE(\nu) = 1$ for light of all frequencies ν. These are rare and most materials have an internal quantum efficiency less than one although the better ones, for use in extracting solar energy, approach 1 at least in the frequency range of most use.

3.4 Photovoltaics

Photovoltaic devices or solar cells work by absorbing light to produce electrons at a higher energy (remember this means a less negative number as discussed in note 8). The electrons are able to drive an external circuit at a voltage which is unfortunately almost always significantly less than the energy difference E_g, the band gap energy, due to non-idealities. This appears to be a nonsensical statement, how can one compare a voltage (units of volts) to a band gap energy (units of Joules or volt-amps)?

In a photovoltaic device the excited electron of charge q is pulled to one electrode (anode) which operates at a potential V to the counter-electrode (cathode). This yields an energy of qV. However, a photovoltaic device can only provide, at most, an energy of E_g since this is the photon energy value that the material can absorb as discussed above. So, one finds $V = E_g/q$ making the maximum obtainable voltage *numerically* equal to the band gap energy in the units eV. Remember an electron-volt is the kinetic energy gained by an electron when accelerated through a potential of 1 V. This is why statements comparing the maximum voltage in a photovoltaic device and the band gap energy in units of eV are made.

To understand how a photovoltaic device works we must take the discussion from Section 3.3 and consider Fig. 3.6. An electron in the valence band (or HOMO) absorbs radiation and is promoted to the conduction band (or LUMO) leaving behind a hole. A *hole* is not a physical particle like an electron, rather it is where an electron is not and has opposite charge. It can move through the crystal lattice just like a physical particle though. In a photovoltaic device the electron goes to one electrode while the hole moves to the other.

One could imagine a material absorbing an energy greater than E_g, why isn't the maximum voltage greater? For example, an electron in the middle of the valence band could be excited to an energy level well within the conduction band by absorbing a photon with an energy greater than E_g, as shown in Fig. 3.7. However, the electron and hole both relax to their respective band edges by giving off heat, or they *thermalize*, so the maximum energy difference becomes equal to the band gap. This is because electrons quickly move to the lowest (more negative) energy value, if possible, while holes do the opposite.[10]

[10] Electrons always go to the lowest (most negative) energy level as they are attracted to it while holes go to the highest (least negative) energy level. Think of electrons as marbles rolling downhill and holes as balloons rising upwards.

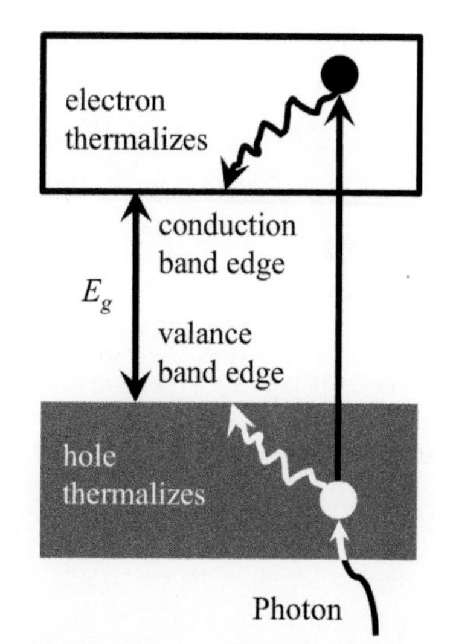

Fig. 3.7 When a photon with greater energy than the band gap energy is absorbed it can produce an electron with an energy higher than that at the conduction band edge and a hole with an energy lower than that at the valence band edge. However, both particles thermalize by giving off heat and relax to the band edge limiting the amount of extractable energy to the band gap.

This effect limits the maximum power one can obtain from a photovoltaic device as discussed below.

The power density or irradiance, P, impinging on the Earth, has already been determined in eqn (2.13)

$$P = \int_0^\infty E \times N_P(E)dE \equiv \int_0^\infty I(E)dE \tag{3.19}$$

which resulted in the Stefan-Boltzmann law for extraterrestrial radiation, eqn (2.4). One finds an irradiance of 1366 $\mathrm{W/m^2}$ just outside the Earth's atmosphere. A standard, AM1.5G, was established to account for reflection, absorption and scattering of photons by the atmosphere, so, the irradiance that would strike a photovoltaic device is less, ≈ 1000 $\mathrm{W/m^2}$, as discussed in the previous chapter. One can use the Stefan-Boltzmann law with a solar temperature less than the actual, which was successfully used in Fig. 2.3, T_s = 5359 K, to represent the AM1.5G spectrum. This is what we do here to find the power density that can go *in* a photovoltaic device, P_{in}

$$P_{in} = \frac{2\pi^4}{15} g_s \frac{k_B^4}{h^3c^2} T_s^4 \tag{3.20}$$

The device efficiency (η) is the power density one obtains from the photovoltaic device, P_{out}, divided by P_{in} or

$$\eta = \frac{P_{out}}{P_{in}} \tag{3.21}$$

Determining P_{out} is reasonably simple and can be found with information given in Chapter 2. Power is *voltage* times *current* and the strategy is to find the current generated in a photovoltaic device since we already know the (maximum possible) voltage, E_g/q.

Assume that every photon is absorbed above an energy of E_g creating an excited electron and the associated hole. So, the integrated photon number rate, see eqn (2.2), N_p, is integrated between E_g and infinity to obtain the number of electrons generated. Remember, only photons with energy greater than E_g will be absorbed and one finds the number of electrons that can go *out* of the device, N_{out}

$$\begin{aligned} N_{out} &= \int_{E_g}^{\infty} N_p(E)\, dE \, [=] \, \frac{\#}{\text{s-m}^2} \\ &= \frac{2g_s}{h^3c^2} [k_B T_S]^3 \left[2 \sum_{n=1}^{\infty} \frac{n x_g + 1}{n^3} e^{-n x_g} - x_g^2 \ln(1 - e^{-x_g}) \right] \end{aligned} \tag{3.22}$$

where x_g is $E_g/k_B T_S$. The current density (J_{out} with units of amp/m^2) is merely N_{out} multiplied by q and so the power out of the photovoltaic device is the current density times the voltage[11]

$$P_{out} = [N_{out} \times q] \times [E_g/q] = N_{out} E_g \tag{3.23}$$

allowing us to write the efficiency as

$$\eta = \frac{15}{\pi^4} x_g \left[2\{x_g + 1\} e^{-x_g} - x_g^2 \ln(1 - e^{-x_g}) \right] \tag{3.24}$$

where we have kept only the first term in the summation in eqn (3.22) (a good assumption). If we consider Silicon with a band gap energy of 1.1 eV one obtains $x_g = 2.4$ and an efficiency of approximately 40%.

This is a rough estimate and several factors have not been taken into account when deriving eqn (3.24), such as recombination of the electron and hole as well as the fact that not all photons will be absorbed. A more accurate calculation of the efficiency was done by Shockley and Queisser in 1961 and takes into account the *detailed balance* of charge generation. They obtained that the maximum efficiency a photovoltaic device could obtain is 31% which occurs at a band gap energy of 1.2–1.4 eV. Yet, after many years of refining processing conditions and architecture of photovoltaic devices, single crystal Silicon photovoltaic devices in laboratory conditions have efficiencies of at most 23%. This highlights the challenges to optimizing device architecture and material properties.

Why is there an optimum band gap energy? We don't need to do a detailed calculation like Shockley and Queisser, rather this comes through inspection of the terms constituting P_{out}; the maximum voltage, V_{out}, and the maximum current density, J_{out}

[11] One can write

$$P_{out} = E_g \times \int_{E_g}^{\infty} N_p(E)\, dE$$

and

$$P_{in} = \int_0^{\infty} E \times N_p(E)\, dE$$

to demonstrate how these two quantities are related. The upper equation will always be less than the lower since the energy below E_g is not harvested plus thermalization occurs reducing energy levels above E_g to E_g.

$$V_{out} = \frac{E_g}{q}\ , \quad J_{out} = q \int_{E_g}^{\infty} N_p(E)\, dE\ , \quad P_{out} = V_{out} \times J_{out} \qquad (3.25)$$

Consider raising E_g to increase V_{out}, however, J_{out} will eventually fall, decreasing P_{out}. Now lowering E_g to raise J_{out} has V_{out} taking a smaller value and again P_{out} will eventually decrease suggesting an optimum which can be predicted via eqn (3.24) as shown in Fig. 3.8.

A metal can have electrons easily excited at optical frequencies, their band gap energy is essentially zero so no power can be produced. Thus, they can make free electrons, however, there is no potential developed and so they do not make a good photovoltaic device. The fact that electrons are so mobile in metals is good though as it makes them appear shiny and they are good conductors.

The curves labeled T_S = 5359 K and T_S = 6000 K in Fig. 3.8 are calculated from eqn (3.24) and an optimum efficiency of over 40% is predicted at a band gap energy of approximately 1 eV. This agrees reasonably well with the *detailed balance* prediction of Shockley and Queisser, given the approximations used. Note the black body temperature used in our prediction was that for the AM1.5G spectrum, 5359 K, while Shockley and Queisser used 6000 K accounting for why their detailed balance curve can become larger than assuming all photons are converted to free electron-hole pairs for band gap energies around 3 eV.

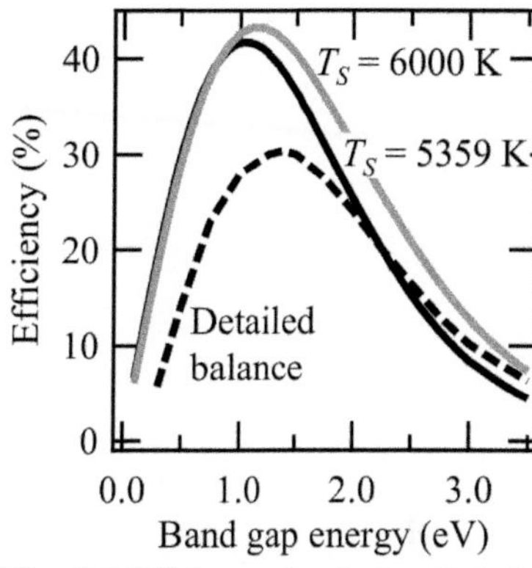

Fig. 3.8 Efficiency of a photovoltaic device as a function of band gap energy. The curves labeled T_S = 5359 K and T_S = 6000 K represents eqn (3.24), with the corresponding black body temperature, and the ultimate limit in efficiency where all photons are absorbed and converted to free electrons and holes. The curve labeled *Detailed balance* is from Shockley and Queisser who took non-idealities into account and is a more accurate calculation. Shockley and Queisser assumed 6000 K for the solar black body temperature in their calculation.

The discussion in Example 3.4 indicated that Germanium may make a better photovoltaic device than Silicon due to its band gap energy being lower: 0.64 eV compared to 1.1 eV. Now with reference to Fig. 3.8 this statement makes no sense. Germanium does not make a large potential despite absorbing more of the solar spectrum. The optimum band gap energy is 1.2–1.4 eV and Silicon has an energy very near this by balancing the maximum possible current that can be generated with development of a potential. A metal, with zero band gap energy, or an insulator, with very large band gap energy, simply do not work very well as photovoltaic devices since they produce either no potential or no current.

The results in Fig. 3.8 show a maximum possible efficiency of 31% when considering all the details in a photovoltaic device. Two strategies that increase the efficiency above the Shockley-Queisser limit are to use concentrators to increase the intensity of incident light, and to use multi-gap materials to capture the incident light more efficiently. A multi-gap material means a material with a large band gap energy is placed closest to the Sun and layered on top of a lower band gap energy material. So, the upper layer absorbs the higher energy photons and the material below it absorbs the lower energy photons. Devices with three gaps under concentrations of 500 times that of the Sun have now achieved efficiencies of 43%. It seems likely that further progress is possible and that a 50% photovoltaic device will be achieved by further developing these approaches. However, this is still much lower than the theoretical limit of having infinite layers on top of each other progressively changing from high to low band gap energy which would yield an efficiency in excess of 90%!

3.5 Solar thermal systems

Solar thermal systems range from (passive) building design of residential dwellings to (active) solar concentrator systems for generating electricity that effectively operate like a coal-fired power plant. Of course, the most cost effective and simplest strategy to reduce energy needs is to use less energy. Most individuals in our society are not amenable to this and solar thermal systems are good replacements for contemporary energy sources due to their relatively low cost and high efficiency.

Solar thermal systems depend on material properties and how they absorb solar radiation. Quantized radiation absorption was discussed above, which is important, especially when considering photovoltaic devices where the absorber must return a potential or voltage. In solar thermal systems the absorbed radiation will be turned into heat by the excited electrons falling back to their original energy level through non-radiative energy transfer to the surrounding atoms.[12]

[12] Non-radiative energy transfer means the electron transfers it energy to the surrounding atoms as it falls to a lower energy level. Light emitting diodes or LEDs emit the energy as light rather than heat and this is an example of radiative energy transfer.

The material properties for this application are very different to photovoltaic devices. They must absorb at all frequencies and then lose their energy to atomic lattice vibrations called phonons. This is a complicated process and an important one, however, here we take a macroscopic view when considering absorption and describe absorption in a single parameter $a(\lambda)$ which may be a function of the radiation wavelength λ.

In Example 3.3 we assumed all the energy from the Sun was absorbed, which is not completely true for real materials. The radiation can be reflected by, absorbed into or transmitted through a material and the sum of all these processes must be equal to the incident radiation. This can be written mathematically as

$$\rho(\lambda) + a(\lambda) + \tau(\lambda) = 1 \tag{3.26}$$

where $\rho(\lambda)$ is the reflectance, $a(\lambda)$, the absorbance and $\tau(\lambda)$, the transmittance or the transmission coefficient (also see eqn (2.16)), and they represent the fraction of the incident radiation that is reflected, absorbed or transmitted through a material. If there is no reflection ($\rho(\lambda) = 0$) then a familiar relation between absorbance and transmittance is found, $a(\lambda) = 1 - \tau(\lambda)$, frequently used in analysis of data from spectrophotometers.

Table 3.2 Absorbance classification.

Black body	$a(\lambda) \equiv 1$ for all λ
Gray body	$a(\lambda) \equiv a < 1$ for all λ
Real body	$a(\lambda) < 1$

All the parameters have been written as a function of wavelength which is a reasonable assumption for real materials and can be graded according to their wavelength dependence as shown in Table 3.2. A black body and gray body are similar in their wavelength dependence, there is none, however, a gray body has an absorbance less than 1. A real body can have a complicated absorbance spectrum and it is not constant at all.

We know that a hot material will emit radiation, as discussed in Chapter 2, to arrive at the Stefan-Boltzmann law in eqn (2.4). However, not all systems are perfect emitters like the Sun and we can write for gray bodies at temperature T_g

$$e = \frac{P_{gb}}{P_{bb}} \equiv \frac{P_{gb}}{\sigma_S T_g^4} < 1 \tag{3.27}$$

where e is the emissivity which is less than one, P_{gb}, the irradiance of the gray body, P_{bb}, the irradiance of a black body at the same temperature as the gray body, σ_S, Stefan's constant and T_g, the gray body temperature. Since we have assumed a gray body e is a constant and not a function of wavelength (or energy) of the emitted radiation. The emissivity is actually easily related to the absorbance, in fact they are equal. To prove this requires a simple derivation.

Consider putting a black body in an enclosed system, which also acts as a black body, as shown in Fig. 3.9, and have them at equal temperature and in equilibrium. Then whatever radiation is emitted by the system P_{sys} is absorbed by the black body and vice versa. So, we can write

$$P_{bb} = P_{sys}$$

Now put a gray body in the same system under similar conditions and we will have

$$P_{gb} = aP_{sys}$$

by the definition of the absorbance since the incident radiation on the gray body is P_{sys}. Taking the ratio of these two equations results in

$$a = \frac{P_{gb}}{P_{bb}} \tag{3.28}$$

which is exactly equal to eqn (3.27) and one obtains

$$e \equiv a \tag{3.29}$$

This is called *Kirchhoff's law* and states that at thermodynamic equilibrium (*i.e.* steady state) whatever is absorbed by a body is emitted at steady state, which makes intrinsic sense. So, the emissivity is equal to the absorbance and with a more thorough derivation one would obtain for a real body $e(\lambda) \equiv a(\lambda)$ at every wavelength as long as the body and source are at thermal equilibrium. One must write the following equation for a real body though

$$L_{rb}(\lambda) = e(\lambda)L_{bb}(\lambda)\ [=]\ \frac{\mathrm{W}}{\mathrm{m}^2\text{-nm}} \tag{3.30}$$

where $L_{bb}(\lambda)$ is obtained from eqn (2.6), and is the spectral irradiance for a black body, and $L_{rb}(\lambda)$ is the spectral irradiance emitted from a real body. One would find the total irradiance emanating from a real body by integrating over all wavelengths, which may be difficult to do since the emissivity could be a complicated function of wavelength.

Now consider an opaque material which does not transmit radiation ($\tau(\lambda) = 0$), this could happen if it has a high absorbance and/or it is

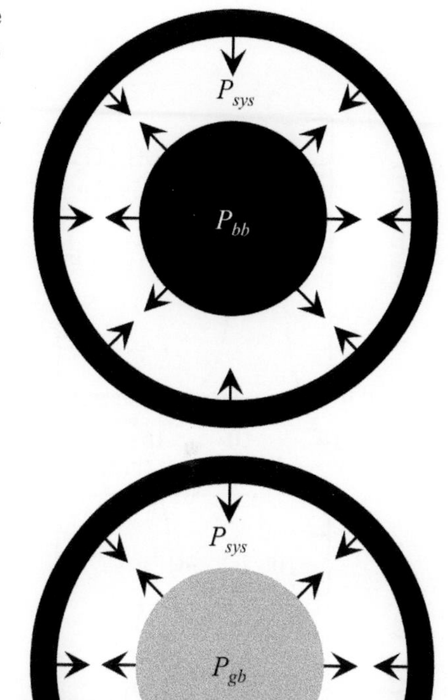

Fig. 3.9 A black body (top) and gray body (bottom) are placed in a system that acts as a black body and allowed to come to thermal equilibrium.

very thick. Equation 3.26 can be written as $1 = \rho(\lambda) + a(\lambda) \equiv \rho(\lambda) + e(\lambda)$ and so

$$e(\lambda) = 1 - \rho(\lambda) \tag{3.31}$$

which is a useful relation since one can measure the amount of radiation reflected from a material allowing one to find its emissivity.

In general, for an absorber in a solar thermal application, one wants a low reflectivity ($\rho(\lambda) \to 0$), a high absorbance ($a(\lambda) \to 1$) and a low emissivity ($e(\lambda) \to 0$), however, the fact that the emissivity equals the absorbance ($e(\lambda) = a(\lambda)$) and $e(\lambda) = 1 - \rho(\lambda)$ complicates device performance and analysis.

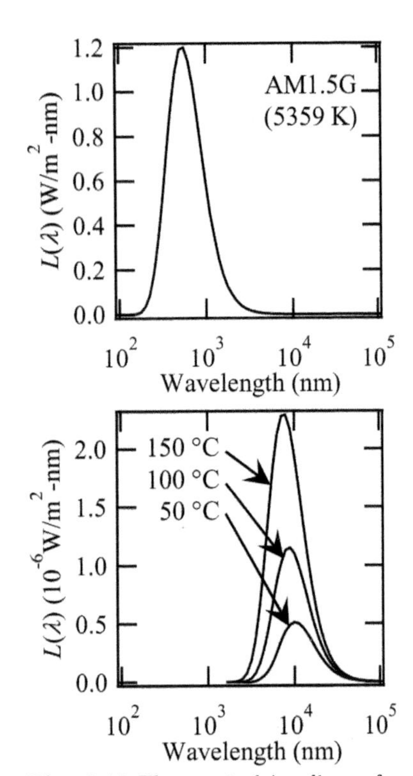

Fig. 3.10 The spectral irradiance for AM1.5G radiation, as representative of the radiation impacting an absorber (top), and for a black body radiating at 50, 100 and 150 °C (bottom). Note the large change in vertical axis magnitude.

This seems an impossible relation, how does one maximize absorption while minimizing emission and reflectance? The answer revolves around two concepts. The first is surface texturing whose size and spacing is related to the wavelength of the radiation to make emitted radiation be absorbed back into the material. This also reduces reflected radiation and is obvious to any who have observed matte compared to gloss finishes.

The second is done by making a *selective surface* to minimize radiative energy loss. This is a surface that absorbs most of the radiation from the Sun, yet, does not re-emit it at that wavelength range. Instead it will emit at much longer wavelengths which follows naturally according to Planck's law or the Wien displacement law since the solar powered device will be at a much lower temperature than the Sun, as shown in Fig. 3.10. This shows the principle of a selective surface.

However, the material design should be more sophisticated than this for optimal performance. The reflectance should be very small for the wavelength range of the solar spectrum, say for wavelengths below 2 μm. Call this the *short wavelength regime*. Then it should have a high reflectance for longer wavelengths, above 2 μm, call this the *long wavelength regime*. Remember, according to eqn (3.31), this means the emissivity will be low. This type of behavior fits in well with the spectral irradiance shown in Fig. 3.10.

Pure materials that fulfill this requirement are typically semiconductors like Silicon and Germanium with band gap energies of 1.1 eV (will absorb for wavelengths below 1200 nm) and 0.67 (will absorb for wavelengths less than 1900 nm). However, they have a large reflectance in the small wavelength regime ($\rho \approx 0.35$) which limits their applicability. Instead coatings are put on their surface, and on the surface of most absorbers for solar thermal applications, to make what is called a *tandem*.

There are many different types of tandems such as black nickel and black chrome which have short wavelength absorbances of 0.95 and long wavelength emissivities of 0.1. This is a great combination of properties. The coating is typically of order 1 μm thick and has high short wavelength absorbance with high long wavelength transmittance. Thus, the tandem promotes solar radiation absorbance while simultaneously allowing the substrate material to use its superior reflectance at long wavelengths thereby reducing its emissivity.

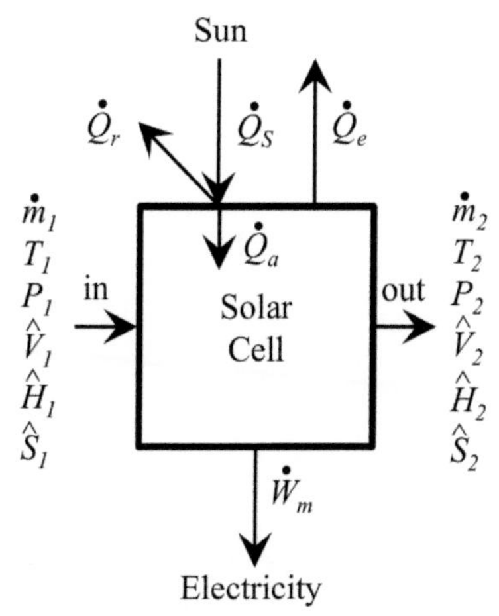

Fig. 3.11 Schematic of a photovoltaic device or solar cell used to heat water as well as make electricity.

Example 3.5

Consider Example 3.3 except in this case let the absorber have a short wavelength reflectance ρ of 0.05 and absorbance a of 0.90 while the long wavelength emissivity e is 0.1. Determine the water temperature, again assuming the photovoltaic device efficiency η is 15%.

In this case, as shown in Fig. 3.11, some of the radiation (heat) from the Sun $\dot{Q}_S$ is reflected at a rate $\dot{Q}_r$ and then absorbed at a rate $\dot{Q}_a$ in the short wavelength regime. Some radiation is emitted in the long wavelength regime and is $\dot{Q}_e$ that we do need to consider.

We first use the energy balance given in eqn (3.12) and arrive at

$$\dot{m}\left[\hat{H}_2 - \hat{H}_1\right] = \dot{Q}_a - \dot{Q}_e - \dot{W}_m$$

or

$$\dot{m}C_p\Delta T = \dot{Q}_a - \dot{Q}_e - \dot{W}_m$$

using the relation between enthalpy and the heat capacity. We don't need to consider $\dot{Q}_r$ since this heat does not enter the system. To find $\dot{Q}_a$ we require the definition of the absorbance

$$\dot{Q}_a = a\ P_{bb}(T_S)A$$

where A represents the photovoltaic device area (1.28 m^2) and the black body irradiance from the Sun is evaluated at the Sun temperature T_S. Now radiation will be emitted at the rate

$$\dot{Q}_e = e\ P_{bb}(T_{sc})A$$

where T_{sc} is the solar cell or photovoltaic device temperature. The efficiency of the photovoltaic device is defined (here) as the fraction of the solar radiation that is converted into electricity and so we can write

$$\dot{W}_m = \eta \times P_{bb}(T_S)A$$

The energy balance can now be written as

$$\Delta T = \frac{[a-\eta]\ P_{bb}(T_S) - e\ P_{bb}(T_{sc})}{\dot{m}C_p} \times A$$

The difficulty is we do not know the photovoltaic device temperature unless further assumptions are made. Indeed to solve this problem completely a more sophisticated energy balance would have to be made and will be performed in Chapter 8. Here, the upper limit on Q_e will be used and the temperature rise from Example 3.3 will be used 4.3 K, assuming the initial system temperature was 20°C then T_{sc} = 297.4 K. Assuming AM1.5G solar radiation is incident on the photovoltaic device, making $[a-\eta]\ P_{bb}(T_S) = 750\text{W/m}^2$, we can write

$$\Delta T = \frac{750\ \text{W/m}^2 - 0.1 \times 5.670 \times 10^{-8}\ \text{W/m}^2\text{-K}^4 \times [277.5\ \text{K}]^4}{6 \times 10^{-2}\ \text{kg/s} \times 4187\ \text{J/kg-K}} \times 1.28\ \text{m}^2$$

$$= 3.7\ \text{K}$$

It is clear that radiation emitted by the photovoltaic device, or hot water heater, can be significant, especially at higher temperatures, and should be taken into account since even at this low temperature radiative energy emission decreased the temperature rise by 0.5 K. To do a more detailed analysis is challenging since a more thorough consideration of the heat transfer to the surroundings has to be done, and this will be considered in Chapter 8.

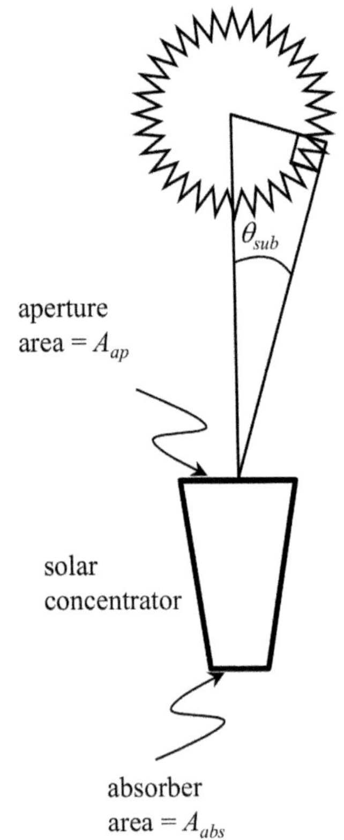

Fig. 3.12 The ideal solar concentrator that allows all the solar radiation impinging onto an aperture to be concentrated onto an absorber. The Earth and solar concentrator are both considered points compared to the Sun making the amount of radiation reaching them scale with the angle subtended between the Earth and Sun, θ_{sub} which is 0.265°. The aperture and absorber areas are taken as A_{ap} and A_{abs}, respectively.

The device in the above example used whatever solar radiation that impacts onto its surface to heat water and to drive a photovoltaic device. This limits its performance to some degree and is dictated by the area presented to the Sun. If hotter water or more current in a photovoltaic device is desired then a larger area must be used. There is another way to increase performance, concentrate sunlight, which effectively increases the device area.

The concentration ratio, C_R, is defined as the ratio of the aperture area A_{ap} to the absorber area A_{abs}, see Fig. 3.12

$$C_R = \frac{A_{ap}}{A_{abs}} \tag{3.32}$$

and is a measure of how much radiation will reach the absorber. The maximum value it can obtain is determined by the SLOT. First consider radiation coming from the Sun to the aperture. Using eqn (2.4) we can find the irradiance P_{ext} and the heat transfer rate $\dot{Q}_S$ coming from the Sun after multiplying by the area

$$\dot{Q}_S = \sin(\theta_{sub})^2 \sigma_S T_S^4 A_{ap} \; [=] \; W$$

where σ_S is the Stefan-Boltzmann constant and T_S is the Sun's temperature. Here it is assumed that all the radiation from the Sun can reach the aperture and it is not attenuated by the atmosphere.

The *maximum* radiation that the absorber can produce (emit) is

$$\dot{Q}_{abs} = E_{a-S} \sigma_S T_{abs}^4 A_{abs} \tag{3.33}$$

where E_{a-S} is the specular exchange factor between the absorber and Sun that indicates the amount of radiation that can go from the absorber to the Sun by all specular means, or that reflected like a mirror, and the usual radiative emission. If all the radiation goes to the Sun then it is equal to 1 and this is what we will assume below. Combining the above equations yields

$$C_R = \frac{1}{\sin(\theta_{sub})^2} \frac{\dot{Q}_S}{\dot{Q}_{abs}} \frac{T_{abs}^4}{T_S^4}$$

The SLOT for an open system, eqn (3.13), can be written for a reversible (isentropic) process of exchanging radiation between the two systems with no mass transfer

$$0 = \frac{\dot{Q}_S}{T_S} - \frac{\dot{Q}_{abs}}{T_{abs}} \Rightarrow \frac{\dot{Q}_S}{\dot{Q}_{abs}} = \frac{T_S}{T_{abs}}$$

If $T_{abs} = T_S$ then $\dot{Q}_{abs} = \dot{Q}_S$ and combining the above two equations we find

$$C_{R,2i} = \frac{1}{\sin(\theta_{sub})^2} \quad \text{(two dimensions)} \tag{3.34}$$

which is valid for a two dimensional absorber surface and the subscript i indicates *ideal* due to the assumption of reversibility.[13] Since $\theta_{sub} = 0.265°$ the maximum concentration ratio is 46,700! Of course, there are many losses and the ratio never reaches this value. If the absorber is linear or one-dimensional, such as a pipe filled with water, then the maximum concentration ratio is

$$C_{R,i} = \frac{1}{\sin(\theta_{sub})} \quad \text{(one dimension)} \tag{3.35}$$

and has a maximum value of 216, still quite large although it is rarely, if ever, reached.

[13] The reader may suggest to put the emissivity of the absorber into eqn (3.34), however, we assumed $T_{abs} \rightarrow T_S$ which is very hot. The likelihood of $e < 1$ in this limit is low and it is assumed to be one. In reality this limit is a fantasy only...

Example 3.6

Determine the length of a pipe in a one-dimensional concentrator used to heat water by 30 °C. Assume the water mass flow rate is 0.06 kg/s, the absorber (pipe) external diameter is 5 cm and it has an absorbance of 0.9.

The device set-up is shown in Fig. 3.13 where water flows in a pipe that allows concentrated solar radiation to heat it. We need to find the length L and use the energy balance with the ideal concentration ratio given in eqn (3.35), so, from eqn (3.12)

$$\dot{m}\left[\hat{H}_2 - \hat{H}_1\right] = \dot{Q}_a$$

where Q_a is the absorbed, concentrated insolation. This equation can be written

$$\dot{m}C_p\Delta T = C_R \times aP_{bb}(T_S) \times \pi DL$$

where C_R is the concentration ratio, a is the absorbance of the absorber pipe in the concentrator, $P_{bb}(T_S)$ is the black body irradiance from the Sun at temperature T_S, D is the absorber (pipe) external diameter and L is the absorber length. Note the entire pipe area was assumed to absorb radiation which may not always occur and is an approximation used here. We will assume AM1.5G radiation from the Sun with $T_S = 5359$ K, making $P_{bb}(T_S) = 1000\ \text{W/m}^2$ and can write

$$L = \frac{\dot{m}C_p\Delta T}{C_R a P_{bb}(T_S)\pi D} = \frac{0.06\ \text{kg/s} \times 4187\ \text{J/kg-K} \times 30\ \text{K}}{C_R \times 0.9 \times 1000\ \text{W/m}^2 \times \pi \times 0.05\ \text{m}}$$

and find $L = 53.3\ \text{m}/C_R$. If this is an ideal system then $C_R = 216$ and the pipe length is a mere 0.25 m! Of course, it is difficult to realize such a concentration ratio and it is typically of order 25 (in some cases for a tracking system up to 50–80) making the length 2 m. This is still quite short although we have not taken into account all the heat losses and heat transfer considerations which will significantly increase the length.

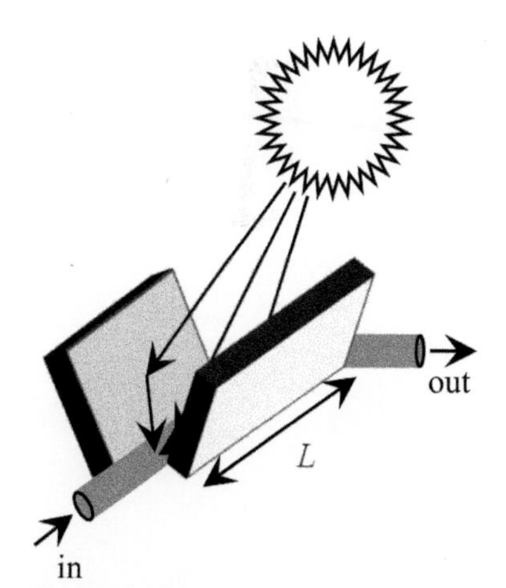

Fig. 3.13 Ideal solar concentrator used to heat water.

3.6 Conclusion

A discussion of the basic principles required to develop design strategies of solar devices was given in this chapter. We began with the most basic principles of all, the First and Second Laws of Thermodynamics. These were modified for open systems allowing the reader to use them to full effect in the future when flow processes are considered.

Light absorption was then considered. This is the basic mechanism in solar energy utilization where energy from the Sun is absorbed by the material in a solar device for a photovoltaic or solar thermal application. In the former application a free excited electron and the accompanying hole are required to make electrical power. In the later application the electron falls back to its lower energy state and transfers that energy to the surrounding atoms in the form of a lattice vibration or phonon. This more enthusiastic lattice vibration is equivalent to a higher temperature allowing this energy to be used for a variety of applications.

An optimum material for a photovoltaic device must generate a potential (voltage) and current, thus, a metal or an insulator are not suitable. A metal cannot produce a potential while an insulator has too high a band gap energy to make much current. The optimum material is a semiconductor which can make the most power that is the product of voltage with current.

A material for a solar thermal application must have minimal emissivity and high absorbance which is an impossibility when Kirchhoff's law is considered. The emissivity equals the absorbance at thermal equilibrium! However, some thought allows an optimal material to be designed. The Sun emits radiation with a peak irradiance near a wavelength of 500 nm while a material in a solar thermal device will heat to a temperature of order 100 °C with a peak irradiance near 10,000 nm. So, the absorbance is maximized at optical wavelengths which has the added benefit of reducing the reflectivity. Then the emissivity is minimized at long wavelengths to retain as much energy as possible. This combination of properties is denoted as a selective surface to maximize the temperature rise in the solar thermal device.

So, one can see for both applications, photovoltaic or solar thermal, optimal performance is found by a trade off of physical principles. This is typical in designing any system since a material is rarely found which maximizes all desired parameters. Furthermore, if the Sun's energy is not enough, solar concentrators were considered to increase the irradiance available to the device.

3.7 General references

J.A. Duffie and W.A. Beckman, 'Solar energy of thermal processes,' John Wiley and Sons, 3rd Edition (2006)

M. Fox, 'Optical properties of solids,' Oxford University Press (2001)

T.S. Hutchinson and D.C. Baird, 'The physics of engineering solids,' John Wiley and Sons (1963)

A.B. Meinel and M.P. Meinel, 'Applied solar energy, An introduction,' Addison-Wesley Publishing Company, Reading, Mass. (1976)

V. Quaschning, 'Renewable energy and climate change,' J. Wiley & Sons (2010).

S. Sandler, 'Chemical, Biochemical, and Engineering Thermodynamics,' 4th Edition, J. Wiley & Sons (2006).

W. Shockley and H.J. Queisser, 'Detailed Balance Limit of Efficiency of p-n Junction Solar Cells,' J. App. Phys. Volume 32 (1961) 510.

Exercises

(3.1) A closed, well sealed 5 m^3 volume water tank contains 1000 kg of water and has air above it at atmospheric pressure. An additional 1000 kg of water is forced into the tank, determine the work done in compressing the air, in kJ.

(3.2) The average household in the USA uses 940 kW-h of electricity per month. If this energy were not lost to the surroundings, how much would your house heat up in one day? Assume the house has a volume of 300 m^3.

(3.3) A human has a heat transfer rate from their body of order 100 W (this is for a 2000 kcal diet over 24 h). Assume there are 100 people in a room, how much would the enthalpy in the room change in one hour? What would the temperature rise be?

(3.4) Estimate how much the Sun would heat up your car in one day if it did not lose heat, it can only gain heat from the Sun.

(3.5) Estimate how much heat comes through the windows and roof of your black-colored car. Your heat transfer model for the car should use the fact that the interior temperature of the car on a hot summer's day will be about 60 °C at steady state. This can be used to determine how much of the heat absorbed by the roof makes it to the interior. Black paint has a high energy (short wavelength) absorbance of 0.96 and a low energy (long wavelength) emissivity of 0.88. Now if you put a reflective shade inside the front windshield, how much will this reduce the car's interior temperature assuming it is 100% effective? *Hint:* Until we know more about heat transfer assume all heat transfer is through radiative processes, as discussed in this chapter, and use the roof area as an adjustable parameter until you obtain the correct interior temperature.

(3.6) A shell and tube heat exchanger has 1 kg/s of hot water coming into the tube side at 70 °C where it is cooled to 40 °C by thermal contact with river water that enters and exits the shell side at 20 °C and 30 °C, respectively. Draw a picture of a shell and tube heat exchanger, label the control volume and determine the flow rate of the river water required to cool the hot water. The student should learn what a shell and tube heat exchanger is, since it is used in many solar energy applications. You have probably seen the simplest version in a chemistry laboratory and it is called a *condenser* used to reflux reactions or condense certain vapors.

(3.7) Assume the cooling water in Problem 3.6 is the Rhine river which has a volumetric flow rate of 2200 m^3/s, determine the temperature rise caused by the discharge from the heat exchanger immediately downstream of the discharge assuming well mixed conditions.

(3.8) A moderately sized power plant continuously produces 50 MW of electricity and operates at 35% efficiency; determine the amount of coal required for its operation in kg/day assuming the energy content of coal is 35 MJ/kg. Efficiency is defined as power output divided by energy (heat) input. If the coal furnace was replaced by a nuclear reactor, how much nuclear fuel would be required, remember $E = mc^2$. Note that nuclear fission is not 100% efficient and Uranium breaks down into smaller elements, not into pure energy, so more mass than you calculate is required.

(3.9) It is proposed to use the variation of ocean water temperature with depth as a heat source in a Carnot heat engine. Assume the surface water has a temperature of 20°C and well below the surface it is 5 °C; determine the maximum possible efficiency of the engine.

(3.10) Geothermal energy is proposed to be used in a Carnot heat engine. Hot, high pressure water, well below the surface of the Earth, is available at 150 °C; determine the maximum possible efficiency of this engine if it rejects heat to the atmosphere at 20 °C.

(3.11) Assume a Carnot heat engine can reject heat from the low temperature reservoir at temperature T_L only by radiative emission (the form of the heat transfer is given in eqn (3.33)). For a given amount of work output and a given temperature for the high temperature reservoir T_H, see Fig. 3.1, determine that the ratio T_L/T_H which minimizes the low temperature reservoir's area is 3/4. *Hint*: Use eqns (3.4) to write the work divided by T_H^4 in terms of the reservoir area A and T_L/T_H, then minimize the equation by taking the total derivative to find the minimum area.

(3.12) Write eqn (3.16) in energy units to find the equivalent Rydberg constant (don't include the reduced mass in the constant). In other words, write the electronic transition as a change of energy, in Joules, and determine the proportionality constant as done in eqn (3.17). Compare this energy to that for thermal energy, $k_B T$, where k_B is Boltzmann's constant and T is temperature. Is thermal energy large enough to cause a transition in the Lyman series?

(3.13) Estimate the frequency required to cause a transition in the Lyman series using eqn (3.16). Can optical frequencies produce a transition?

(3.14) Plot the maximum voltage and the maximum current that can be obtained from a photovoltaic device as a function of band gap energy. Then overlay the maximum efficiency over these curves and determine what would be the maximum voltage and current one could obtain for a device with the optimum efficiency. Translate this to power and indicate how many 100 W light bulbs one could power with a 1 m^2 area exposed to the Sun.

(3.15) Plot power per unit area versus band gap energy for a photovoltaic device that produces the maximum possible voltage and current. Determine if it would be possible to operate a car covered with a photovoltaic device that produces the maximum power.

(3.16) Assume the band gap energy is zero in eqn (3.22) and find the maximum possible current possible for a photovoltaic device. In this case one can write an analytical form for the summation as it is a zeta function.

(3.17) Consider a selective surface which has a short wavelength absorbance of 0.95 and a long wavelength emissivity of 0.1. Assuming the solar device is at 75 °C, determine the net heat transfer to the system when placed in the Sun on the Earth's surface. Only heat transfer through radiative processes is possible.

(3.18) A gray body has a emissivity of 0.3, determine its irradiance if it is at a temperature of 50 °C.

(3.19) The reflectance (ρ) of Black Nickel as a function of wavelength (λ) is given in the table below. This is a selective surface coating material. Plot the absorbance or emissivity as a function of wavelength on a linear–log graph and assess this material as a selective surface for a solar thermal device operating at 300 °C. Data source: J.R. Lowery. 'Solar absorption characteristics of several coatings and surface finishes,' NASA Report Number TM X-3509 (1977).

λ (nm)	ρ	λ (nm)	ρ
270	0.063	550	0.014
310	0.102	640	0.023
360	0.087	740	0.114
380	0.047	850	0.242
420	0.030	1140	0.393
450	0.039	1500	0.497
480	0.039	2230	0.628

(3.20) The irradiance P of an *opaque* material can be written

$$P = \int_0^\infty [1 - \rho(E)]\, I(E)\, dE$$

where $\rho(E)$ is the reflectance that is a function of energy E and $I(E)$ is the energetic irradiance. A selective surface has a reflectance of 0.05 for wavelengths less than 700 nm and 0.6 for wavelengths greater than 700 nm; determine the rate of energy emitted per unit area if the material is at a temperature of 80 °C.

(3.21) Do Example 3.6 assuming the reflectance of the absorber material is 0.05 in the short wavelength region and that the efficiency is one-third of that given by the concentration ratio. In other words the concentration ratio is one-third the theoretical maximum.

Electrons in solids

4

The advent of quantum mechanics made the description of electron physics relatively simple and furthermore presented predictions that led to an explanation of the periodic table of the elements. For the purpose of this monograph, quantum mechanics demonstrates how an electron, or hole, which is where an electron *is not*, moves through a crystalline lattice. This is very important to understanding how a solar cell works and how a solid becomes a photovoltaic device.

The initial part of this chapter is devoted to finding the quantum mechanical wave function that is related to the probability an electron is at a certain position. This leads to the prediction of quantum mechanics to determine the energy–momentum relation needed to determine the band structure of semiconductors. Here it is shown how band structure naturally arises when an electron is placed in a periodic array of potential energy with position. In other words, a potential is present that can pull an electron towards it and is arranged periodically with position. It is this periodic arrangement that gives rise to a band structure. This model was developed by Kronig and Penney in 1930 and it is useful to see how a band structure evolves due to allowed energy levels.

More detailed aspects of the energy band are then considered. The energy dependence of charge carriers within a band with respect to the wave vector is considered to find it is parabolic, at least near the energy minimum (conduction band) or maximum (valence band). This result eventually leads to calculation of the number of free electrons in the conduction band and holes in the valence band for a semiconductor. Without these relations it would be difficult to understand how a photovoltaic device works and their derivation is necessary for this reason.

4.1 The nature of radiation

We mentioned in deriving eqn (2.6) that energy E is related to the wavelength of radiation λ *via*

$$E = \frac{hc}{\lambda} = hc\tilde{\nu} = h\nu = \frac{h\omega}{2\pi} \tag{4.1}$$

where h is Planck's constant, c, the speed of light, $\tilde{\nu}$, the wave number, ν, the frequency in cycles per second and ω, the frequency in radians per second. This equation is quite important and was proposed by Planck as a model to emphasize the particle–like nature of electromagnetic radiation implying that it can act as discrete energy packets rather than a

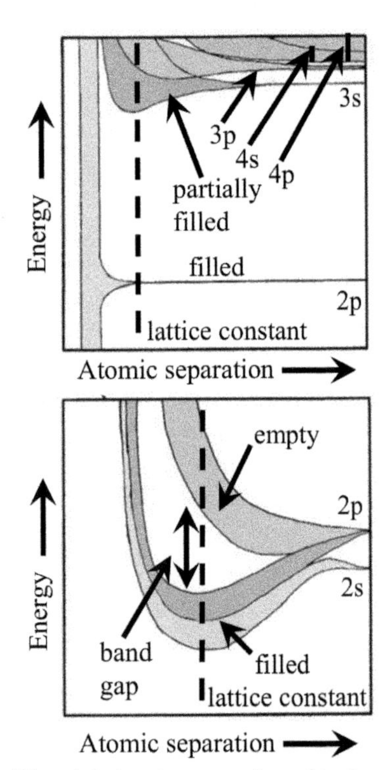

Fig. 4.1 As atoms are brought closer together the Pauli exclusion principle promotes a broadening of energy levels when the atoms are close enough. The upper figure is representative of a metal (Sodium) while the lower figure that of an insulator (Diamond). If the band gap were smaller Diamond would be a semiconductor. The lattice constant represents the atomic separation distance of the pure solid at equilibrium.

wave. This proposal found immediate application in description of the photoelectric effect for which Einstein won the Nobel Prize.

de Broglie subsequently proposed that a particle can also be represented as a wave, now called a de Broglie wave, with momentum p given by

$$p = mv = \frac{h}{\lambda} \tag{4.2}$$

where m is the particle's mass, v its velocity and λ the radiation wavelength. This relation was quite useful in describing electron diffraction by crystals and highlights the wave - particle duality many particles exhibit. For example, if a stream of electrons with charge q is subject to a potential V_a then each electron will have the following kinetic energy:

$$\frac{1}{2}mv^2 = q \times V_a \tag{4.3}$$

Combining eqns (4.2) and (4.3) has the electron wavelength equal to $\lambda(\text{nm}) = \sqrt{1.5/V_a(\text{kV})}$. Electrons are typically accelerated through potentials of 10-100 kV and so have quite small wavelengths, much smaller than those in the optical spectrum, and are readily absorbed by most materials. Interestingly, higher energy particle beams are more easily absorbed while lower energy are not which appears counterintuitive. Hopefully after completion of this chapter one should have a firmer grasp of this and appreciate the mechanism that allows light absorption.

4.2 Band structure

Atoms have specific energy levels, which is nicely described by quantum mechanics discussed below. Accepting only these energy levels are available and that no electron can have the same set of quantum numbers due to the Pauli exclusion principle, then the following scenario occurs.

Consider a Sodium atom which has *1s*, *2s*, *2p* and *3s* orbitals filled with 2, 2, 6 and 1 electrons, respectively. The *s* orbitals can include 2 electrons, each with the same energy, yet, a different spin, being clockwise or anti-clockwise, fulfilling exclusion through the different spin quantum numbers.

Imagine taking two Sodium atoms and gradually moving them closer and closer to each other then the *3s* electrons could form a covalent bond with two electrons in one atom or two in the other, as long as the spin quantum number requirement is not ignored, or they could be divided between the two atoms. Sodium is two rows below Hydrogen and H_2 is a stable compound that follows precisely the above scenario with electrons shared between the *1s* orbitals of the two atoms. This does not happen with Sodium though and the *3p* orbital changes, as does the *3s*, to create a hybrid orbital, shown in Fig. 4.1, allowing for the bonding of other Sodium atoms. Na_2 doesn't exist in Nature.

Instead Sodium metal exists as a solid with many atoms close together separated by a distance characterized by the lattice constant or parameter, or the equilibrium center-to-center distance between atoms.

The *3s* electrons then roam in the hybridized orbitals which are called a *band.* If there are N atoms then the band could contain $2N$ electrons since there can be pairs of electrons with opposite spin. However, there are only N electrons and the band is not filled suggesting high electron mobility with only a minute energy change required to move an electron within a potential. This defines a metal.

Yet, for other elements in the periodic table the situation is different. Diamond does not have much electron conductivity at all and is in fact an insulator. Reference to Fig. 4.1 shows in this case the *2p* orbital hybridizes in a strange manner and blends with the *2s* to create a filled band. There are 2 × *2s* electrons and 2 × *2p* electrons with the entire electronic structure written as: $1s^2 2s^2 2p^2$. In this band, though, one could possibly fit $2 \times 2N$ electrons when N atoms are brought in close proximity. Each atom brings 4 electrons and so the band is full leaving no energy level for an excited electron to achieve should it experience a moderate potential. Instead the other p orbitals hybridize to a higher energy level and a large band gap energy results between the highest energy level in the sp^3 hybrid band and the vacant *2p* band.

The band gap is an important quantity in all solar energy applications dictating material choice. Now we consider quantum mechanics as a way to show how bands develop. After the theory is presented and two simple analyses are performed a more complicated model, the Kronig–Penney model, is discussed that shows how energy bands result when an electron is in a periodic array of atoms. A much more complicated model than the Kronig–Penney model is required to predict orbital hybridization shown in Fig. 4.1, however, it is the purpose of the discussion below to introduce the mechanism by which energy bands appear.

4.3 Schrödinger's equation

In this section the wave function is introduced, since it is hypothesized that particles, or specifically electrons, can have wave-like and particle-like properties. Of course the wave-like properties come immediately from the wave function. The strategy is to assume the wave function exists and then use it in an equation which sums the kinetic and potential energies of a particle together to give the total energy. Since the wave-like nature only admits certain solutions to the equation, such as solutions that are periodic in some way, then the allowed results will be quantized which leads to quantum mechanics.

The de Broglie wave can represent a particle with its wave function Ψ generally written as a function of distance x and time t and the wavelength and frequency

$$\Psi = A\cos\left(2\pi\left[\frac{x}{\lambda} - \nu t\right]\right) + B\sin\left(2\pi\left[\frac{x}{\lambda} - \nu t\right]\right) \tag{4.4a}$$

which is more commonly written

$$\Psi = A\cos(\kappa x - \omega t) + B\sin(\kappa x - \omega t) \tag{4.4b}$$

where A and B are constants and κ is referred to as the wave vector and is equal to $2\pi/\lambda$.[1] The wave vector is an extremely useful quantity and should be appreciated as representing a number of phenomena. First among them is that reference to eqn (4.2) shows κ is related to the particle momentum ($p = \kappa\hbar$, where $\hbar \equiv h/2\pi$) and will frequently be referred to as the wave's *momentum* vector. The term vector is used since the wave can propagate in any direction; above we have chosen the x–direction. Strictly speaking κ should be written in vector notation which will not be done here since this does not add to the basic principles we wish to convey.

[1] A comment on the argument of the cosine and sine should be made. Why is it written $\kappa x - \omega t$? Why is there a negative sign? This convention merely assures us that as distance x or time t is increased then the wave will travel in the same direction. Consider the wave maximum for the sine function, this occurs when the argument equals $\pi/2$ or, $\kappa x - \omega t = \pi/2$. Thus, $\kappa x = \pi/2 + \omega t$, showing that as time increases so too does distance.

We have accepted the wave nature of a particle and written Ψ in the above equation as a function of a cosine and sine which appears reasonable, however, what does Ψ represent? Treated properly it represents the probability of finding a particle in a given volume, or in mathematical terms, $|\Psi|^2 dV$ is the probability of finding the particle in the volume dV. The operator $|\cdots|^2$ means magnitude of and is written in this way since we will find that Ψ is a complex number and should be treated accordingly. Now that it is known what the wave function is, how can it be used to solve actual problems?

Equations (4.2) and (4.3) were combined to find the wavelength of an electron accelerated in a potential. The particle has momentum ($p = mv$) and can be influenced by a potential suggesting it can have kinetic[2] and potential energy equal to its total energy or

[2] The kinetic energy is frequently written as $\frac{1}{2}p^2/m$ rather than $\frac{1}{2}mv^2$ for a couple of reasons. A primary reason is that momentum is a conserved quantity, so, similar momentum in opposite directions will cancel, which does not necessarily apply to velocity.

$$\frac{p^2}{2m} + V = E \tag{4.5a}$$

which can be written in terms of wave parameters as

$$\frac{h^2\kappa^2}{8\pi^2 m} + V = \frac{h\omega}{2\pi} \tag{4.5b}$$

The potential energy V can be a complicated function of position as well as time. This equation is merely a sum of the kinetic and potential energies for a particle, that is all. Now consider the wave-like nature of the particle and introduce the wave function discussed above to discover its utility.

Derivation of Schrödinger's equation is actually quite simple from here, although admittedly not obvious. Take the following derivatives: $\partial\Psi/\partial t$ and $\partial^2\Psi/\partial x^2$ and solve for ω and κ, respectively. Substitute into eqn (4.5b) and rearrange and everything will look great except there will be a term that is $A\cos(\kappa x - \omega t) - B\sin(\kappa x - \omega t)$ which isn't equal to Ψ, as written above. Set this term equal to $C \times \Psi$, where C is a constant, and one finds that if $A = 1$, $B = i$ and $C = -i$ then the term does equal Ψ which is written as, to within a multiplicative constant, $\cos(\kappa x - \omega t) + i\sin(\kappa x - \omega t)$.

Now one has the time dependent Schrödinger equation

$$-\frac{h^2}{8\pi^2 m}\frac{\partial^2\Psi}{\partial x^2} + V\Psi = \frac{ih}{2\pi}\frac{\partial\Psi}{\partial t} \tag{4.6}$$

This is a useful equation, however, the number of time dependent problems that can be solved is limited and not useful to this treatment of solar energy. Instead we use separation of variables and propose that $\Psi(x,t) = \psi(x)\phi(t)$ and arrive at the time independent Schrödinger equation

$$\frac{\partial^2\psi}{\partial x^2} + \frac{8\pi^2 m}{h^2}[E - V]\psi = 0 \tag{4.7}$$

where we recognize in the derivation that $\phi(t) = \exp(-[2\pi i/h]Et)$. Although this equation could merely be given, it is important the reader realizes its origin derives from several hypotheses on the wave-like and particle-like nature of light or radiation in general. The wave-like nature is particularly important as we will find only certain wavelengths are admissible to some systems, this means only some momenta are admissible (eqn (4.2)) and hence energy levels. This is illustrated in the next section to begin an understanding of why only certain energy levels are allowed.

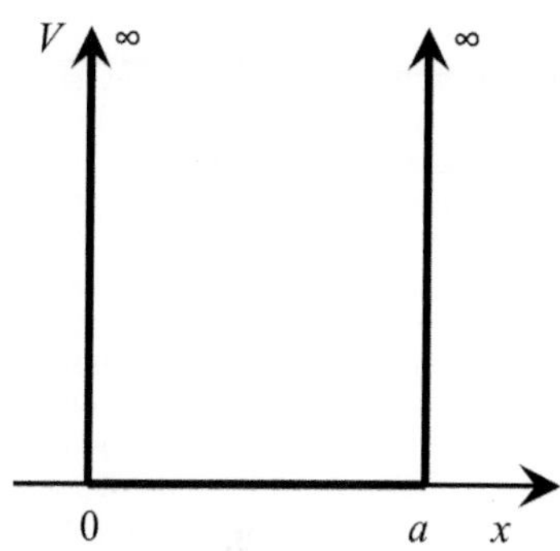

Fig. 4.2 The potential in a one-dimensional box of length a along the x-axis that will not allow an electron to pass through its boundaries.

4.4 Electron in a box

So far, a *generic* particle has been considered in the Schrödinger wave equation. Consider an electron trapped between two infinite potentials separated by a distance a, as shown in Fig. 4.2. In this case eqn (4.7) simplifies to

$$\frac{\partial^2\psi}{\partial x^2} + \frac{8\pi^2 m}{h^2}E\psi = 0$$

which can be written

$$\frac{\partial^2\psi}{\partial x^2} + \alpha^2\psi = 0$$

where α is an energy parameter.[3] This equation is valid only between the two infinite potential lines on either side of the box, in other words for $0 < x < a$. We expect the solution to have a form $\psi = A\cos(\alpha x) + B\sin(\alpha x)$ with boundary conditions

$$\begin{aligned} &\text{for } x = 0, \quad &\text{then,} \quad &\psi = 0 = A \\ &\text{and for } x = a, \quad &\text{then,} \quad &\psi = 0 = B\sin(\alpha a) \end{aligned}$$

Remember ψ is related to the probability an electron will be at a given position and it can't be located at the infinite potential position making $\psi = 0$ at the two boundaries. The second boundary condition can only be true if $\alpha a = n\pi$ or

$$\sqrt{8\pi^2 mE/h^2}\; a = n\pi, \quad \text{for} \quad n = 1, 2, 3, \ldots$$

which means

[3] α is a useful parameter and is defined as

$$\alpha^2 = \frac{8\pi^2 m}{h^2}E = \frac{E}{6.103 \times 10^{-39}\,\text{J-m}^2}$$

the later equation is for an electron.

$$E = \frac{n^2 h^2}{8ma^2}, \qquad n = 1, 2, 3, \ldots \tag{4.8}$$

As suggested above, only certain energy levels are available to the electron. However, if a is large, as would be true in a macroscopically sized device, then the energy levels will be separated by a fairly small amount since they scale with $[n/a]^2$.[4] This represents a *free* electron, that is confined in a device by its edges, which are the infinite potentials, and one can see that energy changes quadratically with the quantum number n.

[4] Nanoparticles can now be made that are so small, on the order of 5–10 nm in diameter, where the size of the particle no longer yields an *infinite* sized box. It is found that different sized particles with the same chemical composition have different colors when suspended in a solvent. What happens is this: a single atom has only certain discrete energy levels available to electrons and as more and more atoms are brought together, to make a nanoparticle, these levels gradually spread out to accommodate more electrons due to the Pauli exclusion principle. The distance between the available energy levels then is reduced. This means the band gap energy is reduced creating differently colored particles for different sized particles when they absorb and emit light.

4.5 Electron in a periodic potential

Now consider an electron which is in a periodic potential such as an array of atoms in a crystal. Of course the potential is provided by the attraction of an electron to the nuclei of the atoms making up the crystal lattice as shown in Fig. 4.3. The challenge is to solve for the energy levels available to the electron which will dictate what energy levels can be absorbed from external radiation. Here we follow the Kronig–Penney model which is simple enough to allow a solution to the wave equation and demonstrate the reason why a band gap energy would develop in a periodic array.

Given the periodic potential one can write the wave equation, in eqn (4.7), as

$$\frac{\partial^2 \psi}{\partial x^2} + \frac{8\pi^2 m}{h^2} E\,\psi = 0 \quad \text{for} \quad 0 < x < a \tag{4.9a}$$

$$\frac{\partial^2 \psi}{\partial x^2} + \frac{8\pi^2 m}{h^2} [E - V_0]\,\psi \quad \text{for} \quad b < x < 0 \tag{4.9b}$$

where we now define

$$\alpha^2 = \frac{8\pi^2 m}{h^2} E \qquad \beta^2 = \frac{8\pi^2 m}{h^2} [V_0 - E]$$

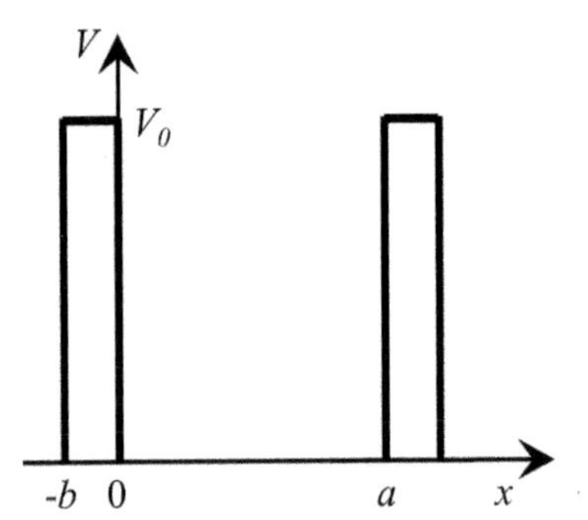

Fig. 4.3 The potential in a one-dimensional periodic array along the x-axis. A finite potential of magnitude V_0 is of width b and separated from adjacent similar potentials by the distance a.

The solution to these equations is not difficult, yet it isn't straightforward. One assumes a solution form for ψ and finds coefficients to the wave function that must be satisfied, similarly to what had to be done in considering a single electron in a box. From this one obtains the following criterion:

$$\frac{\beta^2 - \alpha^2}{2\alpha\beta} \sinh(\beta b) \sin(\alpha a) + \cosh(\beta b) \cos(\alpha a) = \cos(\kappa[a + b])$$

which looks quite complicated and of course it is. To find more useful information let V_0 tend to infinity while b tends to zero, however, their product is kept finite. In this case one finds

$$P \frac{\sin(\alpha a)}{\alpha a} + \cos(\alpha a) = \cos(\kappa a) \tag{4.10}$$

where P is $4\pi^2 m a V_0 b/h^2$.

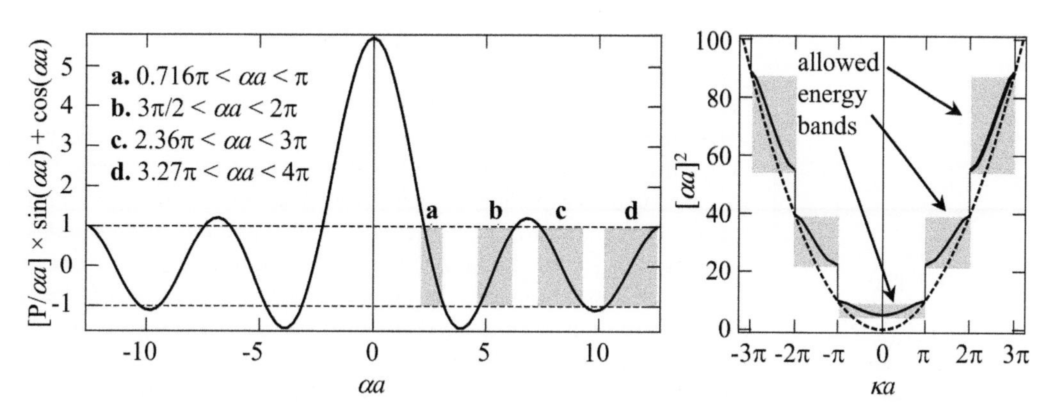

Fig. 4.4 (left) The left-hand side of eqn (4.10) plotted as a function of αa. Note that only certain values of αa are allowed to keep the function within the bounds of ± 1 and are labeled **a** through **d**. (right) The allowed dimensionless energy levels $[\alpha a]^2$ plotted as a function of the dimensionless wave vector κa clearly demonstrating the energy band levels that are allowed. The dotted curve is the energy for a completely free electron in no potential which agrees remarkably well in energy magnitude with the more complicated result. The value of P assumed was $3\pi/2$ in both graphs.

Firstly, if the dimensionless potential P tends to zero, as when there are no potential barriers, one finds

$$\cos(\alpha a) = \cos(\kappa a)$$

meaning that $\alpha = \kappa$ or equivalently $\alpha^2 = \kappa^2$. Rearranging and using the definition of the wave momentum vector κ and momentum p one finds $E = p^2/2m$, which is merely the kinetic energy of a free electron (particle) in no potential. Secondly, if the potential tends to infinity ($P \to \infty$) then $\sin(\alpha a) \to 0$, or there is no solution to eqn (4.10), which can only be true if $\alpha a = n\pi$, where n is an integer. Using the definition of α we would find the result in eqn (4.8). This is expected since at large potential the electron would be confined to a box of size a.

The interesting prediction comes when P is neither very small nor very large. Regardless of the value of P, the right-hand side of eqn (4.10) must be within the bounds of ± 1. However, the left-hand side can take on any value depending on the magnitude of P. This restricts the values that αa can take so the right-hand side remains within ± 1. So, only *certain* energy levels (α) are allowed while some are not! Remember α^2 is directly proportional to energy E so an energy *band* structure forms according to the allowed values of α.

The left-hand side of eqn (4.10) is shown in Fig. 4.4 and plotted as a function of αa with horizontal lines drawn at ± 1. So, only some energy values can be allowed. Also in the figure are the allowed energy levels $[\alpha a]^2$ as a function of the wave vector κa where distinct jumps in energy levels can be seen at integer multiples of π. The free electron model agrees in magnitude with the Kronig–Penney model, however, it does not allow predictions of energy bands that are available to the electrons only when they have certain values of κa.

Remembering that the momentum wave vector κ is related to the wavelength then only certain wavelengths λ of radiation will promote an electron to a given energy level $[\alpha a]^2$. Note that as κ increases then the energy of the absorbed radiation increases, since λ decreases, and one can potentially promote an electron to a higher energy band.[5]

We can also consider the definition of the kinetic energy for an electron given in eqn (4.5b) $(h^2\kappa^2/8\pi^2 m)$. Then define a dimensionless kinetic energy, $[\kappa a]^2$, in the same units we used for the total energy, $[\alpha a]^2$ (see note 3 and multiply the kinetic energy by $8\pi^2 ma^2/h^2$). Subtracting $[\kappa a]^2$ from $[\alpha a]^2$ yields the potential energy. This energy represents the pull of the nuclei on the electrons and clearly as the electron's dimensionless wave vector κa approaches an integer multiple of π the electron has only kinetic energy and not potential according to the Kronig–Penney model (see Fig. 4.4).

[5] Remember, $E = hc/\lambda$ (eqn (4.1)) and $\kappa \equiv 2\pi/\lambda$ (eqn (4.4b)) allowing one to arrive at $E = \hbar c\kappa$.

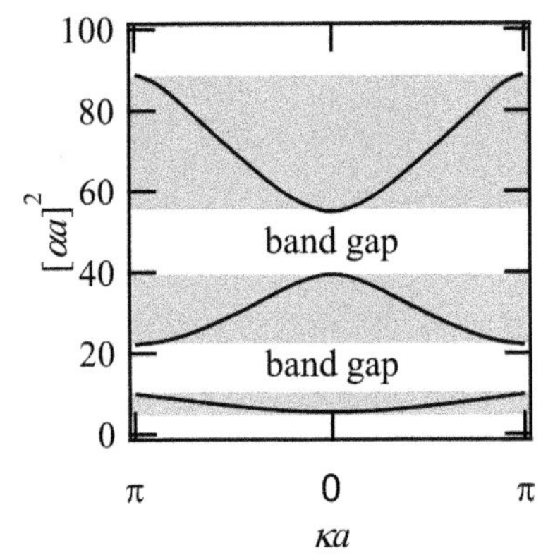

Fig. 4.5 The band structure for the Kronig–Penney model with $P = 3\pi/2$.

4.6 Band diagrams

It is clear the energy bands can be considered only for certain values of κ and these regions are called *Brillouin zones.* The first Brillouin zone in Fig. 4.4 occurs for $-\pi < \kappa a < \pi$ and the second for $-2\pi < \kappa a < -\pi$ and $\pi < \kappa a < 2\pi$ and so on. There is a clear energy discontinuity between them and this defines the different zones. A three-dimensional lattice of atoms will have a much more complicated Brillouin zone structure where the momentum wave vector κ must be considered in various directions.

We don't consider complicated two- or three-dimensional band diagrams here. Instead the reduced band diagram is used, as shown in Fig. 4.5, which is a useful representation. Since the first Brillouin zone occurs for κa in the range of $\pm\pi$ and $\cos(\kappa a)$ is periodic, one can subtract multiples of 2π from κa in Fig. 4.4 to obtain the reduced band diagram.

This diagram is clearly symmetric about zero momentum vector κ and so a shorthand method to look at different directions is often made. The positive κ axis will represent one direction in the crystal and the negative κ axis will represent another. There are two directions of most importance for a cubic crystal. Along the cube edges are the [100]-directions while from the center to the cube vertices are the [111]-directions. There are many more, in fact an infinite number, of unique directions in a crystal, however, these are two of importance. Thus, as shown in Fig. 4.6, the positive and negative κ axes represent completely different directions in the crystal. The symbols; L, Γ and X, represent certain positions in the crystal where direct and indirect band gap transitions occur and VB and CB represent valence and conduction bands, respectively.

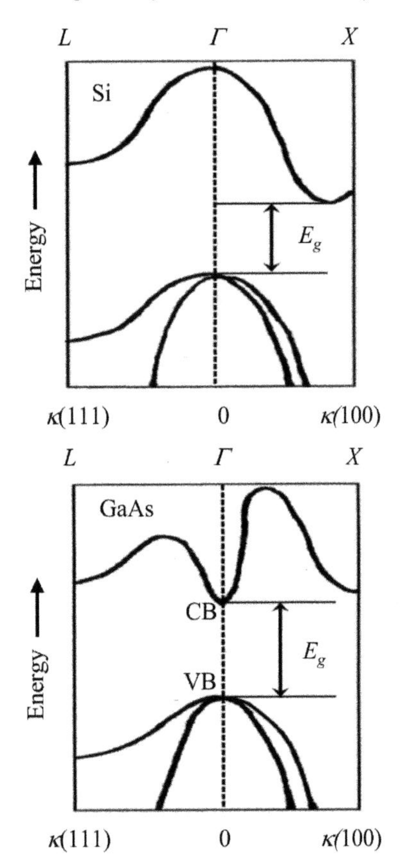

Fig. 4.6 The band structure of Silicon (Si) and Gallium Arsenide (GaAs). Silicon has an indirect band gap energy of 1.1 eV while Gallium Arsenide has a direct energy of 1.4 eV.

This shorthand method conveys much information in a small graph. It is clear that the band gap energy depends on direction, which may be expected to be due to the different packing of atoms (*i.e.* the period of the potential energy array changes). Also, one can see that Silicon has a minimum band gap (E_g) where an electron on the top of one band would have to change momentum ($\kappa a \equiv pa/\hbar$) to reach the bottom of the other. This is called an *indirect* band gap energy and greatly influences the performance of photovoltaic devices that use materials like Silicon. In general, materials with an indirect band gap suffer in ability to absorb light. Specifically the electron must gain momentum from the atomic lattice if the photon has an energy less than E_g, or if the photon has more energy then it must lose momentum. This two stage process is much less efficient than for a direct band gap material like GaAs, see Fig. 4.6, and will be considered in more detail in the next section.

What causes an indirect band gap? The true answer lies in complex quantum mechanical calculations, however, we try and simplify it here. The origin of the energy band comes from the Pauli exclusion principle, as discussed above. Consider two atoms being brought together, then a bonding and an anti-bonding orbital are formed. As more and more atoms are forced together then many orbitals are formed with the lowest energy orbital[6] being fully bonding and the highest energy orbital being fully anti-bonding. This is a complicated phenomenon and imagine calculating the empty orbitals in the conduction band, as shown in Fig. 4.1 for Diamond, let alone the filled orbitals discussed above in the valence band.

[6]Remember this is the most negative energy. Have you noticed that the energy axis in Fig. 4.5 is positive? We merely redefined what zero energy is ... and this can be corrected by subtracting the potential due to the nucleus which is attracting the electron to it. Since we didn't specifically state what the nucleus is, we kept everything general.

This is what must be done and the procedure is to calculate the electron charge density as atoms are brought together, as well as the potential presented by the nuclei in the crystal. Then Schrödinger's wave equation is solved leading to an indirect band gap in Silicon. Now in order for the electron to leave the valence band it must gain momentum as κ must change value; it increases from zero to a given value as shown in Fig. 4.6 along the (100) direction.

What does a momentum change imply? Firstly, remember $p = \hbar\kappa$ and so momentum is directly proportional to the wave vector. If the wave vector increases then the wavelength decreases since $\kappa = 2\pi/\lambda$. This also implies an increase in energy according to eqn (4.1), $E = hc/\lambda$. However, this energy increase does not exactly correspond to the band gap energy E_g since the total energy increase for the electron is due to *both* kinetic and potential energies. The potential energy can change too when the electron resides in the conduction band even though the kinetic energy certainly changes, since it is equal to $p^2/2m = \hbar^2\kappa^2/2m$.

Why does the momentum have to change? Consider the simplest solution to the Schrödinger wave equation we have, the energy levels for a particle in a box given in eqn (4.8), $E = n^2h^2/8ma^2$. One could imagine the *box* size changing in the conduction band by the effective a decreasing which will increase the overall energy levels available to the electron. Subsequently, this will reduce the wavelengths that the electron must possess and so an increase in the wave vector (momentum) is required.

This is an involved process and the above discussion merely highlights factors that could be involved in the requirement for an energy and momentum change when an electron goes from the valence to conduction band.[7] We have tried to lend some *concreteness* to the discussion of a quantum mechanical phenomenon. In summary, when an electron is

[7]Although it is difficult to give a rule-of-thumb as to when an indirect or direct band gap is present, one can give an estimate of its value. The Moss rule is: $n^4E_g(\text{eV}) = 77$, where n is the refractive index, which is fairly accurate.

excited into the conduction band it goes to a higher energy level and it may be required that its momentum (kinetic energy) must also change independently. This will be dictated by subtle changes in the potential energy that the electron experiences in the conduction band compared to the total energy that it had in the valence band. Since energy and momentum must be conserved the momentum change that the electron must obtain is completely prescribed. How the electron gains momentum is discussed later.

Finally, to demonstrate that it is not necessarily the chemical make-up of a material that dictates its band structure consider crystalline and amorphous Silicon. The former is the typical form of Silicon used in household and industrial solar photovoltaic installations, while the latter is used to power calculators and the like. Crystalline Silicon has an indirect band gap with an energy of 1.1 eV while amorphous (*i.e.* not crystalline) Silicon has a direct band gap with an energy of 1.7 eV! This is the same chemical compound and when going from a crystalline to an amorphous structure atomic packing affects the band structure. In the spirit of full disclosure though, amorphous Silicon must be *passivated* with Hydrogen to terminate unfulfilled bonds and has 5–10% Hydrogen within the network.

4.7 Dynamics

The conduction and valence bands have been labeled in Fig. 4.6. For semiconductors and insulators, that is for materials with a very high band gap energy like diamond, the valence band is completely full and an electron must gain an energy equivalent to the band gap energy to escape the valence band. Consider an electron excited to the bottom of the conduction band. While it resides in this excited state it will feel a potential due to the shape of the energy band with respect to the wave vector. Reference to Fig. 4.4 shows that the free electron model represents the energy relation with wave vector, at least for the Kronig–Penney model.

Regardless, the free electron model is used to represent the potential an electron would feel near the band edge. Looking at this in another way, very near the band edge the first nonzero term in an expansion of energy with momentum κ would be of order κ^2 since a linear term would change sign depending on whether κ is positive or negative. Because of this and the results given by the Kronig–Penney model, one can realize that the free electron model is a reasonable approximation to the energy–wave vector relation. Now that the functional form is known it is used to determine the effective force on an electron or hole in this potential.

This can be quantified by considering the acceleration a an electron or hole possesses in the potential

$$a = \frac{dv}{dt} \tag{4.11}$$

where v is velocity and t, time. The challenge is to determine what velocity we use since the electron will experience different values of energy as it explores E with κ.

If there were a single energy then the wave describing it is unique and follows

$$\Psi = \sin(\kappa x - \omega t)$$

Following the crest of the wave we can determine its velocity by noting $\pi/2 = \kappa x - \omega t$. Velocity is given by dx/dt so it is equal to ω/κ. This is denoted as the phase velocity v_p and is unique for a single wave. If there are many waves though with different wavelengths, what average velocity do we use?

Let the wave functions be additive and for simplicity assume there are only two to write

$$\Psi = \sin(\kappa_1 x - \omega_1 t) + \sin(\kappa_2 x - \omega_2 t)$$

which can be written

$$\Psi = 2\sin(\bar{\kappa} x - \bar{\omega} t) \times \cos(\Delta\kappa x - \Delta\omega t)$$

where $\bar{\kappa} = [\kappa_1 + \kappa_2]/2$ and $\Delta\kappa = [\kappa_1 - \kappa_2]/2$, with similar relations for ω. This is as if the overall wave function given by the sine function is modulated by the cosine. The modulated wave will move at a velocity given by the cosine function or

$$v_g = \frac{dx}{dt} = \frac{\Delta\omega}{\Delta\kappa} \to \lim_{\Delta\to 0} v_g = \frac{d\omega}{d\kappa} \tag{4.12}$$

Thus the *packet* of waves will move at the group velocity v_g. The group velocity can be rewritten in terms of energy and momentum by using eqns (4.1) and (4.2), as well as the definition of the wave vector, to find $v_g = dE/dp$.

Now the velocity in eqn (4.11) is taken as the group velocity

$$a = \frac{d}{dt}\left(\frac{dE}{dp}\right) = \frac{dp}{dt}\frac{d}{dp}\left(\frac{dE}{dp}\right) = \frac{dp}{dt}\left(\frac{d^2E}{dp^2}\right)$$

The rate of change of momentum dp/dt is force, making the second derivative of the energy with momentum equivalent to a mass, actually it's the inverse of a mass and it is called the effective mass m^*

$$m^* = \left(\frac{d^2E}{dp^2}\right)^{-1} = \hbar^2\left(\frac{d^2E}{d\kappa^2}\right)^{-1} \tag{4.13}$$

Since the very bottom of the conduction band is a local minimum the second derivative will be a positive number, and the more curved the band then the lower will be the effective mass. Generally an electron's effective mass is less than the mass of a free electron and in some cases is much lower. For example, Silicon has a relative electron effective mass (the effective mass in units of the actual electron mass) of 0.97 while Gallium Arsenide has a relative mass of 0.07 which is much smaller than the electron's physical mass. The curvature of energy with wave vector is much greater at the conduction band minimum for Gallium Arsenide.

A free electron has its total energy equal to its kinetic energy, $p^2/2m$, and in semiconductors it is usually assumed that this holds true with the caveat that the electron's mass is given by the effective mass as discussed above. With reference to Fig. 4.7 we can write

Direct band gap

$$\text{Conduction band:}\quad E - E_c = \frac{\hbar^2\kappa^2}{2m_e^*} \tag{4.14a}$$

$$\text{Valence band:}\quad E_v - E = \frac{\hbar^2\kappa^2}{2m_h^*} \tag{4.14b}$$

for a direct band gap material where E_c and E_v are the energies at the very bottom of the conduction band and the very top of the valence band, respectively. Here we have used the variables m_e^* and m_h^* to represent the electron and hole effective masses, respectively.

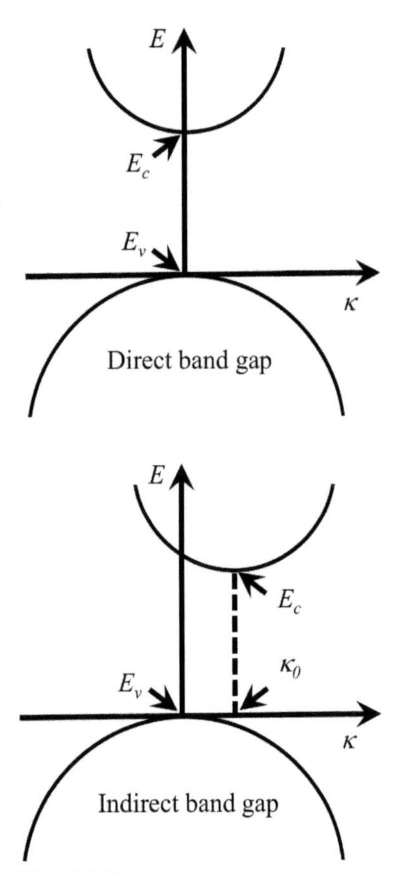

Fig. 4.7 Energy–wave momentum vector diagram showing the difference between a direct and indirect band gap energy.

The above equations deserve some explanation. Firstly, the charge on an electron and hole are negative and positive, respectively. Based on the curves in Fig. 4.7 one would expect an electron to have a positive effective mass (upward curvature) and a hole a negative effective mass (downward curvature). This is not true and by convention the energy for a hole is reversed and increases downward in Fig. 4.7 thus making a hole's effective mass positive. This is a reasonable convention since the hole will try to find the lowest energy possible and by reversing the direction that energy increases will allow this to be possible. With this combination of signs for mass and charge the two particles will move in opposite directions when in an electric field, as expected. The acceleration a particle with charge q will feel in an electric field E is qE/m^* where the sign of q and m^* must be included to get the proper sign of the acceleration.

Finally, eqn (4.14b) can also be applied to an electron in the valence band. In this case the direction energy increases is not changed and an electron in the valence band will have a negative charge and mass, so, it will move in the same direction as a hole when a potential is applied! Regardless, charge transport is determined by movement of electrons in the conduction band and holes in the valence band.

Now, for a material, like Silicon, which has an indirect band gap one can write (see Fig. 4.7)

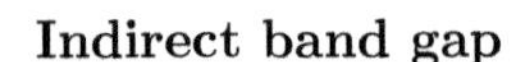

Indirect band gap

$$\text{Conduction band:}\quad E - E_c = \frac{\hbar^2[\kappa - \kappa_0]^2}{2m_e^*} \tag{4.15a}$$

$$\text{Valence band:}\quad E_v - E = \frac{\hbar^2\kappa^2}{2m_h^*} \tag{4.15b}$$

where κ_0 is the wave vector offset between the top of the valence band and the bottom of the conduction band. In order for the electron to be excited into the conduction band it must absorb an energy equivalent to the band gap energy, plus change its momentum by $\hbar\kappa_0$. This can actually occur in two ways, if the absorbed energy is less than the band gap energy then the electron can gain energy from vibrations in the crystalline lattice. These vibrations are called phonons and the electron will gain energy and change its momentum from phonon absorption. Now if the energy is greater than the band gap the electron can lose energy to the lattice and similarly change its momentum to be excited up to the conduction band. Both of these are two-step processes and this accounts for why the absorption of light is much less for indirect band gap materials since its likelihood is much lower than for a direct band gap material.

4.8 Density of states

The density of states is the number of states available to electrons in both the valence and conduction bands. In order to find this we multiply two functions together; one function indicates what states are available and the other finds the probability that they are occupied.[8] Consider the latter function first.

The probability that an electron will be at a certain energy level $F(E)$ is given by

$$F(E) = \frac{1}{\exp([E - \mu]/k_B T) + 1} \tag{4.16}$$

where μ is the chemical potential.[9] This relation is valid for all fermions, that is particles that have half integer spin, such as electrons and holes and is called Fermi–Dirac statistics. Consider a metal at 0 K, absolute zero temperature, as shown in Fig. 4.8. The chemical potential is defined as the energy level a system would achieve if a single electron were added or removed. Since there are so many electrons in the solid material the energy levels available to an electron are very close together (*c.f.* eqn (4.8)) so the chemical potential is essentially the energy of the highest energy electron.

Now consider a semiconductor under a similar condition. Here the highest energy electron is right at the top of the valence band. If an electron were added then it would have to move up in energy to the conduction band E_c, while if it were removed it would essentially have an energy of the top of the valence band E_v. Which one is μ?

For now take the top of the valence band as zero energy, *i.e.* $E_v \equiv 0$, and let the temperature be above absolute zero. If an electron is thermally excited into the conduction band it will occur with probability $F(E_g)$. Since it is excited it must leave behind a hole which will have a probability $1 - F(0)$ of occurring. Clearly these two probabilities must be equal, $F(E_g) = 1 - F(0)$. This can only be true if $\mu = E_g/2$, or more properly $\mu = [E_c + E_v]/2$, as shown in the figure.[10]

One comment should be made about the chemical potential and its confusing use in the literature. Strictly speaking at absolute zero tem-

[8] This is like determining the number of rooms in a building you will visit. It is likely you will go to your office or a meeting room, etc. and it is unlikely you will go to the other sex's toilet or to the utilities room, etc. So, the number of rooms that you will visit is the sum of each room multiplied by the probability you will visit it.

[9] The chemical potential is frequently, if not almost all the time, referred to as the Fermi level. This is confusing as there is the Fermi energy, discussed below, and so here, chemical potential is used. The chemical potential is an energy where half the available states are above it and half below, $F(\mu) \equiv 0.5$.

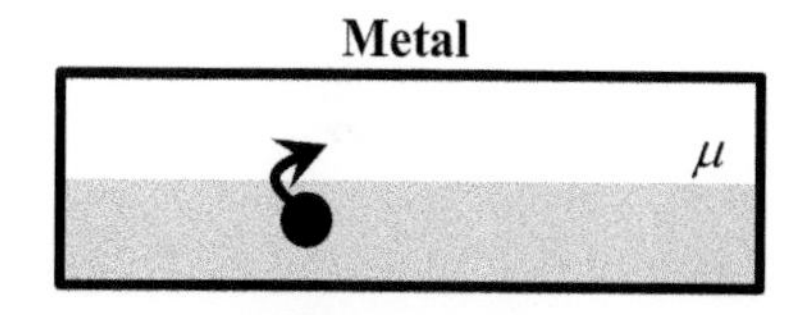

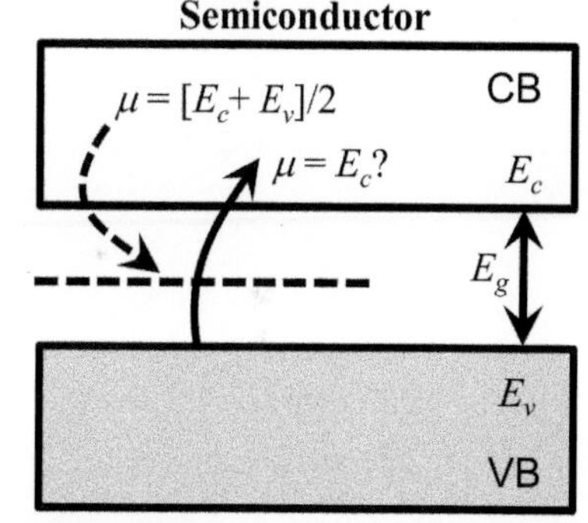

Fig. 4.8 Energy diagram showing the partially filled band of a metal where addition of one electron will increase the energy very slightly and defines the chemical potential μ as that energy. In a semiconductor the valence band (VB) is completely filled and the chemical potential is directly in the middle of the valence and conduction bands (CB) and is not the conduction band's lowest energy, E_c.

[10] The chemical potential is actually not exactly in the center of the energy gap since the density of states in the valence and conduction bands are not necessarily equal, see eqn (4.24).

perature the energy of the uppermost (in energy) electron is (frequently) called the Fermi energy E_F. This is a specific definition and is valid only at 0 K. Above this temperature one uses the chemical potential μ instead of E_F. Yet, the reader will find eqn (4.16) written with E_F rather than μ, which is not strictly correct. In fact, the chemical potential is a very slight function of temperature and so changes depending on the temperature of interest, while the Fermi energy does not due to its definition.

Equation (4.16) is used to find the probability an electron will have a certain energy. Now we find the number of energy levels available in the system. To do this we use the model of electrons confined to a device of size a as given in eqn (4.8) and generalize it to three-dimensions

$$E = \frac{h^2}{8m_e^* a^2}\left[n_x^2 + n_y^2 + n_z^2\right] \equiv \frac{h^2}{8m_e^* a^2}\, n^2 \tag{4.17}$$

where the principal quantum numbers are subscripted according to the x, y or z direction within the crystal. One can write n^2 as the sum of the square of the three individual quantum numbers as shown in the equation. Each quantum number will give a unique energy state and as commented before the energy levels are very close together, so, consider a spherical shell at a quantum number n which is of thickness dn whose volume $4\pi n^2 dn$ will give the number of energy levels available to the system. However, all the quantum numbers must be positive and so the volume is an eighth of this by using only the spherical sector where n_x, n_y and n_z are all positive. The number of energy states at energy E in dE, defined as $Z(E)dE$, is

$$Z(E)\, dE = \frac{1}{2}\pi n^2\, dn \tag{4.18}$$

Rearranging eqn (4.17) so it is explicit in n one can substitute for n and dn in this equation to find

$$Z(E)\, dE = \frac{\pi}{2}\left[\frac{8m_e^* a^2}{h^2}\right]^{3/2} E^{1/2}\, dE$$

Actually this equation is twice as large as expected since each energy level can hold two electrons, one with spin up and the other spin down. Redefine $Z(E)$ and write it on a per unit volume basis or

$$Z(E)\, dE = \frac{\pi}{2}\left[\frac{8m_e^*}{h^2}\right]^{3/2} E^{1/2}\, dE \;[=]\; \frac{\#}{\mathrm{m}^3} \tag{4.19}$$

where a^3 is the volume of the crystal.

Now we are in a position to find the number of electrons in the conduction band of a semiconductor. First assume zero energy is at the top of the valence band and write $Z(E) \propto [E - E_g]^{1/2}$. Next multiply $F(E)$ and $Z(E)$ together to find the number of electrons per unit volume $N(E)$ in the energy range dE

$$N(E)\, dE = Z(E) \times F(E)\, dE$$

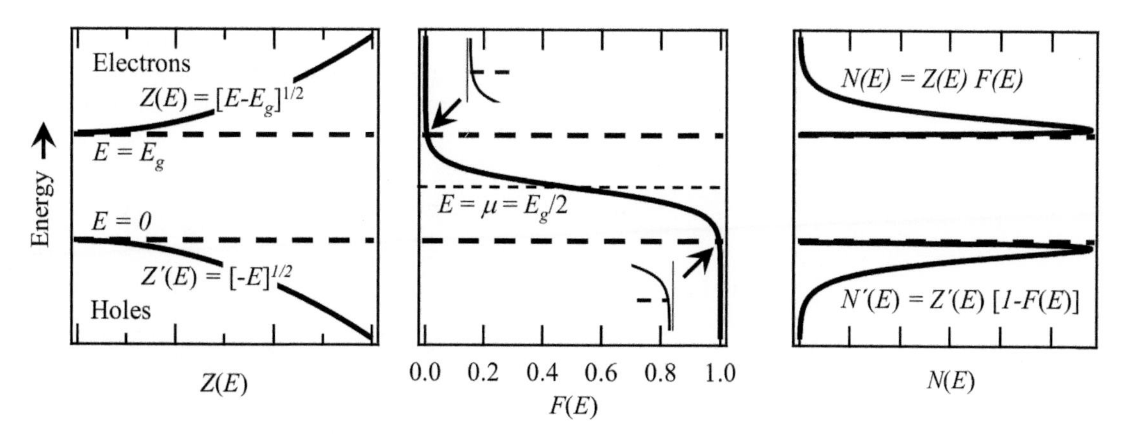

Fig. 4.9 The density of states $Z(E)$, the Fermi function $F(E)$ and the densities of electrons $N(E)$ and holes $N'(E)$ at a temperature above absolute zero. The energy was made equal to zero at the valence band edge. The insets for the Fermi function shows how close the function is to its limits near the band edges. The position for the chemical potential μ is also shown and for an intrinsic semiconductor it lies in the middle of the band gap (not quite in the middle, see eqn (4.24)).

or

$$N(E)\, dE = \frac{\pi}{2}\left[\frac{8m_e^*}{h^2}\right]^{3/2}[E - E_g]^{1/2}\frac{1}{\exp([E-\mu]/k_BT)+1}\, dE \quad (4.20)$$

The number of holes per unit volume $N'(E)$ in the energy range dE is given by

$$N'(E)\, dE = Z'(E) \times [1 - F(E)]\, dE$$

or

$$N'(E)\, dE = \frac{\pi}{2}\left[\frac{8m_h^*}{h^2}\right]^{3/2}[-E]^{1/2}\left[1-\frac{1}{\exp([E-\mu]/k_BT)+1}\right] dE \quad (4.21)$$

where the energy is negative in value for holes since we take the top of the valence band as zero energy. The density of states for holes is given by the symbol $Z'(E)$.

Figure 4.9 shows the result of these calculations. The density of states for electrons follows a parabola and would increase to infinity at infinite energy, as would that for holes at large negative energy. However, the Fermi function tempers this and the densities of electrons and holes peak near the band edges and tend to zero rapidly after these energies.

The total number density of electrons and holes, n and p, respectively, is found by integrating the density of electrons and holes over all energy levels

$$n = \int_{E_c}^{\infty} N(E)\, dE\, [=] \frac{\#}{\text{m}^3} \quad \text{and} \quad p = \int_{-\infty}^{E_v} N'(E)\, dE\, [=] \frac{\#}{\text{m}^3}$$

These integrals are difficult to evaluate and so assume that $F(E) \approx \exp(-[E-\mu]/k_BT)$ and is a good approximation for ambient temperatures. Using this approximation and performing the integrals one finds

$$n = N_c \exp(-[E_c - \mu]/k_BT), \text{ with } N_c = 2\left[\frac{2\pi m_e^* k_B T}{h^2}\right]^{3/2} \tag{4.22a}$$

and

$$p = N_v \exp([E_v - \mu]/k_BT), \text{ with } N_v = 2\left[\frac{2\pi m_h^* k_B T}{h^2}\right]^{3/2} \tag{4.22b}$$

The variables N_c and N_v are the effective density of states in the conduction and valence bands, respectively, and are not equal in general since $m_e^* \neq m_h^*$.

When an electron is promoted up to the conduction band it leaves behind a hole; one must realize that the number density of the two should equal each other or, $n = p$. If eqns (4.22a) and (4.22b) are multiplied together then the number of intrinsic carriers n_i can be found

$$n_i^2 \equiv n \times p = n^2 = p^2 = N_c N_v \exp([E_v - E_c]/k_BT)$$

so

$$n_i^2 = N_c N_v \exp(-E_g/k_BT) \tag{4.23}$$

and is sometimes called the *law of mass action* which will be appreciated more later. If we set them equal to each other then we arrive at

$$N_c \exp(-[E_c - \mu]/k_BT) = N_v \exp([E_v - \mu]/k_BT)$$

which yields

$$\mu = \frac{E_c + E_v}{2} + \frac{k_BT}{2}\ln\left(\frac{N_v}{N_c}\right) \tag{4.24}$$

This is the chemical potential or, as it is frequently called, the Fermi level.

Above we indicated an electron right at the conduction band edge had the probability $F(E_g)$, which left behind a hole in at the valence band edge with the probability $1 - F(0)$, which must equal each other to find the chemical potential. This is not exactly correct and for an intrinsic semiconductor[11] one must use the correction in eqn (4.24) to find that it is not exactly in the center, however, it isn't off by that much. The reason μ is not exactly in the center of the band gap energy is because the number of states differs between the valence and conduction bands. If they are exactly equal, meaning $m_e^* = m_h^*$, then the chemical potential is exactly in the middle.

[11] The material described here is called an *intrinsic semiconductor* since its properties are *intrinsic* to it. If impurity atoms or *dopants* are added then it is a *doped* or *extrinsic semiconductor*, which is used in solar cells. Doped semiconductors are considered in Chapter 6 and it is found that eqn (4.24) does not apply. This is extremely important to the operation of a solar cell and the discussion in Chapter 6 is centered on describing this phenomenon.

4.9 Conclusion

Discussion in this chapter centered around electrons and holes is solids. A brief introduction to quantum mechanics was given so that simple problems could be solved that naturally yield the band structure seen in semiconductors. The electron in a box showed that allowed energy levels were quantized, however, since the *box* is macroscopic in size the electron essentially sees a continuum of energy levels available to it. Then the electron was exposed to a periodic array of potentials and when the potential was increased to infinity the solution to the Schrödinger wave equation naturally arrived at the previous solution. Yet, when it was decreased so that the electron could experience the entire periodic array, *energy bands* became apparent dictating where the electron can reside. Although quantum mechanical calculations on a real material are more involved, the Kronig–Penney model serves to show that it is this periodic array which is responsible for band gap energies.

After consideration of electron and hole dynamics, details of the energy levels experienced by these particles in the potential could be established. Ultimately this allowed us to calculate the density of states near the band edge and the number density of electrons and holes. These values are required since they set the number of intrinsic carriers within the material and also the chemical potential; both important quantities to the operation of a photovoltaic device. Although this chapter represents a cursory introduction to semiconductor physics this will allow the reader to progress within this book should they not be familiar with the topic and to fully appreciate how a photovoltaic device operates.

4.10 General references

M.A. Green, 'Solar cells: Operating principles, technology and system applications,' Prentice-Hall Inc. (1982).

A.S. Grove, 'Physics and technology of semiconductor devices,' John Wiley & Sons (1967) (This is an excellent text to learn semiconductor physics and was written by the eventual Chief Executive Officer and President of Intel Corporation. He did his PhD in chemical engineering at the University of California - Berkeley with Professor Andreas Acrivos in the area of fluid dynamics! So, please read the Preface of this book again to realize that you really can't know the type of job you will have in the future!)

T.S. Hutchinson and D.C. Baird, 'The physics of engineering solids,' John Wiley and Sons (1963).

D.A. McQuarrie, 'The Kronig–Penney model: A single lecture illustrating the band structure of solids,' Chem. Educator **1** (1996) 01003-5.

B.G. Streetman, 'Solid state electronic devices,' Prentice-Hall Inc. 7th Edition (2014).

Exercises

(4.1) Find the momentum of an electron travelling at the speed of light. Compare this to the momentum a car would have travelling at 100 kph. Ignore relativistic effects and use the rest mass of an electron.

(4.2) Determine the normalized wave function for an electron in a one-dimensional box and plot the probability of finding the electron at certain positions in the box for the first three principal quantum numbers (*i.e.* for $n = 1, 2, 3$).

(4.3) Estimate the magnitude of V_0 in the Kronig–Penney model and calculate the value of P to the best of your ability. Is this value of P of an order to produce bands in a crystalline array? If so what is the lowest band gap energy? Based on the above, do you think the Kronig–Penney model is a reasonable model for semiconductor materials? Hint: Let V_0 be equal to the ionization potential of Silicon as a first approximation. What is the ionization potential?

(4.4) The following values are available for the lowest energy band in Fig. 4.5. Determine the dimensionless potential energy in units used to make the total energy dimensionless, *i.e.* dimensionless total energy $\equiv [\alpha a]^2$.

κa	$[\alpha a]^2$
0.254	5.12
1.27	6.32
1.88	7.64
2.49	9.10
3.14	9.87

(4.5) Determine the effective mass of a free electron that only has kinetic energy.

(4.6) Determine the number of intrinsic carriers n_i in Silicon at room temperature using eqn (4.23) in dimensions of $\#/\text{cm}^3$. The effective mass of the electron and hole are 1.08 m_e and 0.81 m_e, respectively, and assume the temperature is 300 K.

(4.7) Calculate the correction to the chemical potential for Silicon given in eqn (4.24) at 300 K and compare its magnitude to the approximate value of the chemical potential. Use the effective masses for the electron and hole given in Exercise 4.6. Is it significant?

(4.8) What is the probability an electron will occupy an energy level equal to the chemical potential and what is the probability for an energy 0.5 eV higher than the chemical potential?

Light absorption

5

The optical properties of materials is a deep and complex subject of study covering many aspects of the interaction of radiation with materials. There are many textbooks written on this topic and here a concise, and more or less macroscopic, discussion is given. Specifically, consideration of how much radiation absorbed by a device is performed.

The absorbance is the key parameter in the performance of any solar powered device. It dictates the optical properties of materials used in solar devices and indicates how much radiation is absorbed within a given distance as it penetrates the device. Since every material of interest will absorb at least a small amount of radiation, a very thick sample will not allow radiation to escape. Materials used in solar devices of course cannot be infinitely thick to absorb all the radiation, and the absorbance is of a magnitude to allow most of the radiation to be captured while balancing transport properties, like electrical and thermal conductivity, to make an efficient, cost effective device. Thus, this material property is a key component in product optimization.

The absorption of radiation is unusual in that, for a material which has a band gap energy (*i.e.* not a metal), low energy radiation will not be absorbed. This is equivalent to stating that it is easier to catch a ball that is thrown to you very fast rather than one gently rolled to you. The reason is that an electron must be excited to a higher energy level and if the radiation is not of sufficient energy then it is not absorbed. This is why insulators are optically clear, they have a high band gap energy and so cannot absorb much visible radiation at all. It is the impurities in a diamond that give it color since Carbon in a tetrahedral structure has a very large band gap energy.

The quantum world is very different to the physical world around us and the remarkable optical properties of materials are controlled by it. The electron is the primary part of a material that interacts with radiation, after all what else is there? There are electrons and the atomic nuclei; whether atoms are covalently bound, ionically coupled or just interact through van der Waals forces is not a primary concern. The electrons are the part of a material that can interact with light, or radiation in general, and (usually) not the nucleus made of protons and neutrons. The fact that electrons live in discrete energy levels, which become somewhat blurred when atoms are forced to be together in condensed matter, dictates what radiation, if any, will be absorbed.

To gain some understanding of radiation absorption, a discussion of how much radiation is absorbed within a given distance is presented first.

Solar Energy, An Introduction, First Edition, Michael E. Mackay

This is followed by a brief discussion of the absorption coefficient and how the band gap energy is determined. Finally, how much heat can be generated within a material as it absorbs radiation from the Sun is discussed.

5.1 Absorption of radiation

The integrated photon number rate, $N_p(E)$, from Planck's law discussed in Chapter 2, provides the number of photons available to the solar device at its surface

$$N_p(E) = \frac{2g_s}{h^3c^2}\frac{E^2}{\exp(E/k_BT_S) - 1}[=]\frac{\#}{\text{s-m}^2\text{-J}} \tag{2.2}$$

where $g_s \equiv \sin(\theta_{sub})^2$ (θ_{sub} is the angle subtended between the Sun and Earth), h, Planck's constant, c, the speed of light, E, energy, k_B, Boltzmann's constant and T_S, the (black body) temperature of the Sun. The symbol [=] means 'has dimensions of.' We want to determine whether the photon is absorbed and use the absorption coefficient, $\alpha(E)$, usually given in units of cm^{-1}, as a measure for the probability a photon is absorbed in a unit length. The absorption coefficient is a function of the radiation energy E and for semiconductors is zero below a certain energy level representing the band gap energy.

The product of α with $N_p(E)$ yields the generation rate of excited states per unit volume per unit energy, assuming each photon generates an excited state, $G(E, x)$

$$G(E,x) = \alpha(E) \times N_p(E,x)[=]\frac{\#}{\text{s-m}^3\text{-J}} \tag{5.1}$$

where we have written $N_p(E, x)$ since the number of photons will decrease with distance through the sample, x, as they are absorbed. The generation rate will also change accordingly. In a photovoltaic device or solar cell the excited state will (hopefully) be a free electron and hole to produce power while in a solar thermal application the electron will fall back to its base energy level while releasing the excess energy to the crystalline lattice (*i.e.* the material) to increase its temperature.

At steady state the continuity equation for the photons is[1]

$$\frac{dN_p(E,x)}{dx} = -G(E,x) = -\alpha(E)N_p(E,x) \tag{5.2}$$

where the final equation results by substitution of eqn (5.1). This equation is easily solved to yield

$$N_p(E,x) = N_p(E)\exp(-\alpha(E)x) \tag{5.3}$$

where $N_p(E)$ is the number rate at the sample surface and we have assumed that none of the photons were reflected from the surface. This relation can be cast in a more familiar form if both sides are multiplied by E to yield the energetic irradiance and this may be called the *intensity*, I.

[1] The continuity equation merely means a balance and as stated in eqn (5.2) means as the photons move into the sample they are lost when they are absorbed.

Consider a sample with thickness L, then the intensity of light $I(E)$ that will propagate through it is given by the *Beer–Lambert–Bouguer law*

$$I(E) = I_0(E)\exp(-\alpha(E) \times L) \quad (5.4)$$

where $I_0(E)$ is the incident intensity on the sample and $I(E)/I_0(E)$ is the transmittance τ (see (2.16)).[2] In many cases α can be written as a function of frequency ν rather than energy. Furthermore, many physical chemistry textbooks have α written in other forms particularly if the absorbing species is dissolved in solution at concentration c: $\alpha(\nu, c) = \epsilon(\nu) \times c$, where ϵ is the molar absorption coefficient. Since we are considering absorption by bulk materials we will not use this formalism and exclusively use $\alpha(\nu)$, or $\alpha(E)$.

[2] Careful with the definition of τ as it is sometimes defined as

$$\tau = 10^{-\alpha(E)\times L}$$

One limitation of this law is that scattering could influence the results. For example, particulates that are present in a sample could scatter the radiation making a false absorption coefficient and erroneous results are obtained. So, care should be taken to ensure the absorption properties are being measured and not data affected by spurious effects. Furthermore, the surface of the material could reflect the radiation. Consider the very surface of the sample, if the fraction of light that is reflected is $\rho(E)$ then $1 - \rho(E)$ will pass into the material. This will modify eqn (5.3) to be

$$N_p(E, x) = [1 - \rho(E)]N_p(E)\exp(-\alpha(E)x) \quad (5.5)$$

which may be a sizable effect. In this case $N_p(E)$ is related to the number of incident photons *not* those absorbed by the material. The amount of radiation reflected can be measured with certain instruments and is a valuable material property to determine since, in general, it should be minimized in solar photovoltaic and thermal applications.

An example for the absorption properties of a material is given in Fig. 5.1. Monocrystalline Silicon, in this case, shows a complicated absorption spectrum and absorbs light to varying degrees depending on the radiation energy level or wavelength. This dictates the sample thickness required to absorb most of the radiation or really light since much of the energy magnitudes in the figure are for optical wavelengths. At a 500 nm wavelength Silicon has an absorption coefficient of approximately 10^4 cm^{-1}, which seems quite high. According to eqn (5.4) though, a Silicon sample must be about 2 μm thick to absorb 90% of the light, which is quite thick compared to the absorptive properties of many materials. This relatively poor absorption coefficient is partly due to the fact that crystalline Silicon has an indirect band gap requiring generation of an excited state plus a change in electron momentum.

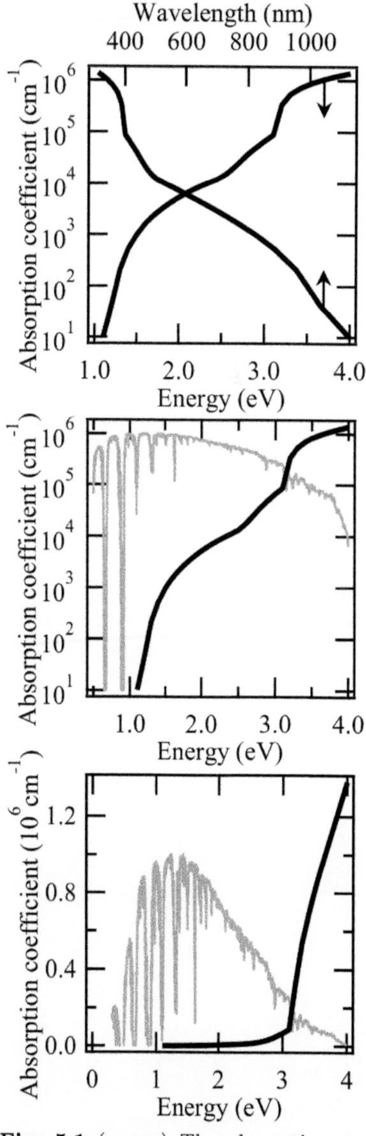

Fig. 5.1 (upper) The absorption coefficient for crystalline Silicon demonstrating how it varies with the radiation energy or wavelength. The absorption spectrum compared to the normalized AM1.5G energetic irradiance on a log - linear graph (middle) and linear-linear graph (lower). Source: http://www.ioffe.ru/SVA/NSM/Semicond/Si/index.html.

Moving to the lower energy region of the optical spectrum one finds $\alpha(E)$ approaches 10^3 cm^{-1} suggesting that the sample should be 10 times thicker to absorb a similar amount of the light. So, if the sample or device were made to absorb most of the radiation at a given frequency it may not absorb much at another as it may be too thin. Device design is truly dictated by the absorption spectrum and must be matched to the terrestrial solar spectrum on the Earth's surface as given in Fig. 2.3. Suffice to say at this point the absorption spectrum is key to defining the device performance. Also shown in Fig. 5.1 is a comparison of the absorption spectrum to the energetic irradiance. A linear - linear scale clearly demonstrates Silicon strongly absorbs at higher energy levels when the energetic irradiance is quite small.

Closer inspection of Fig. 5.1 reveals that not much radiation is absorbed for energies below 1.1 eV. The reason for the limited absorption below this energy level is due to the *band gap energy* of Silicon which is approximately 1.1 eV, so, radiation below this energy level cannot be absorbed.

What is a band gap energy? Conceptually, it represents an energy level which must be achieved for absorption to occur, as discussed in the previous chapter. A more detailed understanding demands consideration of the energy levels available to an electron in a material since electrons are the excited entities when radiation is absorbed, as discussed in Chapter 4. Clearly, this must be so since a photovoltaic device produces electrical current when exposed to light.

Consider the generation rate of electron - hole pairs that can be written as

$$G(E,x) = N_p(E)\alpha(E)\exp(-\alpha(E)\,x) \tag{5.6}$$

by combination of eqns (5.1) and (5.3). The generation rate can be integrated over the sample thickness to find

$$G_L(E) = \int_0^L G(E,x)dx = N_p(E) \times [1 - \exp(-\alpha(E) \times L)] \tag{5.7}$$

Integrating $G_L(E)$ over all energy levels (that can be absorbed) yields the generation rate of excited states G

$$G = N_{out} - \int_{E_g}^{\infty} N_p(E)\exp(-\alpha(E) \times L)\; dE \;[=]\; \frac{\#}{\text{s-m}^2} \tag{5.8}$$

If the absorption coefficient tends to infinity for all energy levels above the band gap, then we would find the same result as given in eqn (3.22), N_{out}.

The absorption coefficient is a complicated function of energy, as seen in Fig. 5.1, so the integral in eqn (5.8) is a complicated one. Assuming it is a constant value above the band gap energy one can determine a first-order correction to the maximum current density expected from a photovoltaic device

$$J = J_{out}\,[1 - \exp(-\langle\alpha\rangle \times L)] \tag{5.9}$$

where $\langle\alpha\rangle$ is a constant, J is equal to $G \times q$, where q is the charge of an electron, and the maximum current density J_{out} is given in eqn (3.25). This is the first correction to the maximum current a photovoltaic device can provide and is related to the absorptive properties of the material.

Equation (5.8) is more complicated that one realizes. In addition to $\alpha(E)$ being a function of energy E so too is $N_p(E)$. In fact, for a real spectrum it is a very complicated function of energy as seen in Fig. 2.3. So, the integral will most likely have to be integrated numerically and an analytical solution is not in general possible. However, eqn (5.9) gives, to first order, the effect of a non-infinite absorption coefficient on the current that can be generated.

Example 5.1

How thick does a Silicon photovoltaic device have to be so that 99% of the theoretical maximum current is provided by the cell?

Reference to eqn (5.9) dictates that $\exp(-\langle\alpha\rangle \times L)$ must be equal to 0.01. This equation was derived by assuming the absorption coefficient was constant and not a function of the radiation energy level. As seen in Fig. 5.1 this is certainly not the case for Silicon. However, if it is assumed that it is of order $10^3 - 10^4$ cm^{-1} for the energy levels within the solar spectrum (see Fig. 2.3) one can write

$$L = -\frac{\ln(0.01)}{\alpha} \approx 5 - 50\ \mu\text{m}$$

Since most Silicon photovoltaic devices are 200 - 500 μm thick it is clear that much of the solar radiation is absorbed.

The above example demonstrates, in a simple way, how thickness affects the current by assuming a constant absorption coefficient. The following example shows how to perform a more accurate calculation by letting the absorption coefficient vary over the solar spectrum. Rather than a numerical integration a piecewise continuous method is used allowing a rapid and reasonably accurate calculation to be made.

Example 5.2

Determine the current from a Silicon photovoltaic device when exposed to the AM1.5G spectrum. Assume the photovoltaic device is 300 μm thick.

The absorption coefficient for Silicon is given in Table 5.1 by assuming it is constant over small energy increments. One must now evaluate the integral in eqn (5.8) over the selected energy increments by assuming α is a constant. In particular one must solve

$$J = q \times \int_{E_g}^{\infty} N_p(E)\,[1 - \exp(-\alpha(E) \times L)]\ dE \qquad (5.10)$$

Assume black body radiation applies for the energy of the Sun at the Earth's surface allowing one to write

Table 5.1 The absorption coefficient for Silicon at certain energy levels.

E(eV)	α(cm^{-1})
1.2	4.24×10^1
1.5	9.60×10^2
1.9	3.11×10^3
2.3	9.42×10^3
2.7	2.51×10^4
3.1	8.74×10^4
3.5	8.19×10^5
4.1	1.49×10^6
$>$4.1	1.79×10^6

$$J = q \times \int_{E_g}^{\infty} \frac{2\pi \sin(\theta_{sub})^2}{h^3 c^2} \frac{E^2}{\exp(E/k_B T_S) - 1} [1 - \exp(-\alpha(E) \times L)] \; dE$$

A change of variables is warranted by allowing $x \equiv E/k_B T_S$ and write

$$J = K_J \int_{x_g}^{\infty} \frac{x^2}{e^x - 1} [1 - \exp(-\alpha(x) \times L)] \; dx$$

where $K_J \equiv 2\pi q \sin(\theta_{sub})^2 [k_B T_S]^3 / [h^3 c^2] = 333.4\ \text{A/m}^2$ ($33.34\ \text{mA/cm}^2$). To solve this integral one must first assume α is constant over small energy ranges, represented by x_1 and x_2, and then integrate by parts

$$\int_{x_1}^{x_2} \frac{x^2}{e^x - 1} \; dx = x^2 \ln(1 - e^{-x}) \Big|_{x=x_1}^{x_2} - \int_{x_1}^{x_2} 2x \ln(1 - e^{-x}) \; dx$$

then the natural logarithm is expanded in a Maclauren series $\ln(1-e^{-x}) = -[e^{-x} + e^{-2x}/2 + e^{-3x}/3 + \cdots] = -\sum_{n=1}^{\infty} e^{-nx}/n$ which will allow the integral to be easily evaluated. Now eqn (5.10) can be written

$$\frac{J}{K_J} = \sum_{i=1}^{N} \left[1 - e^{-\alpha_i L}\right] \times \left[x_{i+1}^2 \ln(1 - e^{-x_{i+1}}) - x_i^2 \ln(1 - e^{-x_i}) - 2 \sum_{n=1}^{\infty} \frac{n x_{i+1} + 1}{n^3} e^{-n x_{i+i}} + 2 \sum_{n=1}^{\infty} \frac{n x_i + 1}{n^3} e^{-n x_i}\right]$$

where α_i is the (constant) absorption coefficient used over the dimensionless energy range of x_i to x_{i+1} and N ranges have been assumed. For the values of x of interest, the infinite series can be truncated after the first term, as was done for the series used to evaluate N_{out}, so

$$\frac{J}{K_J} \approx \sum_{i=1}^{N} \left[1 - e^{-\alpha_i L}\right] \times \left[x_{i+1}^2 \ln(1 - e^{-x_{i+1}}) - x_i^2 \ln(1 - e^{-x_i}) - 2[x_{i+1} + 1] e^{-x_{i+i}} + 2[x_i + 1] e^{-x_i}\right] \equiv \sum_{i=1}^{N} \left[1 - e^{-\alpha_i L}\right] f_i \tag{5.11}$$

where the definition of f_i is determined from the equation.

The black body equivalent of the AM1.5G spectrum (T_S = 5359 K) was assumed and the results of the calculation are shown in Table 5.2. The photovoltaic device is thick enough to essentially absorb all the radiation for this spectrum and the sum of $[1 - \exp(-\alpha_i L)] f_i$ is 1.111 making J equal to 37.03 mA/cm^2, which is quite large (if α were assumed to tend to infinity then the sum is 1.175 rather than 1.111). This is a simple calculation and a more refined model integrating the true spectrum with a more complicated absorption coefficient as a function of radiation energy will yield a different value. For the purpose here though this is an accurate enough calculation and the maximum current a Silicon photovoltaic device can make is 37 mA/cm^2.

Table 5.2 Calculation results following eqn (5.11).

i	E_i (eV)	x_i	α_i (cm^{-1})	$1 - \exp(-\alpha_i L)$	f_i
1	1.1	2.382	4.24×10^1	0.720	0.2281
2	1.3	2.815	9.60×10^2	1.00	0.3649
3	1.7	3.681	3.11×10^3	1.00	0.2438
4	2.1	4.548	9.42×10^3	1.00	0.1498
5	2.5	5.414	2.51×10^4	1.00	0.0868
6	2.9	6.280	8.74×10^4	1.00	0.0481
7	3.3	7.146	8.19×10^5	1.00	0.0258
8	3.7	8.012	1.49×10^6	1.00	0.0204
9	4.5	9.745	1.79×10^6	1.00	0.0068
10	10	21.66	–	–	–

If the cell thickness is great enough then all the radiation will be absorbed. This can be quantified by considering the photon mean free path, $\bar{x}$, meaning that if the cell thickness is much greater than this then almost all the radiation would be absorbed

$$\bar{x} = \frac{\int_0^\infty x \exp(-\alpha x)\, dx}{\int_0^\infty \exp(-\alpha x)\, dx} = \frac{1}{\alpha} \tag{5.12}$$

where we have assumed that the cell thickness is infinity.[3] Certainly, Silicon has a very large absorption coefficient for high enough energy radiation (Fig. 5.1) and the photon mean free path would be very small. Near the band gap energy, though, the path length would be very large and not much of this energy radiation would be absorbed. Comparing to the example above with $\alpha = 10^3 - 10^4$ cm^{-1} the range of $\bar{x}$ will be 1 - 10 μm, which is a comparable order of magnitude for absorbing most of the light.

One final comment should be made about the absorption coefficient and its relation to other optical properties. The complex refractive index m is written as $n[1 - ik]$ where nk is the damping factor and k the extinction coefficient. The absorbance a is related to k *via* $4\pi k/\lambda$ demonstrating the complex part of the refractive index is related to absorption.[4]

[3]We wrote following eqn (5.4) that $\alpha(\nu, c) = \epsilon(\nu) \times c$ for a dilute solution of absorbers. The molar absorption coefficient can be written as $\epsilon(\nu) = \sigma \times N_A$ where N_A is Avogadro's number. The constant σ is the absorption cross-section and one can write $\bar{x} = 1/\sigma N_A c$. Equation (5.12) is similar to the mean free path between molecular collisions in an ideal gas $\lambda = 1/[\sqrt{2}\sigma N_A c]$ where σ is the collision cross-section. The factor of $\sqrt{2}$ comes from the molecular movement through Brownian motion that effectively increases the cross-section.

[4]What is the relation between the absorbance and the absorption coefficient? Equation (5.4) gives the transmittance τ as $\tau(E) = \exp(-\alpha(E) \times L)$ so, assuming there is no reflection at the surface, their relation can be written

$$a = 1 - \exp(-\alpha(E) \times L)$$

Some care should be taken here as the absorbance is sometimes defined as $-\log_{10}(I(E)/I_0(E))$.

5.2 Absorption coefficient

Consider absorption of radiation in a direct band gap material shown in Fig. 5.2 as a graph of energy E versus wave vector κ. The electron has initial energy E_i and absorbs enough radiation to have energy E_f above the band gap energy. Assuming parabolic bands, as discussed in Chapter 4, we can write

$$E_f - E_c = \frac{\hbar^2 \kappa^2}{2m_e^*} \tag{5.13a}$$

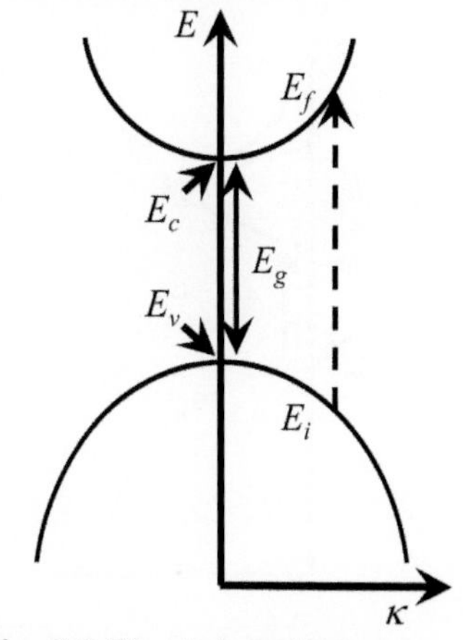

Fig. 5.2 Direct absorption.

$$E_i - E_v = -\frac{\hbar^2\kappa^2}{2m_h^*} \tag{5.13b}$$

where we ignore the sign for the effective mass of a hole. Subtracting these two equations results in

$$E - E_g = \frac{\hbar^2\kappa^2}{2}\left[\frac{1}{m_e^*} + \frac{1}{m_h^*}\right] \tag{5.14}$$

Equation (4.19) provided the number of energy states at energy E in dE ($Z(E)dE$) and is needed to find the likelihood of absorption for a given photon energy. Re-write this in terms of the wave vector since $E = \hbar^2\kappa^2/2m_e^*$ for parabolic bands to find

$$Z(E)\; dE = \frac{1}{\pi^2}\;\kappa^2\; d\kappa \tag{5.15}$$

Equation (5.14) can be written in terms of κ and substituting into eqn (5.15) finds

$$Z(E)\; dE = \frac{\sqrt{2}m_r^{3/2}}{\pi^2\hbar^3}\;[E - E_g]^{1/2}\; dE \tag{5.16}$$

where m_r is the reduced mass, $1/m_r = 1/m_e^* + 1/m_h^*$.[5]

To find the absorption coefficient from the above equation is not difficult, however, the details are beyond the scope of this monograph. The number of states available $Z(E)$ is used to find the overall probability per unit volume per unit time $\mathcal{P}$ that energy E is absorbed. The absorption coefficient can be written as

$$\alpha(E) = \frac{\mathcal{P}E}{\text{incident energy per unit time per unit area}}[=]\frac{1}{\text{cm}}$$

where cm^{-1} is the traditional unit. The denominator is the time averaged Poynting vector, usually written as $\langle S\rangle$, which is the irradiance for an electromagnetic wave. It is equal to $1/2\, nc|E_0|^2$, where n is the real part of the refractive index, c, the speed of light in vacuum and $|E_0|$, the electric field amplitude of the electromagnetic wave.[6] From this one can determine the absorption coefficient

$$\alpha(E) = A^*\;[E - E_g]^{1/2} \tag{5.17}$$

where A^* is a group of constants. This relation applies near the band edge, meaning near the band gap energy, since parabolic bands were assumed and is useful to find E_g as described below.

Now consider an indirect transition, it is known that a momentum change must accompany the electron energy change, and is shown schematically in Fig. 5.3. The momentum change is mediated by *phonons* in the crystalline material and is represented by E_p in the figure. A phonon is a vibration of the atoms making up the crystal and these are in fact coordinated vibrations. The electron can gain or lose momentum from/to the crystalline lattice, which are denoted as $E_g - E_p$ and $E_g + E_p$

[5]We found $Z(E)$ for an electron and hole in the conduction and valence band, respectively, and used m_e^* and m_h^* in each $Z(E)$ expression. Equation (5.16) has m_r, which is smaller than m_e^* and m_h^*, thereby making $Z(E)$ smaller. This is because we are considering absorption which requires generation of a hole and electron with energy difference E which restricts the number of states available.

[6]Have you ever wondered why the irradiance (or what is called *intensity* by many) is proportional to the square of the amplitude of the electric vector $|E_0|^2$? A physical reason can be stated like this. The frequency of light is very large, of order 10^{15} Hz, while processes used to absorb light are many orders of magnitude slower, thus, the material will effectively integrate the light wave over time since it cannot react fast enough resulting in the square of the amplitude!

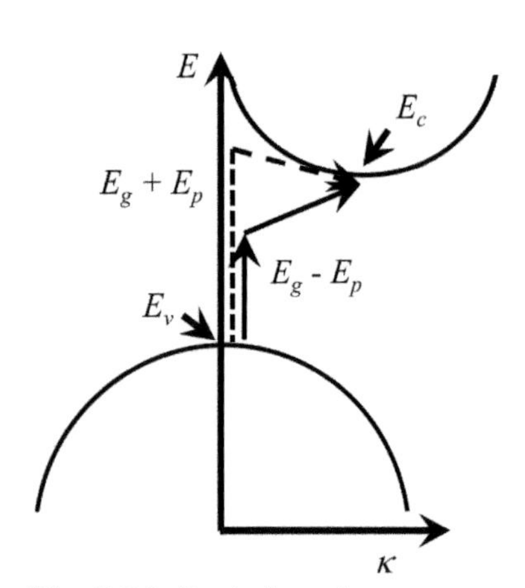

Fig. 5.3 Indirect absorption.

in the figure. Notice that the arrows have a slope since it would be impossible for them to be horizontal. Remember a larger wave vector κ suggests an increase in momentum and hence energy. If the electron gains momentum then it will simultaneously gain energy and the slope is positive, as shown for the arrow labeled $E_g - E_p$, while the opposite is true for the $E_g + E_P$ case (*n.b.* clearly the slope would have to be different in these two cases).

Derivation of the absorption coefficient is more involved and one finds

$$\text{absorption: } \alpha_a(E) = \frac{A[E - E_g + E_p]^2}{\exp(E_p/k_B T) - 1} \quad \text{for} \quad E > E_g - E_p \tag{5.18a}$$

$$\text{emission: } \alpha_e(E) = \frac{A[E - E_g - E_p]^2}{1 - \exp(E_p/k_B T)} \quad \text{for} \quad E > E_g + E_p \tag{5.18b}$$

making

$$\alpha(E) = \alpha_a(E) + \alpha_e(E) \tag{5.18c}$$

One can see that α for an indirect band gap material varies as E^2 while a direct band gap has a $E^{1/2}$ dependence. This is applicable near the band gap energy and the complete absorption spectrum is much more complicated as seen in Fig. 5.1. In fact one can write the Silicon absorption spectrum as a sum of indirect and direct transitions

$$\alpha(E,T) = \sum_{i=1,2}\sum_{j=1,2} A_{ij}\left[\frac{[E - E_{g,j}(T) + E_{p,i}]^2}{\exp(E_{p,i}/k_B T) - 1} + \frac{[E - E_{g,j}(T) - E_{p,i}]^2}{1 - \exp(E_{p,i}/k_B T)}\right] + A_d[E - E_{gd}(T)]^{1/2} \tag{5.19}$$

where the constants are given in Table 5.3. As eqn (5.18) suggests, the various terms in eqn (5.19) can only be applied when the photon energy is greater than a given band gap energy. Reference to Fig. 4.6 shows three possible transitions with one being direct. Clearly the terms involved with absorption of a photon are much greater than the phonon energy required to move the electron up to a larger energy value for the indirect transitions. Although phonons have low energy they do have relatively large momentum, which is opposite to that of photons, since the particles (atoms) involved have much larger mass.

In summary, the absorption coefficient is a vastly different function of energy for indirect and direct band gap materials. Near the band gap energy they can be written

$$\text{Direct: } \alpha(E) \propto [E - E_g]^{1/2} \tag{5.20a}$$

$$\text{Indirect: } \alpha(E) \propto [E - E_g]^2 \tag{5.20b}$$

Of course, far away from the band gap energy the other transitions come into play and the absorption spectrum becomes more complex (see Fig.

Table 5.3 Constants for eqn (5.19). Note: $E_g(T) = E_g(0) - \beta T^2/[T + 1108\text{ K}]$, $\beta = 7.021 \times 10^{-4}$ eV/K. Source: M.A. Green, 'Solar cells: Operating principles, technology and system applications,' Prentice-Hall Inc. (1982).

Constant	Value
$E_{g,1}(0)$	1.156 eV
$E_{g,2}(0)$	2.5 eV
$E_{gd}(0)$	3.2 eV
$E_{p,1}$	1.827×10^{-2} eV
$E_{p,2}$	5.773×10^{-2} eV
A_{11}	1.777×10^3 cm^{-1}/eV^2
A_{12}	3.980×10^4 cm^{-1}/eV^2
A_{21}	1.292×10^3 cm^{-1}/eV^2
A_{22}	2.895×10^4 cm^{-1}/eV^2
A_d	1.052×10^6 cm^{-1}/eV$^{1/2}$

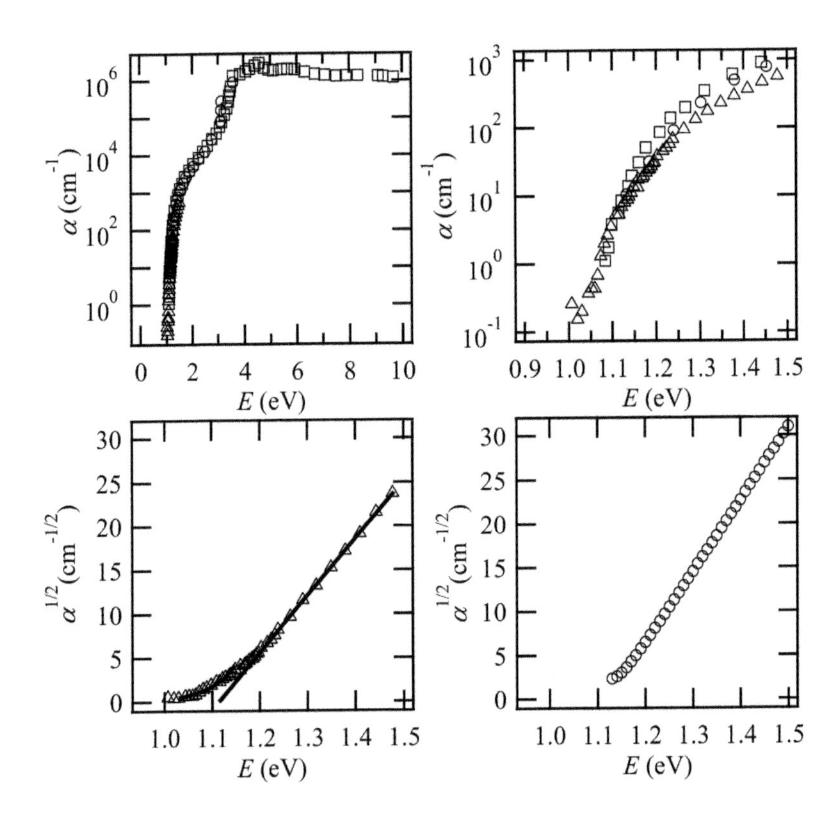

Fig. 5.4 (upper left) The absorption coefficient for Silicon as a function of photon energy comparing three different data sets. (upper right) The same data for lower energy levels. (lower left) The square root of the absorption coefficient as a function of energy for one data set. Extrapolation of the data to the horizontal axis gives a band gap energy of 1.114 ± 0.012 eV which agrees well with the value in Table 5.3 of 1.111 eV for E_g at 300 K. (lower right) Calculated absorption coefficient, using eqn (5.19), plotted as the square root of the absorption coefficient as a function of energy. Some evidence of curvature due to absorption and emission of phonons may be present as shown in Fig. 5.5. Source: http://www.ioffe.ru/SVA/NSM/Semicond/Si/index.html.

4.6). Near E_g it would appear easy to determine its value according to eqns (5.20) and this will be discussed in the next section.

5.3 Band gap energy determination

We begin with an analysis of Silicon to determine its band gap energy. Data from several sources were used and are plotted in Fig. 5.4, and when plotted for the entire energy spectrum appear quite consistent as seen in the upper left graph. However, when the lower energy region is explored some variation can be seen which may affect the analysis (upper right graph). As suggested by eqns (5.18), (5.19) and (5.20) the absorption coefficient should be analyzed as

$$\alpha^{1/2} \propto E - E_g \tag{5.21}$$

to find E_g.

So, a graph of $\alpha^{1/2}$ versus E should generate a straight line. As shown in Fig. 5.4, lower right graph, there is a linear region at energies above the band gap energy. Indeed eqn (5.18) suggests that at lower energies the absorption coefficient will follow the mechanism due to phonon absorption, and at slightly higher energies, that due to phonon emission. This is shown schematically in Fig. 5.5 where extrapolation to the

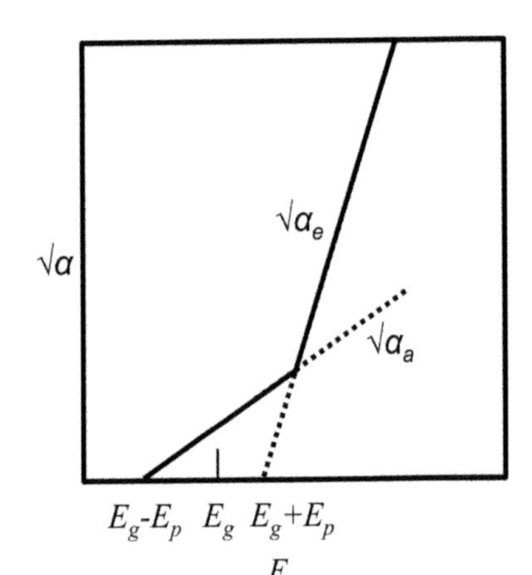

Fig. 5.5 One should be able to ascertain the absorption processes due to absorption and emission of a phonon in an indirect semiconductor.

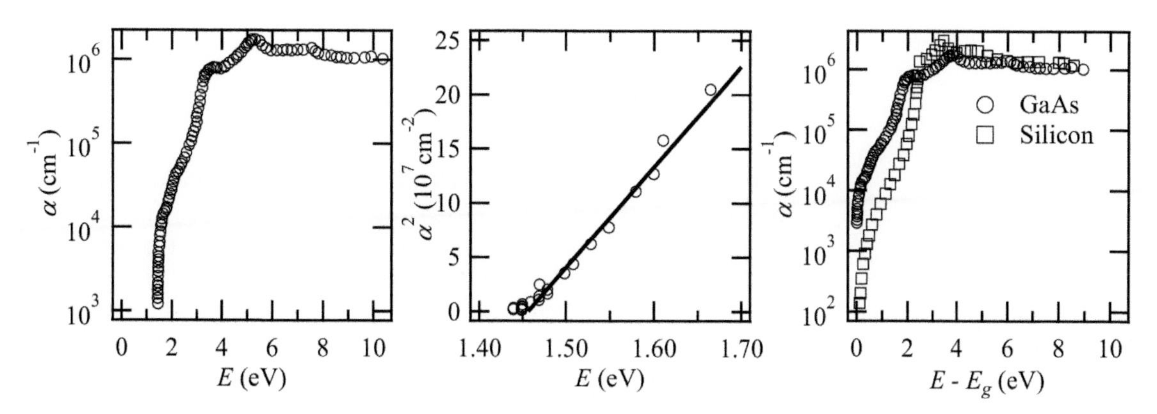

Fig. 5.6 (left) Absorption coefficient as a function of photon energy for GaAs at room temperature. (middle) The square of the absorption coefficient as a function of energy, demonstrating the linear region near the band gap energy, which, from the intercept is 1.456 ± 0.101. Urbach tailing is seen near the band gap and was not included in the linear regression. (right) Comparison of the absorption coefficient of GaAs with Silicon, the abscissa has the band gap energy subtracted from the radiation energy. Source: http://www.ioffe.ru/SVA/NSM/Semicond/Si/index.html.

horizontal axis should yield two different intercepts related to these phenomena. This was tested in Fig. 5.4 in the lower right graph using eqn (5.19), so this is in fact *perfect* data and shows some bending suggestive of the two absorption mechanisms in an indirect semiconductor.

Indeed the data in Fig. 5.4 suggests something that resembles this (lower left graph), however, this *band tailing* is most likely the result of *Urbach tailing*. This is caused by defects in the crystalline structure leading to an exponential band edge rather than following eqns (5.18). Note that the phonon energy is quite small for crystalline Silicon, 20–60 meV, see Table 5.3, and so would be difficult to detect. This together with extrapolation of $\sqrt{\alpha}$ versus E leading to the correct band gap energy, as well as Urbach tailing, suggests that it was not possible to determine the phonon energy with this technique under these conditions.

For a direct band gap material the analysis is simpler and demonstrated in Fig. 5.6. The absorption coefficient rises quickly above the band gap energy since the material strongly absorbs the radiation. Analysis of the absorption coefficient for a direct band gap material reveals that the data should be plotted as

$$\alpha^2 \propto E - E_g \tag{5.22}$$

which is shown in the figure. At moderate photon energies eqn (5.22) follows the data, yet, again there appears to be an Urbach tail since the absorption coefficient curves at lower energies. This is a common phenomenon at lower photon energy levels for many semiconductors.

The Urbach absorption tail results from disorder in a perfect crystal introduced by doping, inherent defects and fluctuation of the energy bands from lattice vibrations. A relation frequently used to model the Urbach tail is to write the absorption coefficient as

$$\alpha = \alpha_U \exp\left(\frac{E - E_g}{E_U}\right) \tag{5.23}$$

where α_U and E_U are Urbach tail absorption coefficient and energy parameters, respectively. The magnitude of E_U is approximately 10 meV although this can vary depending on the material and temperature and determines the broadness of the exponential region.

Finally, a direct band gap material is fairly efficient at absorbing radiation. An indirect band gap material has two processes which must occur, moving an electron to a higher energy level then changing its momentum; it is typically a much less efficient process than that for a direct band gap material. One can see in Fig. 5.6 the absorption coefficient for GaAs is much larger near the band gap energy than for Silicon and is the result of much more efficient absorption of the radiation. At higher energy levels the absorption coefficient for Silicon eventually becomes larger than or equal to that for GaAs. However, inspection of the solar spectrum shows the number of available photons is greatly reduced at these energies and the absorption becomes almost insignificant in terms of power generation (see Fig. 2.3 or Fig. 5.7).

Thus, the performance of Silicon can suffer somewhat if thinner active regions are used. The *active region* is the region in the device where photon absorption occurs. Contemporary manufacturing techniques for crystalline Silicon have thicker active regions merely because it is difficult to handle thinner ones and as shown in Example 5.2 almost all the radiation is absorbed. Yet, if new manufacturing techniques were developed to generate thinner films then radiation absorption could limit device performance. The absorption coefficient of GaAs is 2–10 times larger than Silicon in the energy region where the most number of photons are generated by the Sun requiring a much thinner active region.

5.4 Generation of heat

Heat generation comes about by the material absorbing radiation for use in solar thermal applications. In this case, rather than producing electricity as would occur in a photovoltaic device, the electron and hole recombine to form heat. Heat is produced by the electrons losing energy to the surrounding crystalline matrix by inducing lattice vibrations to the atoms, which are called phonons. There are a variety of mechanisms that heat, or phonons, can be generated and it is not important to outline them here.

We take a macroscopic view and use the absorbance as a gauge to heat production. Equation (5.7) gives the generation rate of excited electrons and is related to the absorption coefficient of the material. To find the rate of heat generation we multiply $G_L(E)$ with energy E and integrate over all energy levels

$$
\begin{aligned}
\dot{Q}_{gen} &= \int_0^\infty G_L(E)E\,dE \\
&= \int_0^\infty N_p(E)E \times [1 - \exp(-\alpha(E) \times L)]dE
\end{aligned}
\tag{5.24}
$$

where $\dot{Q}_{gen}$ is the rate of energy production through absorption of radiation. This will yield the heat generation rate in W/m^2 where we have assumed 100% efficient conversion of radiation to heat. An efficiency factor could be inserted in the integral to account for imperfect conversion and would be a function of the radiation's energy.

This equation is somewhat misleading though. Most absorbers will *absorb* all the radiation within a very short distance, however, not all the radiation is converted to heat, as discussed above by mention of the efficiency factor. There are inefficiencies from internal processes within the material at the atomic level. For a photovoltaic device this is characterized by the internal quantum efficiency or IQE(E), which is a function of energy E. So the theoretical potential current that could be generated would be multiplied by IQE(E) at each energy value to determine the total current.

In the case of an absorber material this will be done by noting the absorbance dependence on energy. Consider the absorbance of Black Nickel on an Aluminum substrate as shown in Fig. 5.7. An alloy of zinc and nickel is electrodeposited to make a Black Nickel (selective) coating which also has small amounts of carbon, nitrogen and sulfur within it. The coating is a reasonable selective coating, although its absorbance at the spectrum maximum is somewhat lacking, and has been used for years in solar thermal applications due to its price and durability.

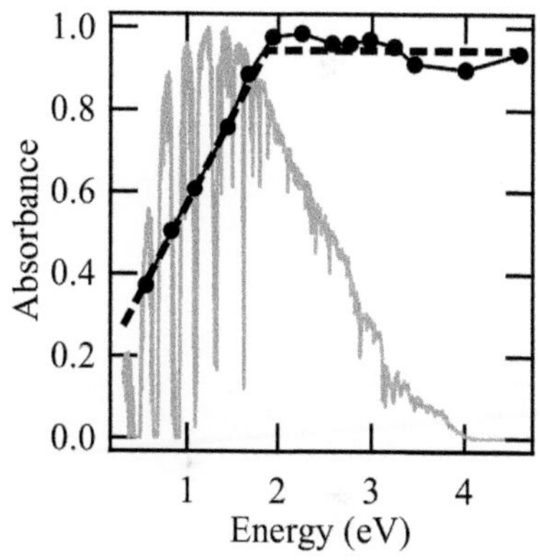

Fig. 5.7 Absorbance of Black Nickel on an Aluminum substrate as a function of radiation energy. A linear fit of absorbance with energy at low energy levels and a constant value at higher energies is shown in the graph. Also shown in the figure is the AM1.5G solar spectrum where the irradiance has been arbitrarily scaled to have a maximum of one.

The absorbance can be fitted to the energy with a linear relation at low energies and is constant at higher. The result of this fit is given by

$$a = 0.140 + 0.429E(\text{eV}), \text{ for } E < 1.88\text{eV} \tag{5.25a}$$

$$a = 0.944, \text{ for } E > 1.88\text{eV} \tag{5.25b}$$

Now the heat generated can be written as

$$\dot{Q}_{gen} = \sum_i a_i P_{\Delta,i} \tag{5.26}$$

where $P_{\Delta,i}$ is the useful irradiance over an energy range and a_i is assumed to be constant over it. The subscript i is used to denote different energy ranges. The useful irradiance was introduced in Chapter 2 and is given by

$$\frac{P_\Delta}{K} = \left[x_1^3 + 3x_1^2 + x_1 + 6\right]e^{-x_1} - \left[x_2^3 + 3x_2^2 + x_2 + 6\right]e^{-x_2} \tag{2.41}$$

where $K = 15\sin(\theta_{sub})^2\sigma_S T_S^4/\pi^4$ and $x_i = E_i/k_B T_S$ with $i = 1$ or 2. In the above equation $E_1 < E_2$. Thus, it is possible to determine how much heat is generated in a solar thermal device as long as the absorbance is known as a function of energy. The use of the above equations is best demonstrated with an example.

Example 5.3

Determine the rate of heat generation for a material coated with Black Nickel when exposed to the standard AM1.5G solar spectrum.

Assume the material absorbs light in a similar manner to the coated Aluminum samples characterized in Fig. 5.7. This allows the absorbance to be approximated by eqn (5.25). To determine the rate of heat generation first define P_i as

$$P_i = K\left[x_i^3 + 3x_i^2 + x_i + 6\right]e^{-x_i} \tag{5.27}$$

where $K = 154.0$ W/m^2 for AM1.5G radiation ($T_S = 5359$ K). Use eqn (5.26) and let $P_{\Delta,i} \equiv P_i - P_{i+1}$ and the absorbance a_i will be calculated at the middle of this energy range $\langle E_i \rangle$. The result of the calculation is given in the table below.

E_i (eV)	x_i	P_i (W/m^2)	$\langle E_i \rangle$ (eV)	a_i	$a_i P_{\Delta,i}$ (W/m^2)
0.00	0.00	924			
			0.25	0.247	75.3
0.50	1.08	619			
			0.75	0.461	21.7
1.00	2.17	572			
			1.25	0.676	82.7
1.50	3.25	450			
			1.69	0.864	99.7
1.88	4.07	334			
			–	0.944	316
∞	∞	0			
				$\sum a_i P_{\Delta,i}$	595

This solar device will produce 595 W/m^2 of power which is substantial. Of course if it were a perfect absorber over the entire energy range then it would produce 1000 W/m^2 since it would be exposed to the AM1.5G spectrum. This calculation is not perfect since eqn (2.41) utilizes the first term in an infinite series, however, the error in using this truncated series is acceptable. For example, the total power determined for the AM1.5G spectrum between energies of $E_1 = 0$ eV and $E_2 \to \infty$ eV is found to be 924 W/m^2 rather than 1000 W/m^2. This is an 8% error over the true AM1.5G total power which is acceptable for many calculations.

The utility of using a selective absorber can be seen when considering the energy region where one wants to minimize thermal loss. Assuming this absorber operates at 50°C then according to the Wein displacement law (eqn (2.7)) the spectrum maximum for radiation emission from the absorber will occur at a wavelength of approximately 9000 nm. This corresponds to an energy of 0.14 eV ($E(\text{eV}) = 1240/\lambda(\text{nm})$) which is more or less in the middle of the first energy region given in the above table. Thus, although it is desirable to maximize energy absorption at this energy level, and there is of order 300 W/m^2 available in the energy region between 0.0–0.5 eV, this is precisely the energy region where radiation emission is to be minimized.

The absorbance at an energy of 0.14 eV is 0.2 which is small although a much smaller value would be desired to minimize radiation emission. Other more expensive and newly developed selective coatings give better results although Black Nickel, and similar selective coatings, have shown their utility over the past decades.

5.5 Conclusion

The absorption of radiation was considered in this chapter. This process can generate current in a photovoltaic device or heat energy in solar thermal devices, with both processes dictated by the material's electronic structure. A band gap energy is desired for photovoltaic applications to ensure generation of current with the presence of a potential to produce power. The lifetime of a charge carrier, that is the electron or hole, will be discussed in the next chapter and is a key parameter to the operation of a photovoltaic device. The lifetime is related to the distance a charge carrier can diffuse before it recombines with an oppositely charged particle to produce useless heat. Thus, if the device is too thick many of the charge carriers will recombine and performance suffers, so the thickness should be limited.

The absorption coefficient mediates how much radiation is absorbed and dictates how thick the device's *active region* should be so the maximum current can be generated. Thus, a tradeoff between absorbing radiation and allowing the charge carriers to reach the appropriate electrode exists. If the device is too thick then the charge carriers recombine, since, in their lifetime, they cannot reach the electrode. On the contrary, a thin device will not efficiently absorb the radiation to generate sufficient charge carriers. As with many devices or processes an optimization must be performed to make the best device. In this chapter the basics of how much radiation is absorbed and the possible current that is concomitantly generated was presented allowing this optimization to be performed.

The absorption coefficient can also be used to find the band gap energy, the fundamental variable of a photovoltaic device. The dependence of the absorption coefficient on photon energy can be used to determine its value and the functional form can be used to ascertain whether a direct or indirect transition is present. Some care should be used since non-idealities in the crystal structure can develop an exponential *tail* in the absorption coefficient with energy near the band gap energy through Urbach tails to complicate the analysis. However, being aware that they are most likely present alerts one to ensure proper analysis.

No band gap energy is required for a material used in solar thermal applications since all the energy available to it should be absorbed. The *caveat* to this is that as the material increases in temperature it becomes a radiator itself and this is to be minimized. Ironically, something akin to a band gap energy is desired. The absorbance should be zero for lower energy levels, as dictated by the Wien displacement law, to minimize energy loss to the surroundings (Kirchoff's law states that the absorbance equals the emissivity at thermal equilibrium). So, in the ideal case, energy is absorbed only above a certain energy level to maximize energy gain, remarkably similar to what is expected of a photovoltaic device.

Of course, no potential is developed in a solar thermal device one merely desires the energy lost by the system to be minimized and yet again an optimization has to be performed. The properties of the selective absorber should be engineered to have a minimal absorbance at low photon energies (long wavelength) since this minimizes the emissivity. Yet, the energy range over which this minimal absorbance occurs should be carefully designed so one does not affect absorption of energy from the solar spectrum. Thus, an optimization has to be performed which is also dependent on the available materials (*i.e.* the selective surface) and cost.

The two equations used to perform this optimization are eqns (5.10) and (5.24) for the photovoltaic and solar thermal device, respectively. Their simplicity is deceiving though as the true solar spectrum is complicated by absorption bands for chemicals in the atmosphere. Here a black body radiator model was used and it is expected for a first design that this assumption will allow the optimization to proceed.

5.6 General references

M. Fox, 'Optical properties of solids,' Oxford University Press (2001)

M.A. Green, 'Solar cells: Operating principles, technology and system applications,' Prentice-Hall Inc. (1982)

J.I. Pankove, 'Optical processes in semiconductors,' Dover Publications (2010)

Exercises

(5.1) A material looks perfectly clear to the human eye; is it a metal, semiconductor or insulator? Estimate the lowest possible value for its band gap energy.

(5.2) Calculate the transmittance of radiation at an energy of 1.2 eV and 4.1 eV through Silicon as a function of depth and graph your results.

(5.3) Determine the maximum current from a Silicon photovoltaic device that is only 5 μm thick. How does this compare to a 300 μm thick device?

(5.4) Use eqn (5.19) to calculate the absorption spectrum for Silicon and compare it (as best you can) to Fig. 5.4 if the temperature is 300 K. *Hint:* Terms in eqn (5.19) can only be used when the radiation energy is above a certain value.

(5.5) Amorphous Silicon is a direct band gap semiconductor. Data for the absorption coefficient as a function of radiation energy near the band gap energy is given in the table below. Determine the band gap energy and compare it to the accepted literature value as well as to the band gap energy for crystalline Silicon. Provide a valid reference to the literature value.

E (eV)	α (cm^{-1})
1.74	1.35×10^2
1.75	6.39×10^2
1.79	2.19×10^3
1.89	7.53×10^3
1.95	1.19×10^4
1.99	1.71×10^4

(5.6) Data for the absorption coefficient near the band gap energy of Gallium Arsenide are given in the table below. Analyze these data as either a direct band gap or indirect band gap semiconductor and, based on the correlation coefficient, determine what type it is. What is the band gap energy in eV?

E (eV)	α (cm^{-1})	E (eV)	α (cm^{-1})
1.50	5.94×10^3	1.60	1.13×10^4
1.51	6.63×10^3	1.61	1.26×10^4
1.53	7.91×10^3	1.66	1.43×10^4
1.55	8.83×10^3	1.72	1.53×10^4
1.58	1.05×10^4		

(5.7) Additional data for the absorption coefficient of Gallium Arsenide, to that in Example 5.6, at higher radiation energies are (given as data points of (E (eV), α (cm^{-1})): $(1.90, 2.22 \times 10^4)$, $(1.93, 2.48 \times 10^4)$ and $(1.98, 2.77 \times 10^4)$. Graph all the data together and explain why the higher energy data do not lie on the trend line fitted to the lower energy data. Assume Gallium Arsenide is a direct band gap semiconductor and graph the data so that they should be linear.

(5.8) Silicon and Gallium Arsenide have absorption coefficients of 1.78×10^3 cm^{-1} and 1.13×10^4 cm^{-1}, respectively, for radiation of energy 1.60 eV. Calculate the sample thickness in μm for each material so that 90% of the radiation is absorbed. What makes one material a better absorber than the other? What is the wavelength of the radiation and is it at an optical frequency?

(5.9) Glass is frequently used in a photovoltaic or solar thermal device and has an absorption coefficient of 0.04 cm^{-1}. If the glass is 3.2 mm thick what is the maximum amount of radiation that can be transmitted through it? How thick would it have to be to absorb 90% of the radiation?

(5.10) A material acts as a perfect selective surface and has an absorbance, or equivalently, emissivity, of identically zero for low energy radiation and an absorbance of one for high energy radiation. Assume the energy where the transition from low to high energy absorption occurs is 0.1, 0.5 1, 2, 3 and 4 eV to determine the energy production rate through absorption or $\dot{Q}_{gen}$ in W/m^2. Graph $\dot{Q}_{gen}$ versus the cut-off energy and comment on the trend. Assume the Sun is a perfect black body radiator at a temperature of 5359 K.

(5.11) Assume you wear perfectly black clothing, meaning it absorbs all radiation that comes to it, and that you cannot emit radiation back to the surroundings. Assume AM1.5G conditions, the Sun is a perfect black body radiator and your body's thermal properties are equivalent to water; determine the rate of temperature rise you will experience due to absorption of radiation, in °C per minute.

(5.12) The Bedouins, who live in a desert environment, wear black robes, does this make sense? *Hint:* Read this manuscript to find out why: A. Shkolnik *et al.* 'Why do Bedouins wear black robes in hot deserts?' Nature **283** (1980) 373.

The photovoltaic device

6

A brief introduction to the physics dictating how a photovoltaic device or solar cell works is presented. This is truly a cursory introduction as we skirt around the details of semiconductor physics, yet, is a required exercise. The reader can then appreciate that a typical photovoltaic device is made from a p-n junction, defined below, that has particular properties to generate current when exposed to light. For this ideal photovoltaic device its operation is dictated by two parameters; the dark field current, J_0, which is a property of the junction, and the current generated through light absorption, J_L, which is a property of the individual materials making up the junction (as a first approximation for the ideal photovoltaic device).

These two parameters are integral to the function of the device that can be gauged through its current–voltage graph. The shape of this plot demonstrates much about performance and a section is devoted to its interpretation followed by a section on the discussion of the maximum power one can obtain. Finally, although this is a chapter considering the ideal photovoltaic device, we consider some non-idealities. These are not microscopically derived, being mere empiricisms representative of device imperfections. At the end of this chapter the reader should be able to read a current–voltage graph much like a company balance sheet to ascertain performance.

6.1 What happens inside a photovoltaic device?

The similarity between a Carnot heat engine and a photovoltaic device was discussed in Chapter 3. The heat engine is fictitious while a photovoltaic device is not. In this case heat or radiation from the Sun is absorbed by the material making up the cell through excitation of an electron from its ground state to an excited state. The key to generating electrical current is to make the excited electron move and eventually find its way to the appropriate electrode. To do this requires the formation of a junction between dissimilar materials to form a diode.

In a typical photovoltaic device used today the two materials are p-type, doped with Boron, for example, and n-type, doped with Phosphorous, semiconductors each with unique electrical properties. In many cases the same material, Silicon as shown in Fig. 6.1, is doped to become p-type or n-type. The electrons have more ease of movement in

Solar Energy, An Introduction, First Edition, Michael E. Mackay

the n-type material, and when they do hop around, they leave behind a positively charged atom that is covalently bonded to other atoms and so cannot move. The opposite is true for the p-type material, with holes (defined as where an electron is not present in the valence band) being mobile, and negatively charged atoms are trapped in its matrix, see Fig. 6.1. Of course, each material, by itself, is electrically neutral although the electrons and holes are quite dynamic. When exposed to light this dynamic equilibrium is disturbed and a light induced current is produced.

The two materials are joined together to find the p-type material draws electrons from the n-type and the n-type material draws holes from the p-type. The reason for this is due to the mobile particles, in other words the electrons or holes, being free to diffuse through the composite material and they will naturally move to occupy any larger volume available to them. However, they can only diffuse a certain distance on average, called the diffusion length, before they recombine with their oppositely charged partner. This is an important effect and will be discussed in detail below.

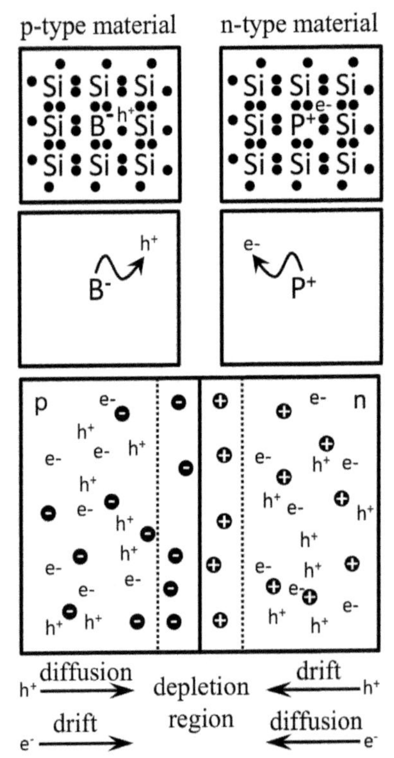

Fig. 6.1 (upper and middle figures) The p-n junction is made from two materials that have different electrical properties, a p-type and n-type material by doping with Boron and Phosphorus, respectively. The holes (h^+) and electrons (e^-) are free to move within each bulk material since its high dielectric constant effectively shields the opposite charges. (lower figure) When the two materials are joined together the holes naturally diffuse to the n-type region while the electrons diffuse to the p-type, this process is called *diffusion* and is merely the result of a greater volume available to the particles. Eventually a depletion region develops which creates an internal electric field that pulls the holes and electron back, this process is called *drift*. At equilibrium a balance exists between the two.

The charged stationary atoms produce an in-built electric field and act as a barrier to holes moving from the p-type material because there are now positively charged atoms in the n-type material, within what is called the *depletion region* which is devoid of mobile charge carriers. This would suggest a total loss of activity within the entire device. However, every once in a while, a hole has enough energy to move through the depletion region and it rapidly diffuses within the n-type semiconductor bouncing from atom to atom. Of course, the opposite occurs for the electrons.

The in-built electric field also pulls the holes from the n-type material back into the p-type in opposition to the diffusion, and a delicate balance results as shown in Fig. 6.1. There is diffusion in one direction and drift caused by the in-built electric field in the other for both the electron and hole. The depletion region's width is dictated by the doping level of, or the number of *impurity atoms* introduced into, the p-type and n-type material as well as any applied potential on the junction. In reality the depletion region width is quite small in relation to other relevant length scales, as we will find later.

If one were to put a wire linking the n-type semiconductor terminus to the p-type, a current would not result. The electrons and holes move about such that their next flux produces zero current. Again this is the result of the balance of drift and diffusion. However, if one were to put an electric field potential across them then an interesting response would happen. A *forward bias* means the positive terminus is applied to the p-type material and a negative to the n-type. This has the effect of reducing the width of the depletion region since the potential increases the diffusion of holes to the n-type semiconductor and vice versa to the p-type material (see Fig. 6.2). This also effectively reduces the internal electric field and the drift component is minimized until, one could imagine, at a sufficiently high potential a large current flows through it.

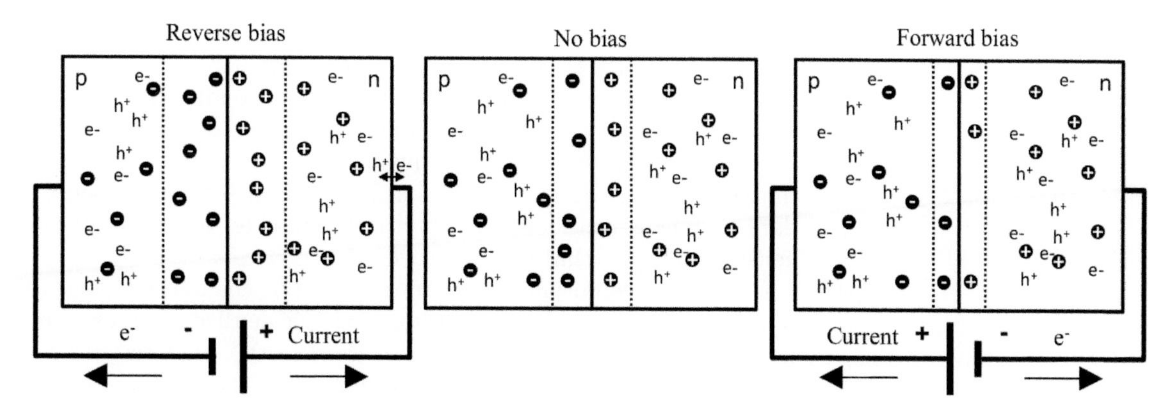

Fig. 6.2 A p-n junction under reverse bias, no bias and forward bias.

Reference to Fig. 6.3 will reinforce that holes flow from the positive terminal to the negative, while the opposite is true for electrons.

If a *reverse bias* is applied the opposite occurs. Now the electrons flow into the p-type material and combine with the mobile holes and holes flow into the n-type material from the p-type material to combine with the mobile electrons. The flow of holes only occurs within the p-n junction for the following reason. The metal wire attached to the n-type material, for example, can be considered to be positively charged nuclei that cannot move since they are confined in a crystalline lattice surrounded by a sea of mobile electrons. Holes would be immediately destroyed if they came into the wire and the concept of holes make no sense in a metal. So, instead what happens is this, viewing the double headed action arrow in the *Reverse bias* part of Fig. 6.2, an electron leaves the n-type material and moves into the wire leaving behind a hole to restore charge neutrality. Holes were introduced to explain charge movement in semiconductors since the electron is promoted to a much higher energy from an almost full valence band to an almost empty conduction band. It leaves behind a hole. A metal has an essentially half empty conduction band and the electron is only promoted to a slightly higher energy level.

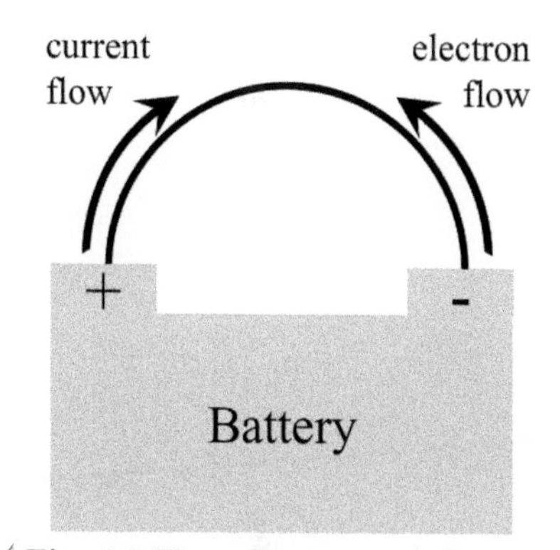

Fig. 6.3 Flow of current and electrons from the terminals of a battery are opposite to each other. This convention originates back to Benjamin Franklin, an American scientist, statesman and entrepreneur, who chose the direction of current flow before electrons were discovered and it was realized the current direction was opposite to that of the electrons.

Reverse bias increases the depletion region's width and the drift component increases, however, diffusion decreases as the number of holes or electrons with enough energy to vault across the wider depletion region becomes smaller. Ultimately the current remains small at large potential since it becomes self-limiting.[1] Operating under this bias requires power to be supplied to the p-n junction and the voltage is negative in sign. Physically one can imagine how this would happen as the electrons are driven into the p-type material and the holes to the n-type to promote much electron-hole recombination and heat.

Under reverse bias the current is dictated by the *minority* carriers within each of the materials.[2] In the p-type material the holes are free to move about, as described above, however, there are also a small amount of freely moving electrons which are the minority carriers. If a current

[1] If a high enough potential is applied under reverse bias the diode can breakdown which forms the basis of a Zener diode.

[2] A minority carrier is exactly what it sounds like, it is the charge carrier in a doped semiconductor that has the lowest concentration. In a p-type material it will be electrons and in a n-type material it will be holes.

is to flow, holes must go one way and electrons the other. Under reverse bias there are only so many minority carriers available in each material, holes are the minority carrier in the n-type semiconductor, and the current saturates under large reverse bias to a value denoted as the *dark current*, or J_0, a very important quantity to photovoltaic device performance and whose details are discussed in the next section. Presently the dark current is considered an empirical parameter as will be the light generated current that makes a successful solar cell.

Under ideal conditions, as we consider in this chapter, the solar cell's p-n junction will generate a constant current density when exposed to light denoted as J_L. Light absorption in the n-type material, for example, raises an electron to a higher energy level, which leaves behind a hole. This is what happens when light is absorbed, the energy must go somewhere and this is most easily done by exciting an electron within an atom to a higher energy state. If the electron falls back to its ground state then heat is commonly produced as the electron and hole recombine, just like we ourselves would experience outside on a sunny day as the Sun warms us.

Fortunately, they don't always recombine and the electron is a *majority* carrier in the n-type material which has plenty of mobile electrons in it, while the hole is the minority carrier. The hole must reach the depletion region and be swept away by the internal electric field where it is taken into the p-type material to then become a majority carrier. If it doesn't make it to the depletion region then it will readily recombine with any one of the numerous free electrons. So, a length scale is introduced called the *diffusion length*, which is related to the diffusion coefficient and lifetime of a charge carrier. If either minority carrier is within a diffusion length of the depletion region then it can be swept over it to become a majority charge carrier. So, excited electrons in the p-type material that are within a diffusion length of the depletion region will move over to the n-type region.

All the processes in an operating solar cell have been described, let's summarize what occurs. Reference to Fig. 6.2, and in particular to the p–n junction under *Forward bias*, has electrons flowing into the n-type material and the depletion region narrows. When radiation or light is absorbed electrons are raised to a higher energy level in either part of the junction. In the p-type material electrons can be swept over the depletion region due to the internal electric field within the depletion region while holes in the n-type material are pulled the other way. Thus, the light generated current flows in the opposite direction to the forward bias generated current and mathematically J_L is negative in value. The applied potential (or voltage) V_a is positive under forward bias and their product, which is the power (work) generated by the solar cell, is negative. Negative work implies that the solar cell is doing work on the surroundings and is exactly what is desired! Power is generated!

Imagine for a minute that a solar cell was operated under reverse bias, as shown in Fig. 6.2. In this case, electrons are forced to enter into the p-type material while the light generated current has holes entering into the same material; they are oppositely charged and they will eliminate each other. Of course, the opposite occurs in the n-type material and light generated electrons are eliminated by the holes. In both cases heat is produced and work is performed on the photovoltaic device (the sign of the applied potential and light generated current are negative, so, their product is positive and work is performed on the device).

The light generated current is the sum of the two rates of minority carrier generation in either doped region. This is because they are randomly generated and independent of each other so the *total* current is their sum. If they were somehow correlated in formation, for example, the given electron and hole formed upon excitation must make it to the appropriate electrode, then a sum would not be taken.

Now, remember the discussion in Chapter 3 and specifically the discussion surrounding eqn (3.25) where it was shown that a potential must be present to create power. The potential is supplied by the band gap energy in the semiconductor and the electron and hole that have made it to their electrodes have a potential difference between them, in other words a voltage difference. Electric power is made!

In summary, a p-n junction photovoltaic device operates in this manner:

- If an electron is put into an excited state by absorption of radiation within the n-type region then the associated hole, *i.e.* the minority carrier, will be swept back over the depletion region by the favorable electric field. It must be within a hole diffusion length, or L_h, of the depletion region for this to occur.
- For an excited electron–hole pair within the p-type region, the minority charge carrier or electron will be pulled over the depletion region again due to the favorable electric field to become a majority carrier. It must be within an electron's diffusion length, or L_e, of the depletion region to generate light generated current. The two lengths, L_p and L_e, are usually much larger than the depletion region width.
- The photovoltaic device must be operated in forward bias to ensure the light generated charge carriers are not eliminated. In addition, forward bias allows extraction of power from the device.

The following example is presented to demonstrate how many minority charge carriers are generated in a Silicon-based photovoltaic device.

Example 6.1

Estimate the number density of minority carriers in a Silicon-based photovoltaic device.

This can be estimated by knowing what light generated current density a Silicon photovoltaic device produces. A typical value is 35 mA/cm^2 or 350 A/m^{2}and if this is divided by the charge of an electron (1.602 ×

10^{-19} C) then one obtains N_{out} defined in eqn (3.22). In reality it is almost N_{out}, however, the operating photovoltaic device is not 100% efficient as was assumed in determining N_{out}. Regardless the value is 2.2×10^{21} #/s-m^2.

This is the number of minority carriers that produce useful work per unit time per unit area. To go further we must assume a lifetime for the carriers which is a function of many factors including the doping level, temperature and what type of minority carrier it is (i.e. electron or hole). An average value of 10^{-6} s is assumed (see Fig. 6.7, 10^{-6} s is a geometric mean of values found in the n-type and p-type regions). In addition, the length scale over which they are formed by light absorption must also be assumed. This is the sum of the width of the depletion region and the minority carrier diffusion length on either side and is called the *active region thickness.* This thickness is also a function of the above three variables and a value of 30 μm is assumed. When N_{out} is multiplied by the lifetime and divided by the active region thickness we obtain 7.3×10^{13} #/cm^3 and is the number density of carriers within the photovoltaic device.

This seems like a very large amount. Yet, the number density of Silicon atoms is approximately 5×10^{22} #/cm^3, so, only one in one billion atoms produces a *usable* electron and hole! Certainly, many more atoms are put in an excited state than this and only a fraction of them become available for use. An excited atom means that the Sun is breaking bonds to do this and so it is fortunate that only a small number of atoms are affected or else the material would be completely degraded at any given time.

A few comments about the p-n junction of an actual photovoltaic device should be made at this point. Firstly, the p-type and n-type materials are not independently manufactured and joined together. The Silicon single crystal is grown as a large ingot (cylinder) with the p-type dopant (usually Boron) within the crystallizing Silicon melt. The ingot is sliced into wafers with a thickness of approximately 300 μm; they could be thinner, however, there are technological issues in processing (holding without cracking) thinner wafers as well as cutting them that thin. Then a very thin layer of n-type chemical (usually Phosphorus) is diffused into the wafer.

The diffusion can be done by exposure to Phosphorus containing gas at high temperature. This is a classic boundary value problem of diffusion into a semi-infinite medium which has an error function solution for the Phosphorous concentration through the wafer. So, there is a high Phosphorus concentration at the surface and it decays rapidly within the bulk material. The concentration is taken as approximately 10^{19} #/cm^3 in the n-type region while the p-type region has a Boron concentration of order 10^{16} #/cm^3, one-thousand times less! Since there are so many more Phosphorus atoms than Boron in the n-type region it is indeed n-type.

Because of the way the cell is made the n-type region has a better surface quality and is the surface that faces the Sun, so it is the negative terminal or *cathode* under forward bias. This region is sometimes called the *emitter* and is very thin, of order 1 μm, yet it is highly doped to generate enough free electrons for successful operation. The thin layer of high concentration ensures that a large number of light generated charge carriers can be swept across the depletion region.[3]

The remaining part of the cell, that is the region which does not include the emitter and depletion region, is the p-type region. This region is sometimes referred to as the *base* since the trapped Boron atoms are missing one valence electron and can accept one. The reason the n-type region is called an emitter is now obvious. A layer of metal is placed at the edge of the base layer to make the *anode* of the cell.

This is a cursory introduction to semiconductor physics, as well as how the p-n junction is made. However, it yields the most important reason as to why a photovoltaic device has a p-n junction central to its operation. The depletion region induces an internal electric field which sweeps away the minority carrier in one of the materials so it doesn't readily recombine with a majority carrier to produce useless heat. The ideal photovoltaic device requires this sort of internal field for its operation. In the next section, details of what happens in an ideal photovoltaic device are considered.

[3]It makes sense to initially lightly dope the material p-type then diffuse a thin layer of highly concentrated n-type material into it. The opposite is certainly not possible! The high concentration is chosen based on the time it takes to diffuse Phosphorus in the semiconductor to make an n-type region with a large enough volume to supply electrons for the junction to work effectively.

6.2 Details of what happens in a photovoltaic device

In the previous section we described in words what happens in a photovoltaic device, while here we present the details. The key aspect of the solar cell is development of the depletion region and the internal potential. To understand how this potential forms we must first determine what happens to the chemical potential in the n-type and p-type regions, since this promotes its formation.

Remember the chemical potential μ, sometimes called the *Fermi level*, which we have discussed before is an imprecise use of this term (*Fermi energy* is certainly not acceptable), and is a measure of the energy change a system will have when an electron is added to or subtracted from a system. So, it lies intermediate among the available energy levels within a given material. The probability that an electron will have an energy greater than μ is 50% and an energy less than μ is 50%, making the Fermi function $F(E) = 0.5$ at its energy value. For an *intrinsic semiconductor*, one that is not doped at all, μ lies right in the middle of the band gap while for a *doped semiconductor* it does not. However, recalling eqn (4.24), the chemical potential is not necessarily exactly in the middle for an intrinsic semiconductor.

To continue, we must first realize how large the degree of doping is. Typical concentrations are on the order of $10^{16}\#/\text{cm}^3$ and $10^{19}\#/\text{cm}^3$ for the p-type (written as N_A) and n-type (N_D) materials, respectively, although these values may be vastly different in some applications, as shown in Table 6.1. Regardless, it is much larger than the number of intrinsic carriers n_i which for Silicon is $1.0 \times 10^{10}\#/\text{cm}^3$ at 300 K. Yet, N_A and N_D are much smaller than the number of atoms in the lattice $5 \times 10^{22}\#/\text{cm}^3$ and so addition of donor or acceptor impurity atoms or dopants greatly increases the number of carriers, yet, at a concentration of only 0.4 ppm for the p-type doped material!

Doping moves the chemical potential near the band edges as shown in Fig. 6.4. To prove this we must realize the space charge density ρ is always zero far from the depletion region to ensure electrical neutrality

$$\rho = q \times [p - n + N_D^+ - N_A^-] \, [=] \, \frac{\text{C}}{\text{m}^3} \tag{6.1}$$

Here q is the magnitude of the charge for an electron, p and n, the number concentration of holes and electrons, respectively, N_D^+ and N_A^-, the number concentration of ionized donors and acceptors, respectively, and the symbol [=] means 'has dimensions of.' In most inorganic semiconductors all the donors and acceptors can be assumed to be ionized because the material's high dielectric constant screens the opposite charges[4] and so it will be assumed that

$$N_D \equiv N_D^+ \quad \text{and} \quad N_A \equiv N_A^-$$

Consider the quasi-neutral region for the p-type material where we can write the space charge density as

$$\rho \equiv 0 = p_p - n_p - N_A$$

where p_p and n_p are the number concentration of holes and electrons in the p-type material. The number of electrons is expected to be very small while the number of holes will be quite large or, $p_p \approx N_A$.

How can we determine how many electrons there are? We answer this by finding out the changes in the material upon doping. The number concentration of holes in the p-type material p_p was given in Chapter 4 as (taking reference to Fig. 6.4 for definition of the variables, N_{vp} is the effective number of states in the valence band of the p-type material)

$$p_p = N_{vp} \exp([E_{vp} - \mu_p]/k_B T)$$

Again this relation is not affected by doping since neither E_{vp} nor N_{vp} is expected to change significantly on doping[5] since the number of (Silicon) atoms is much greater than the number of dopant atoms, however, we know that p_p does change since it is equal to N_A. So, the only variable that can significantly change is the chemical potential μ which must take on the value

$$\text{p-type material: } \mu_p = E_{vp} + k_B T \, \ln(N_{vp}/N_A) \tag{6.2a}$$

and similarly for the n-type material

$$\text{n-type material: } \mu_n = E_{cn} - k_B T \, \ln(N_{cn}/N_D) \tag{6.2b}$$

The effect of doping is to move the chemical potential closer to the valence band in the p-type material and the conduction band in the n-type

Table 6.1 Values of various physical constants for Silicon. N_{atoms} is the number of Silicon atoms in a single crystal and the band gap energy for Silicon E_g is 1.1 eV. The value of n_i is very sensitive to the value of E_g, see eqn (4.23).

Constant	Value ($\#/\text{cm}^3$)
N_{atoms}	5.0×10^{22}
n_i	1.0×10^{10}
N_c	2.8×10^{19}
N_v	1.8×10^{19}
N_A (typical)	2×10^{16}
N_D (typical)	1×10^{19}

[4]This can be justified by the Bohr Hydrogen atom model for an atom, as given in eqn (3.15), which can be written as $E = m_e q^4/8\epsilon_0^2\epsilon^2 h^2$, where we have assumed the principal quantum number n is one and added the dielectric constant of the surrounding medium ϵ. Of course, for a Hydrogen atom the 'dielectric constant' between the nucleus and the electron is one, while for this model of a semiconductor we must take the material's dielectric constant into account. Substituting in values we find $E = 13.6\text{eV}/\epsilon^2$. For most semiconductors $\epsilon \approx 10$ making the binding energy between a nucleus and an electron or hole of order 100 meV according to the Bohr model. In reality, for crystalline Silicon, the binding energy is much lower and equal to 14.7 ± 0.4 meV. Thermal energy or $k_B T$ is 25.8 meV at room temperature and is large enough to separate the electron from the nucleus when accompanied by a random thermal fluctuation.

[5]The definitions of N_c and N_v were determined before

$$N_c = 2\left[\frac{2\pi m_e^* k_B T}{h^2}\right]^{3/2} \tag{4.22a}$$

and

$$N_v = 2\left[\frac{2\pi m_h^* k_B T}{h^2}\right]^{3/2} \tag{4.22b}$$

Since doping occurs on the ppm level, one does not expect m_e^* or m_h^* to change significantly in either the n-type or p-type material. Likewise, the energy levels E_c and E_v are not expected to change.

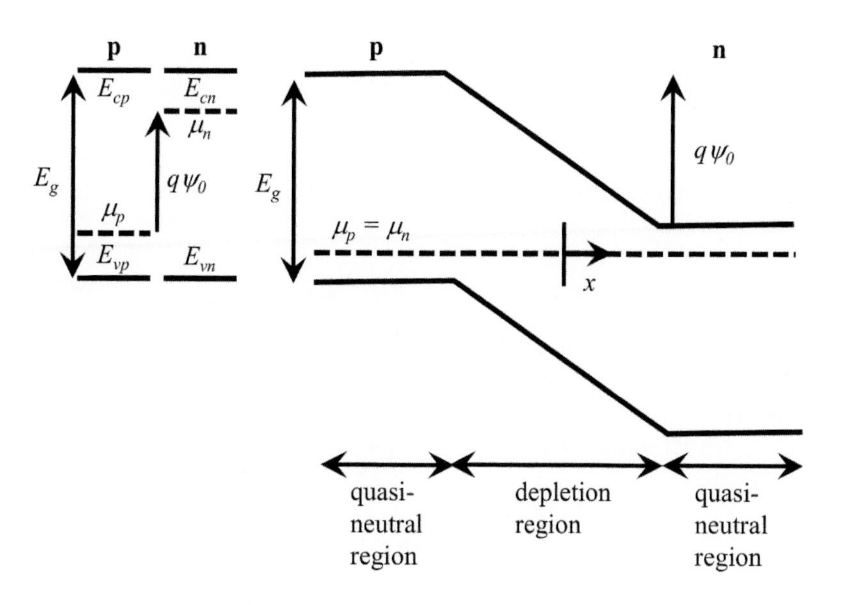

Fig. 6.4 (left) Diagram showing the p-type and n-type materials before joining them together. (right) After joining, the chemical potential must be equal through the heterogenous material producing band bending and an internal built-in voltage $q\psi_0$.

material, as shown in Fig. 6.4. Clearly, as the value of N_A approaches N_{vp} then μ tends towards E_{vp}. Constants for Silicon are given in Table 6.1 and if we take $E_{vp} \equiv 0$, making $\mu \approx 1/2 E_g = 0.55$ eV for undoped Silicon, then at a dopant level of 10^{16} $\#/\text{cm}^3$ for both p-type and n-type material we find $\mu_p = 0.16$ eV and $\mu_n = 0.93$ eV, which is already quite close to the band edges.

What happens when the p-type and n-type materials are joined together to form what is designated a *metallurgical junction*, meaning they are in intimate contact? With reference to Fig. 6.4 one finds a potential ψ_0 will develop since the chemical potential in the p-type material μ_p must equal that for the n-type, μ_n. So, when they are joined the valence and conduction bands must *bend* and the depletion region is formed between the two quasi-neutral regions in the two homogeneous materials.

Based on what has been determined above we can calculate n_p, we know that

$$n_p = N_{cp} \exp(-[E_{cp} - \mu_p]/k_B T)$$

when μ_p from eqn (6.2a) is used one finds

$$n_p = N_{cp} \exp(-E_g/k_B T) \frac{N_{vp}}{N_A} \equiv \frac{n_i^2}{N_A}$$

It was previously determined that

$$np \equiv n_i^2 = N_c N_v \exp(-E_g/k_B T) \tag{4.23}$$

which allowed the substitution for n_i^2 in the above equation. This equation is true for all doped semiconductors as we have not changed the band gap energy or any of the constants in N_c or N_v. This relation is frequently called the *law of mass action*.[6] In a doped semiconductor n_i^2 is merely a number and a shorthand for the very right-hand side of eqn (4.23). So, in general, we find

$$\text{p-type material:} \quad p_p = N_A \quad \text{and} \quad n_p \approx \frac{n_i^2}{N_A} \tag{6.3a}$$

and

$$\text{n-type material:} \quad n_n = N_D \quad \text{and} \quad p_n \approx \frac{n_i^2}{N_D} \tag{6.3b}$$

where n_n and p_n are the number concentrations of electrons and holes in the n-type material, respectively.

Now the energy $q\psi_0$ due to the built-in potential ψ_0 can be calculated. The potential developed when the p-n junction was formed can be written

$$q\psi_0 = \mu_n - \mu_p = E_g + k_BT \ln\left(\frac{N_A N_D}{N_{vp} N_{cn}}\right) \tag{6.4a}$$

with reference to Fig. 6.4. The energy values for the band edges, E_{vp} etc., don't change in value on doping, because the dopant atoms' relative concentration is so low, and we have joined two pieces of Silicon together, for example, so we can write $E_g = E_{cn} - E_{vp}$. If we were to join two different materials together and make a p-n *heterojunction* the analysis becomes a little more involved which we won't be concerned with here. Remembering the definition of n_i one can write eqn (6.4a) in a simpler form,

$$\psi_0 = V_{th} \ln\left(\frac{N_A N_D}{n_i^2}\right) \tag{6.4b}$$

where V_{th} is the *thermal voltage* k_BT/q which equals 25.86 mV at 300 K. Clearly, the potential becomes larger and larger as the level of doping increases which will increase the drift of charges.

The potential across the junction is now known and in order to determine the electric field, and the ability for the photovoltaic device to produce a current, we need to know the thickness of the depletion region, since the field is the change of the potential over a distance. The variables needed for this endeavor are shown in Fig. 6.5 and begin the analysis with Poisson's equation,

$$\frac{d^2\psi}{dx^2} = -\frac{\rho}{\epsilon\epsilon_0} = -\frac{q}{\epsilon\epsilon_0}\left[p - n + N_D - N_A\right] \tag{6.5}$$

where ϵ is the relative permittivity or dielectric constant and ϵ_0 is the permittivity of vacuum (8.854×10^{-12} C/V-m $= 8.854 \times 10^{-12}$ F/m).

In the depletion region we assume all the dopant atoms are ionized and there are no countercharges allowing us to write

$$-L_p < x < 0, \quad \frac{d^2\psi}{dx^2} = \frac{q}{\epsilon\epsilon_0} N_A; \quad 0 < x < L_n, \quad \frac{d^2\psi}{dx^2} = -\frac{q}{\epsilon\epsilon_0} N_D$$

[6] There are a couple of ways of understanding why the law of mass action is true for doped semiconductors, here is a simple one. One must first realize that the rate of electron and hole recombination (R) is proportional to the product of their concentration or, $R \propto np$. This will be true for a second-order reaction and for the typical concentrations used, this is a good approximation. Now, assume a large amount of donor atoms are added to the pure Silicon semiconductor making n initially increase to $N \times n_i$ where N is a large number. The hole density will subsequently decrease to a concentration of $p = n_i/N$ as the recombination rate is vastly increased. This will in return reduce the number of free electrons by a similar amount or, $n = Nn_i - n_i/N$. Take the product of n and p to find, $np = [Nn_i - n_i/N] \times n_i/N = n_i^2[1 - 1/N^2] \approx n_i^2$. Since semiconductors are doped at concentrations much, much greater than n_i the approximation is warranted.

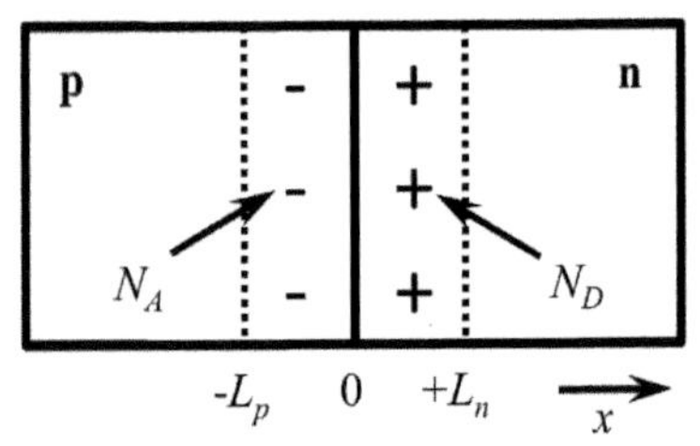

Fig. 6.5 Variables for use in the analysis of the depletion region.

which can be integrated once to find

$$-L_p < x < 0,\ \frac{d\psi}{dx} = \frac{q}{\epsilon\epsilon_0} N_A x + C_1\,;\quad 0 < x < L_n,\ \frac{d\psi}{dx} = -\frac{q}{\epsilon\epsilon_0} N_D x + C_2$$

Realizing the derivative of the potential is zero at $x = -L_p$ and $x = +L_n$ yields

$$-L_p < x < 0,\ \frac{d\psi}{dx} = \frac{q}{\epsilon\epsilon_0} N_A[x + L_p]\,;\quad 0 < x < L_n,\ \frac{d\psi}{dx} = -\frac{q}{\epsilon\epsilon_0} N_D[x - L_n]$$

Integrating again and noting $\psi = 0$ at $x = -L_p$ and $\psi = \psi_0 - V_a$ at $x = L_n$ one has

$$-L_p < x < 0,\ \psi = \frac{q}{2\epsilon\epsilon_0} N_A[x + L_p]^2\,;$$
$$0 < x < L_n,\ \psi = -\frac{q}{2\epsilon\epsilon_0} N_D[x - L_n]^2 + \psi_0 - V_a$$

Here, V_a is the applied potential that may be used to drive an external device. Since ψ should be a continuous function it must have $\psi|_{x=0^-} = \psi|_{x=0^+}$ and $d\psi/dx|_{x=0^-} = d\psi/dx|_{x=0^+}$ where we can find

$$L_p^2 = \frac{2\epsilon\epsilon_0}{q}\frac{\psi_0 - V_a}{N_A^2}\left[\frac{1}{N_A} + \frac{1}{N_D}\right]^{-1} \tag{6.6a}$$

$$L_n^2 = \frac{2\epsilon\epsilon_0}{q}\frac{\psi_0 - V_a}{N_D^2}\left[\frac{1}{N_A} + \frac{1}{N_D}\right]^{-1} \tag{6.6b}$$

and

$$W \equiv L_p + L_n = \left\{\frac{2\epsilon\epsilon_0}{q}[\psi_0 - V_a]\left[\frac{1}{N_A} + \frac{1}{N_D}\right]\right\}^{1/2} \tag{6.6c}$$

The following example is given to show how big the various lengths are.

Example 6.2

Determine L_p, L_n and W for a typically doped Silicon-based photovoltaic device at 300 K with zero applied potential and for a 0.4 V potential at forward bias.

These variables can be determined by noting the typical values for N_A and N_D given in Table 6.1, which are 2×10^{16} #/cc and 1×10^{19} #/cc, respectively. The dielectric constant of Silicon at room temperature is 11.8 while the permittivity of free space is 8.854×10^{-12} C/V-m and the charge of an electron is 1.602×10^{-19} C.

The internal potential ψ_0 needs to be determined that depends on N_A, N_D and n_i, as well as the thermal voltage V_{th} (which is equal to 0.02586 V), see eqn (6.4b). A comment about the value of n_i. The value determined with the equation given by Sproul and Green will be used

Table 6.2 Values of L_p, L_n and W under zero and 0.4 V applied potential.

Variable	Value (nm)
$L_p, V_a = 0$ V	243
$L_p, V_a = 0.4$ V	136
$L_n, V_a = 0$ V	0.486
$L_n, V_a = 0.4$ V	0.272
$W, V_a = 0$ V	243.5
$W, V_a = 0.4$ V	136.3

$$n_i(\#/\text{cc}) = 9.15 \times 10^{19} \left[\frac{T(\text{K})}{300}\right]^2 \exp\left(-\frac{6880}{T(\text{K})}\right) \tag{6.7}$$

which yields the value of 1.00×10^{10} #/cc shown in Table 6.1. Now one can find $\psi_0 = 0.911$ V allowing calculation of the variables given in Table 6.2.

The depletion region in the n-doped material is extremely thin and for practical purposes equal to zero. This is due to the high level of doping that creates numerous free charge carriers. Regardless, as will become evident below, most of the active region in a photovoltaic device is given by the electron and hole diffusion lengths as they are much larger.

In order to progress in understanding what makes a photovoltaic device work we need to determine the concentration of holes and electrons at every position when a potential is applied which will allow us to find the current. Considering the holes first, we know their concentration for $x < -L_p$, as it is assumed all the acceptors are ionized, making $p_p = N_A$, as given in eqn (6.3a). The concentration p_n under zero bias potential is also known by combining eqns (6.3b) and (6.4b) to give

$$p_{n,0} = \frac{n_i^2}{N_D} = N_A \exp(-\psi_0/V_{th}) = p_p \exp(-\psi_0/V_{th}) \tag{6.8a}$$

and it is also simple to find

$$n_{p,0} = \frac{n_i^2}{N_A} = N_D \exp(-\psi_0/V_{th}) = n_n \exp(-\psi_0/V_{th}) \tag{6.8b}$$

with the subscript 0 indicating zero applied potential or $V_a = 0$.

To find the concentration under non-equilibrium conditions, that is when a potential is applied, one must consider the flux of holes J_h (or equivalently for electrons)

$$J_h = -q\mu_h p \frac{d\psi}{dx} - qD_h \frac{dp}{dx} \tag{6.9}$$

The first term is due to drift and the second, diffusion. The variables μ_h and D_h are the mobility and diffusivity for holes, and μ_e and D_e will be used for mobility and diffusivity of electrons below (see Fig. 6.7 for values of D_h and D_e for doped Silicon).

Under equilibrium conditions J_h is identically zero which is what we will assume here under an applied potential. This may seem contradictory at first since the purpose of a photovoltaic device is to make electrical current, however, on inspection of the terms constituting the flux, one realizes that the two individual terms are large in magnitude and sum to a much smaller term. This results in

$$\text{if } J_h \approx 0, \text{ then, } q\mu_h p_a \frac{d\psi}{dx} \approx -qD_h \frac{dp_a}{dx} \tag{6.10}$$

where p_a represents the concentration of holes under an applied potential. Einstein's relation finds $D_h = V_{th} \times \mu_h$ which can be used to give

$$p_{n,a} = p_{p,a} \exp(-\psi_0/V_{th}) \exp(V_a/V_{th}) \tag{6.11}$$

after separation of variables and integrating across the depletion region. The concentration $p_{p,a}$ will not change with applied potential since the hole concentration is dominated by the number of acceptors N_A in the p-type material. Combining the above equation and eqn (6.8a) we find

$$p_{n,a} = p_{n,0} \exp(V_a/V_{th}) \tag{6.12a}$$

and it is also simple to find

$$n_{p,a} = n_{p,0} \exp(V_a/V_{th}) \tag{6.12b}$$

using the definition of the electron flux, $J_e = -q\mu_e n d\psi/dx + qD_e dn/dx$.

The band diagram in Fig. 6.6 is used to show the effect of an applied potential under forward and reverse bias. Since there is a potential we now have the two chemical potentials offset from each other by the amount qV_a. Under forward bias the minority carrier concentrations are increased according to eqn (6.12), which is an exponential dependence. This is extremely important in the operation of photovoltaic devices and is called *injection of minority carriers*, since $p_{n,a}$ and $n_{p,a}$ have much lower concentrations in the n-type and p-type materials' quasi-neutral regions under no bias. Yet, these are the concentrations that dictate the photovoltaic device performance, and the exponential dependence on applied potential is what makes a photovoltaic device perform as it does. Remember, holes must find electrons to complete a circuit and this is controlled by the minority carriers' concentrations since the majority carriers are available in excess.

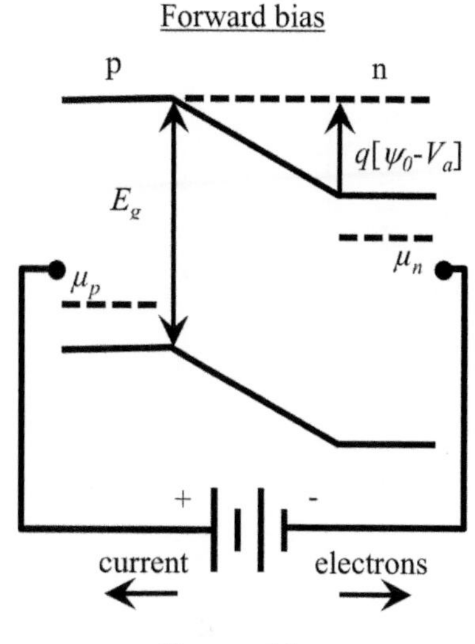

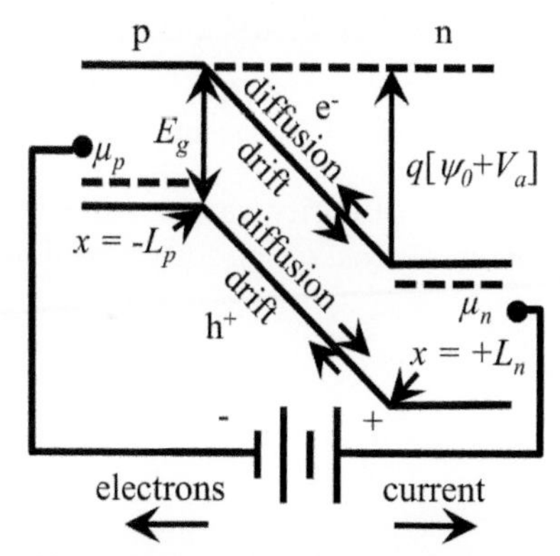

Fig. 6.6 Forward and reverse bias affects the band locations.

Consider movement of the minority carriers. Since there is a large number of majority carriers in the quasi-neutral region, one can assume the potential across the p-n junction does not affect their flux making their movement purely diffusive. The reason is due to the majority carriers *shielding* the minority carriers' charge and the flux is written as

$$\text{n-type quasi-neutral region } (x > L_n),\ J_h \approx -qD_h \frac{dp_n}{dx} \tag{6.13a}$$

$$\text{p-type quasi-neutral region } (x < -L_p),\ J_e \approx qD_e \frac{dn_p}{dx} \tag{6.13b}$$

These expressions are required since the flux will dictate the current. For the present assume the photovoltaic device is operating in the dark and the continuity equations[7] can be written as

$$x > L_n,\ \frac{1}{q}\frac{dJ_h}{dx} = -U_h \tag{6.14a}$$

$$x < -L_p,\ \frac{1}{q}\frac{dJ_e}{dx} = U_e \tag{6.14b}$$

[7] The continuity equation is just like a mass balance and eqns (6.14) merely state mathematically that as the electrons or holes are consumed through recombination there is less flux further into the material.

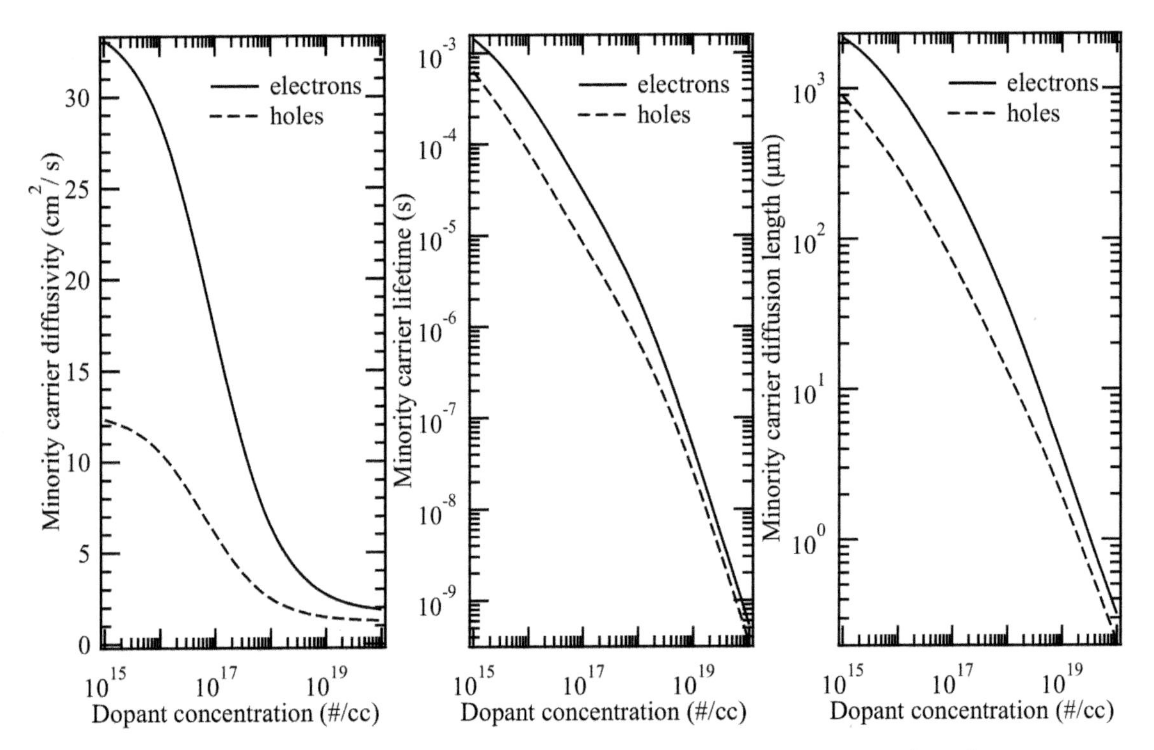

Fig. 6.7 Minority carrier diffusivity (left) lifetime (middle) and diffusion length (right) as a function of dopant concentration for crystalline Silicon.

where U_h and U_e are the recombination rates of holes or electrons, respectively. The simplest model possible to describe recombination is used,

$$x > L_n,\ U_h = \frac{p_{n,a} - p_{n,0}}{\tau_h} \tag{6.15a}$$

$$x < -L_p,\ U_e = \frac{n_{p,a} - n_{p,0}}{\tau_e} \tag{6.15b}$$

where τ_h and τ_e are the minority carrier lifetime for holes or electrons, respectively, before they recombine with the opposite charge (see Fig. 6.7 for values of τ_h and τ_e for doped Silicon). Define excess concentrations above the equilibrium as: $\Delta p_n = p_{n,a} - p_{n,0}$ and $\Delta n_p = n_{p,a} - n_{p,0}$, where $p_{n,0}$ and $n_{p,0}$ are given in eqns (6.8), which are constants for a given p-n junction. Combining eqns (6.13) to (6.15) one finds

$$x > L_n,\ \frac{d^2 \Delta p_n}{dx^2} = \frac{\Delta p_n}{L_h^2}$$

$$x < -L_p,\ \frac{d^2 \Delta n_p}{dx^2} = \frac{\Delta n_p}{L_e^2}$$

where L_h and L_e are the diffusion lengths for holes or electrons, respectively, and equal to $\sqrt{D_h \tau_h}$ for holes and $\sqrt{D_e \tau_e}$ for electrons (see

Fig. 6.7 for values of L_h and L_e for doped Silicon). These are important variables to know as will become clear below. The solution to the above second-order differential equation is of the form $\exp(-[x - L_n]/L_h)$ for the holes and $\exp([x+L_p]/L_e)$ for the electrons. Remember x is negative in the p-type quasi-neutral region. The boundary condition at the edges of the depletion region for each minority carrier is given by eqns (6.12) which we can use to determine

$$x > L_n, \quad \frac{p_{n,a}(x)}{p_{n,0}} - 1 = [\exp(V_a/V_{th}) - 1]\exp(-[x - L_n]/L_h) \qquad (6.17a)$$

$$x < -L_p, \quad \frac{n_{p,a}(x)}{n_{p,0}} - 1 = [\exp(V_a/V_{th}) - 1]\exp([x + L_p]/L_e) \qquad (6.17b)$$

The total current flux of electrons and holes will be constant at each position in the p-n junction since it is assumed to operate under steady state and there can be no charge accumulation anywhere in the device. Thus, if we add the flux of holes to that of electrons at any given position it will be a constant regardless of the position chosen. Assuming there is no recombination of electrons or holes in the depletion region, since it is relatively thin compared to the rest of the device, we can add the flux of holes at $x = L_n$ to that of electrons at $x = -L_p$, which should equal the total current anywhere in the device.[8] In other words, the total current is constant with position and since we assume there is no recombination in the depletion region the sum of these two current fluxes represents the current at any x position. This is simple to do using eqns (6.13) and (6.17) to obtain the following expression:

[8] Formation of minority carriers through radiation absorption occurs randomly and independently allowing this sum to be taken, see Section 6.1.

$$J_{diode} = J_0 [\exp(V_a/V_{th}) - 1] \qquad (6.18a)$$

where

$$J_0 = qn_i^2 \left[\frac{D_h}{L_h N_D} + \frac{D_e}{L_e N_A}\right] \qquad (6.18b)$$

This is the famous *diode equation* and forms the basis of a photovoltaic device's operation. The parameter J_0 is called the dark current and can be determined when there is no solar radiation available (*i.e.* in the dark) at large reverse bias ($V_a \ll 0$).

Example 6.3

Determine the dark current J_0 for crystalline Silicon using the typical dopant concentrations given in Table 6.1 at 300 K.

Equation (6.18b) can be used to determine the dark current, however, the diffusivity and diffusion length are not known for an electron or hole as minority carriers in their respective quasi-neutral region. In fact, one needs to know these variables when the electron or hole are a majority carrier too (although there seems to be only minor differences between minority or majority carriers). The mobility is determined through a relation developed by Caughey and Thomas and is multiplied by the thermal voltage to obtain the diffusivity (Einstein's relation[9]) to find

[9] Einstein's relation is written as

$$D = V_{th} \times \mu$$

where μ is the mobility.

$$D = \frac{D_{max} - D_{min}}{1 + [N_I/N_{ref}]^{\alpha}} + D_{min} \tag{6.19}$$

where N_I is the concentration of *impurity* or dopant atoms and the other parameters are given in Table 6.3 for either electrons or holes. This correlation is only valid at 300 K.

Table 6.3 Values of the parameters for use in eqns (6.19) and (6.20).

Parameter	Electrons	Holes
D_{max} (cm^2/s)	34.4	12.8
D_{min} (cm^2/s)	1.68	1.23
N_{ref} (10^{16} #/cm^3)	8.50	6.30
α (-)	0.72	0.76
τ_0 (ms)	2.50	2.50
C_{SRH} (10^{-13} cm^3/s)	3.00	1.18
C_{Aug} (10^{-31} cm^6/s)	1.83	2.78
γ (-)	1.77	0.570
δ (-)	1.18	0.720

The minority carrier lifetime is determined with a correlation developed by Klaassen, given as

$$\tau^{-1} = \left[\tau_0^{-1} + C_{SRH} N_I\right]\left[\frac{300}{T}\right]^{\gamma} + C_{Aug} c^2 \left[\frac{T}{300}\right]^{\delta} \tag{6.20}$$

where T is in Kelvins and the correlation applies to either electrons or holes, with parameters given in Table 6.3. The two constants C_{SRH} and C_{Aug} are coefficients used to describe the Shockley–Read–Hall and Auger recombination processes of electrons and holes whose physics will not be considered here. The parameter c is the total hole concentration if the minority carrier is an electron or the total electron concentration for a hole minority carrier. This deviates from N_I only at very low doping levels.

All the information is known to determine J_0 using eqn (6.18b) with $N_A = 2 \times 10^{16}$ #/cc and $N_D = 1 \times 10^{19}$ #/cc. From the above correlations or with reference to Fig. 6.7 one finds D_h = 1.47 cm^2/s, L_h = 1.93 μm, D_e = 25.9 cm^2/s and L_e = 633 μm, allowing one to calculate J_0 = 3.40×10^{-10} mA/cm^2, which is quite small.

Two final remarks, firstly, the depletion length was determined in Example (6.2) and was only a fraction of a μm in size. This is much smaller than the diffusion lengths which will constitute most of the active region. Secondly, we see that L_h = 1.93 μm and L_e = 633 μm and it is clear now why a solar cell can work. The physical lengths of the n-type and p-type regions are of order 1 μm and 300 μm, respectively, which is less than their diffusive lengths. So, the minority carriers can survive until the internal field can direct them to the correct electrode to make the light generated current.

Of course, we have not put the diode in the Sun to actually have a photovoltaic device, how does this change eqn (6.18)? Again assuming there is no recombination of holes or electrons in the depletion region and that the diode operates in parallel with the light generated current we can write

$$J = J_0 \left[\exp(V_a/V_{th}) - 1\right] - J_L \qquad (6.21)$$

where J_L is the current produced by absorption of the radiation and can be given by J_{out}, eqns (3.22) and (3.25), as its upper bound. As expected, based on the discussion given in this chapter, the sign of the current through the diode (eqn (6.18)) is opposite to that of the photo-generated current, since J_L is generated by minority charge carriers being forced in the opposite direction to the forward bias *via* the p–n junction's internal potential.

Only electrons and holes generated within the depletion region $L_p + L_n$ and within a diffusion distance L_h and L_e within the depletion region will contribute to the photovoltaic current. This is called the *active* region whose thickness is abbreviated as $L_{act} = L_p + L_n + L_h + L_e$. Furthermore, light will be absorbed if the active region is not exactly at the surface closest to the Sun and will not contribute to the photovoltaic effect. The generation rate within the active region G_{act} is

$$G_{act} = \int_{L}^{L+L_{act}} G(E,x)dx \qquad (6.22)$$

where L is the distance the active region is located below the surface and $G(E,x)$ is given by eqn (5.6). The current generated is thus

$$J_L = q \int_{E_g}^{\infty} \int_{L}^{L+L_{act}} N_p(E)\alpha(E)\exp(-\alpha(E)x)\; dx\; dE \qquad (6.23)$$

Evaluation of this integral depends on details of both the material being used, which may not absorb light in a 100% efficient manner, and the design of the photovoltaic device itself. Of course, the maximum value it can have is J_{out}.

6.3 Current–voltage relation

Before discussing the current–voltage relation, a discussion is given to clarify why a photovoltaic device is operated under forward bias again, knowing the details discussed above. At first forward bias operation seems contradictory; reference to Fig. 6.8 shows that when the excited electron is formed in the n-doped region the resulting hole is swept over to the p-doped region if the photon is absorbed in the active region (for brevity, W is written as the sum of L_p and L_n). Of course, the hole must be within the distance L_h, the diffusion length of a hole, of the depletion region if it is to make it to the p-doped region. If the photon is absorbed within the p-doped region then the electron must be within the distance L_e of the depletion region. Now, as shown in the figure, forward bias has the electron flow opposite to the current flow generated by photon absorption. How can this be?

Consider the photovoltaic device in the dark, which is merely a diode. Under forward bias the depletion region narrows, as shown in Fig. 6.8,

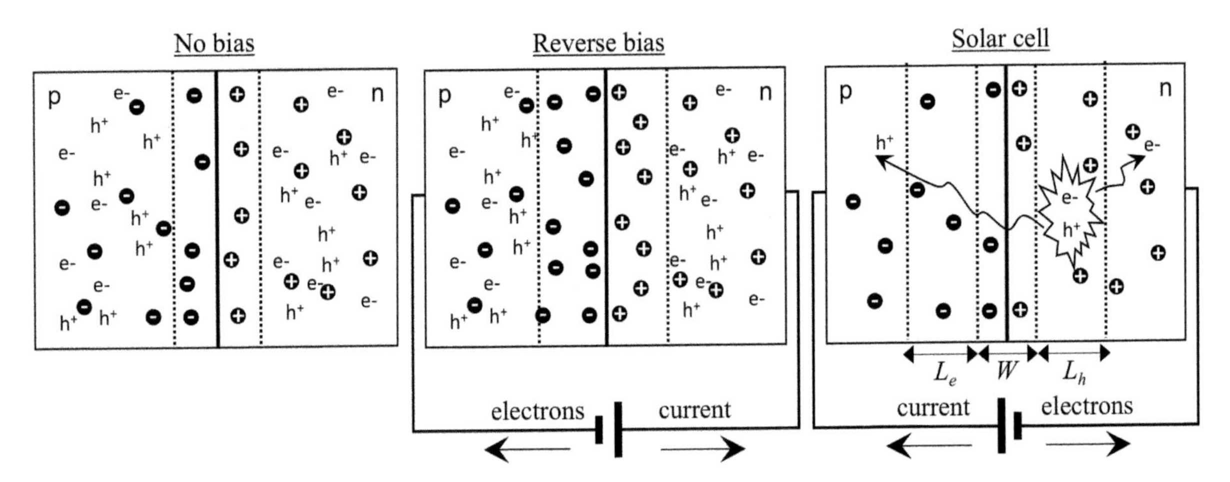

Fig. 6.8 The depletion region has a certain width under no bias (left) and becomes larger under reverse bias (middle) effectively increasing the resistance of the junction. Under illumination an excited electron-hole pair is formed in the n-type material of a solar cell (right) and the hole is swept over to the p-doped region while the electron resides in the n-doped region. A solar cell is operated under forward bias so the effective resistance is lowered and the electrons and holes separate to complete the circuit.

and the current is dictated by diffusion of the majority carriers. Under reverse bias the depletion region widens and now drift of the minority carriers controls the current flow. This, of course, produces a much lower current than forward bias and can be thought of as an increase of the resistance within the diode.

The number of minority carriers is increased under illumination, by a large amount, and the drift current is increased. In other words, the hole generated in the n-doped region would be swept over to the p-doped region to produce this current. If the photovoltaic device were operated under reverse bias the depletion region would increase in size and not aid charge extraction. There would be an accumulation of charge due to the effective increase in the cell resistance. Applying forward bias narrows the region letting the charge escape and one has an operational photovoltaic device.

Previously, we derived the *diode equation* and the sum of this and the light generated current density, J_L, is the operating equation for a photovoltaic device which is written as

$$\begin{aligned} J &= J_0[\exp(qV_a/k_BT) - 1] - J_L \\ &\equiv J_0[\exp(V_a/V_{th}) - 1] - J_L [=] \frac{\text{A}}{\text{m}^2} \end{aligned} \tag{6.21}$$

where, as before, V_{th} is the thermal voltage of 25.86 mV at room temperature (300 K which is actually a warm room ...). A graph of this equation is given in Fig. 6.9. In eqn (6.21) it was assumed that the light generated current operates independently of the diode characteristics, which is a reasonably good assumption. Essentially the assumption is that the electron–hole pairs formed in the active region do not recombine and create the required current regardless of the applied potential. So, the diode equation is merely shifted down by a constant current equal to J_L.

As discussed above, the operating part of a photovoltaic device is in the positive voltage and negative current quadrant. Since power is voltage times current, the power, or work, from a photovoltaic device is negative in value. According to our definition of work in Chapter 2, this indicates work is coming from the device to do work on the surroundings, which is what we desire.

There are several points on the graph worth noting. The first is the *short circuit current*, J_{sc}, which is equal to $-J_L$ at zero applied voltage. Again this is under the condition of the assumption listed above that the light induced current operates independently of the diode characteristic. The second point is the *open circuit voltage*, V_{oc}, which is the voltage when no current flows through the cell. This voltage is dictated by the band gap energy which we discuss below. The last point is that of *maximum power*, (V_m, J_m). This is more readily seen if power, or $J \times V_a$, is plotted as a function of voltage, as shown in Fig. 6.9. Clearly, the point of maximum power is located at the minimum of the curve, as shown in the bottom graph, which is more difficult to see in the top graph.

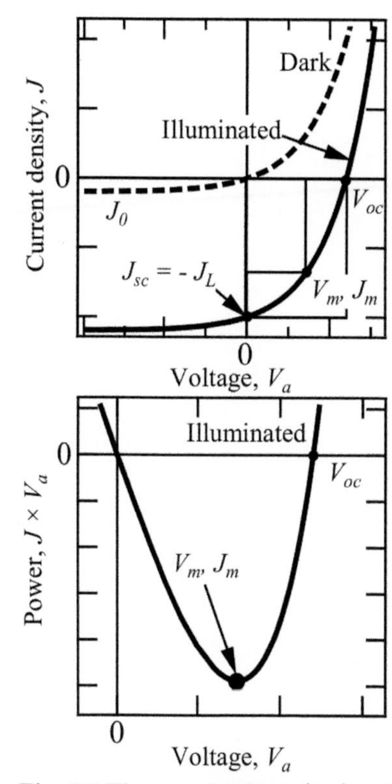

Fig. 6.9 The current-voltage (top) and power-voltage (bottom) curves for an ideal photovoltaic device under dark and illuminated conditions. The symbols are described in the text.

The short-circuit current has already been discussed above since it is equal to J_L. So, the semiconductor properties which affect its value are clear. The electron and hole recombination lengths, as well as the semiconductor absorption characteristics, all play key roles in its magnitude. The parameters that influence the open circuit voltage are a little more obscure. To understand this we take eqn (6.21) and rearrange it

$$V_{oc} = V_{th} \ln\left(\frac{J_L}{J_0} + 1\right) \approx V_{th} \ln\left(\frac{J_L}{J_0}\right) \tag{6.24}$$

where the approximation is used since J_L is much larger than J_0 (this is discussed below). Remembering the definition of J_0 this equation can be re-written as

$$V_{oc} = V_{th} \ln\left(\frac{J_L}{q\left[\dfrac{D_h}{L_h N_D} + \dfrac{D_e}{L_e N_A}\right] N_c N_v}\right) + \frac{E_g}{q} \tag{6.25}$$

The open circuit voltage is primarily dictated by the band gap energy and is modified by the other terms in the logarithm. Thorough consideration of the physics of a photovoltaic device will find a more complicated expression than given in eqn (6.25), yet, this equation reflects the essential dependence on the various parameters.

Another important variable to grade photovoltaic device operation is the fill factor or *FF* given by

$$FF \equiv \frac{J_m V_m}{J_{sc} V_{oc}} \tag{6.26}$$

The fill factor indicates how *square* the current–voltage relation is, the more square the better, since the maximum power that can be obtained is $J_{sc} \times V_{oc}$. The fill factor for a very good photovoltaic device should be greater than 80%.

Example 6.4

Estimate the point of maximum power and the fill factor for this ideally operating photovoltaic device with simple equations that only contain the open circuit and thermal voltages.

Reference to Fig. 6.9 shows the point of maximum power, which is the largest power that can be obtained from a photovoltaic device, and is given by $P_m = J_m \times V_m$, where V_m and J_m are the voltage and current density at the point of maximum power. This point is found by multiplying equation (6.21) by the applied voltage V_a and taking the derivative with respect to V_a to arrive at

$$0 = e^{\hat{V}_m} - 1 - \frac{J_L}{J_0} + \hat{V}_m e^{\hat{V}_m} = e^{\hat{V}_m} - e^{\hat{V}_{oc}} + \hat{V}_m e^{\hat{V}_m}$$

then setting the derivative to zero. Here $\hat{V}_m = V_m/V_{th}$ and $\hat{V}_{oc} = V_{oc}/V_{th}$. The last equation results from using the definition of the open circuit voltage in eqn (6.24). One can now write

$$\hat{V}_m = \hat{V}_{oc} - \ln(1 + \hat{V}_m) \tag{6.27}$$

This is a transcendental equation for $\hat{V}_m$. Of course one can always assume values of $\hat{V}_m$ and calculate $\hat{V}_{oc}$, however, we want $\hat{V}_m$ explicitly in terms of $\hat{V}_{oc}$ so it can be calculated by knowing the band gap energy and other factors, for example. To do this, note the logarithm is a slowly varying function and so we take an approximation scheme like this. As a zeroth-order approximation forget the logarithm term and let $\hat{V}_m^{(0)} = \hat{V}_{oc}$, then subsequent approximations are obtained through this recurrence relation, $\hat{V}_m^{(i)} = \hat{V}_{oc} - \ln(1 + \hat{V}_m^{(i-1)})$ for $i \geq 1$. Following this one finds

$$\hat{V}_m^{(0)} = \hat{V}_{oc} \tag{6.28a}$$

$$\hat{V}_m^{(1)} = \hat{V}_{oc} - \ln(1 + \hat{V}_{oc}) \tag{6.28b}$$

$$\hat{V}_m^{(2)} = \hat{V}_{oc} - \ln(1 + \hat{V}_{oc} - \ln(1 + \hat{V}_{oc})) \tag{6.28c}$$

$$\vdots$$

As it turns out $\hat{V}_m^{(1)}$ is within 3% of the exact relation for $\hat{V}_{oc}$ greater than 10 (at room temperature this is $V_{oc} = 0.25$ V which is a very small open circuit voltage) allowing this approximation to be used. Now the current density at maximum power conditions can be found through direct substitution of eqn (6.28b) into eqn (6.21)

$$\hat{J}_m = \frac{e^{\hat{V}_{oc}}}{1+\hat{V}_{oc}} - 1 - \hat{J}_L = \frac{e^{\hat{V}_{oc}}}{1+\hat{V}_{oc}} - e^{\hat{V}_{oc}} \tag{6.29}$$

where $\hat{J}_m$ and $\hat{J}_L$ are J_m and J_L normalized with J_0. Equation (6.24) was also used to substitute $\hat{J}_L = e^{\hat{V}_{oc}} - 1$. The voltage and current density at maximum power can now both be estimated solely from the open circuit voltage.

We know that $J_{sc} = -J_L$ so the fill factor can be written as

$$FF_0 = \left[\frac{\frac{1}{1+\hat{V}_{oc}} - 1}{e^{-\hat{V}_{oc}} - 1}\right] \times \left[\frac{\hat{V}_{oc} - \ln(1+\hat{V}_{oc})}{\hat{V}_{oc}}\right]$$

where FF_0 is written since this is an approximate relation valid for assumptions given above and that none of the non-idealities described below are considered. One can assume that $e^{-\hat{V}_{oc}}$ is very small and can be neglected, so, after algebraic manipulation one obtains

$$FF_0 = \frac{\hat{V}_{oc} - \ln(\hat{V}_{oc}+1)}{\hat{V}_{oc}+1} \tag{6.30}$$

This equation is almost exactly equal to an empirical equation given by Green (1982) (see also the discussion by Green (1982–1983))

$$FF_0 = \frac{\hat{V}_{oc} - \ln(\hat{V}_{oc}+0.72)}{\hat{V}_{oc}+1} \tag{6.31}$$

Both these equations show that the fill factor approaches 1 in the limit of the open circuit voltage approaching infinity which is desirable. The fill factor indicates how 'square' is the current–voltage graph, the more square the graph the better and the more efficient the photovoltaic device operates. Thus, for a perfect square the maximum power can be extracted from the photovoltaic device which is $V_{oc} \times J_{sc}$ and the fill factor is 1.

6.4 Parameters for an operating photovoltaic device

So far non-idealities have not been considered except with recombination occurring in the quasi-neutral regions through eqns (6.15). However, there can be defects in the crystalline lattice and this can cause recombination in the depletion region. The result of the calculation can be found in most solar photovoltaic textbooks, the key fact to remember is that the width of the depletion region influences whether or not recombination occurs there. There are other recombination mechanisms that also contribute where ultimately one arrives at a current density–voltage relation like this,

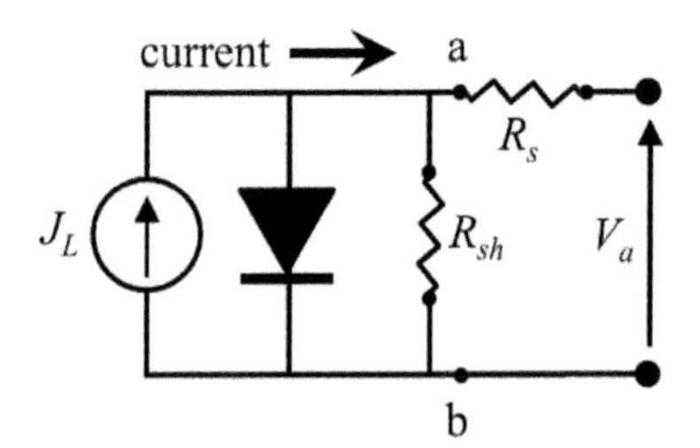

Fig. 6.10 Series and shunt resistances in an operating photovoltaic device showing the direction of current flow and the potential, V_a.

$$J = J_{01}\{\exp(V_a/V_{th}) - 1\} + J_{02}\{\exp(V_a/2V_{th}) - 1\} - J_L \tag{6.32}$$

where J_{01} and J_{02} are very different in value as different physics dictates their occurrence. Charge carrier recombination in the quasi-neutral region produced the J_{01} term as was derived above. If recombination in the depletion region occurs then the J_{02} term results and the factor of 2 in the exponential is present. In general, they are lumped together to this form,

$$J = J_0\{\exp(V_a/nV_{th}) - 1\} - J_L \tag{6.33}$$

where n is the non-ideality factor. Note, one can perform careful measurements in the dark and J_0 will be found to be a function of the applied voltage, as will n, so, it is possible to determine their individual values. However, this is not considered here and the lumped parameters will be used.

A further non-ideality is the appearance of a series and shunt resistance, R_s and R_{sh}, respectively, as shown in Fig. 6.10.[10] The light generated current J_L is shown as is the applied voltage V_a and the diode making up the photovoltaic device. The current–voltage relation can now be written

[10] The shunt resistance frequently comes from defects in the active region (p-n junction) that allow current flow through it. The shunt resistance should be as large as possible and tending to infinity. The series resistance has three major sources; current flow through the p-n junction itself, contact resistance between the active region of the photovoltaic device and the electrodes and resistance of the electrodes (contacts) themselves. Of course, it is desirable to have the series resistance as small as possible.

$$J = J_0\left[\exp\left(\frac{V_a - JR_sA}{nV_{th}}\right) - 1\right] + \frac{V_a - JR_sA}{R_{sh}A} - J_L \tag{6.34}$$

where A is the illuminated area of the photovoltaic device. Derivation of this equation, that accounts for the two resistances, is simple. The potential between the points **a** and **b** in Fig. 6.10 is $V_a - JR_sA$, and this can be substituted into eqn (6.33) for V_a. Since the diode, photogenerated current and shunt resistor operate in parallel their currents are merely added together to yield eqn (6.34). The order of magnitude for the various parameters is shown in Table 6.4 and one can make a reasonable assumption that $V_a \gg JR_sA$, which makes it much easier to analyze eqn (6.34) and so we write

$$J \approx J_0\left[\exp\left(\frac{V_a - JR_sA}{nV_{th}}\right) - 1\right] + \frac{V_a}{R_{sh}A} - J_L \tag{6.35}$$

It is found that R_{sh} affects V_{oc} very little until very small values are used. So, the open circuit voltage can be approximated by

$$V_{oc} = nV_{th}\ln\left(\frac{J_L}{J_0} + 1 - \frac{V_{oc}}{J_0R_{sh}A}\right) \approx nV_{th}\ln\left(\frac{J_L}{J_0}\right) \tag{6.36}$$

Table 6.4 Order of magnitude for parameters in a typical (Silicon) photovoltaic device for eqns (6.34) and (6.35). The thermal voltage V_{th} is 25.86 mV at 300 K. The row that has J_0 (V_{oc} = 0.6 V) is the value of J_0 which makes V_{oc} = 0.6 V. This must be done because eqn (6.36) is not detailed enough to account for losses in V_{oc}.

Parameter	Value
V_{oc}	0.6 V
J_L	35 mA/cm^2
J_0 (typical)	10^{-10} – 10^{-9} mA/cm^2
J_0 (V_{oc} = 0.6 V)	5×10^{-7} mA/cm^2
R_sA	1 Ω-cm^2
$R_{sh}A$	10^3 Ω-cm^2
n	1 – 2

The approximation comes after using typical values for the parameters and keeping only the significant terms. The effect of R_{sh} on V_{oc} is discussed in Exercise 6.17 and an analytical expression is developed.

Even with the above simplification, eqn (6.35) is difficult to fit to data as the parameters are coupled should non-linear regression be used.

Instead the series and shunt resistances are determined by taking the derivative of the applied potential with the current density, then further analysis is required. Since the equation is implicit in the applied potential, one takes the derivative of the current with the potential to arrive at

$$\begin{aligned}\left(\frac{\partial J}{\partial V_a}\right) \times \left[1 - \frac{J_0}{nV_{th}} \exp\left(\frac{V_a - JR_sA}{nV_{th}}\right) R_sA\right] \\ = \frac{J_0}{nV_{th}} \exp\left(\frac{V_a - JR_sA}{nV_{th}}\right) + \frac{1}{R_{sh}A}\end{aligned} \qquad (6.37)$$

The shunt resistance is most easily determined by taking the derivative of voltage with current density near zero voltage (*i.e.* near the short circuit current) and from eqn (6.37) one finds

$$\left(\frac{\partial V_a}{\partial J}\right) \approx R_{sh}A \qquad (\text{for } V_a \approx 0) \qquad (6.38)$$

since the exponential terms will be essentially zero.

The series resistance is frequently estimated by taking the slope of voltage with current near the open circuit voltage. A slightly more involved technique, which yields more information, is to take the derivative of J with respect to V_a in eqn (6.35) to arrive at

$$\left(\frac{\partial V_a}{\partial J}\right) = -R_sA + \frac{nV_{th}}{J + J_L} \qquad (\text{for V} \approx \text{Voc}) \qquad (6.39)$$

since one can write $J_0 \exp\left([V_a - JR_sA]/V_{th}\right) \approx J + J_L$. The effect of the shunt resistance has been ignored at higher voltages and is easily validated by taking orders of magnitude in eqn (6.35). So, a graph of $(\partial V_a/\partial J)$ versus $[J + J_L]^{-1}$ will have an intercept of R_sA and a slope of nV_{th}. Some error is apparent in this technique since a derivative of data must be taken and this will frequently amplify any noise in data (see Example 6.6 below).

It is recommended to use the above techniques to estimate the shunt and series resistance first then if a non-linear regression is used these values can be used as the starting value. Equation (6.34) is highly non-linear and parameter estimation is difficult since they are coupled. The example below shows the accuracy of eqns (6.38) and (6.39) where it is determined that they are reasonably accurate as long as the resistances do not deviate too much from the *normal* values.

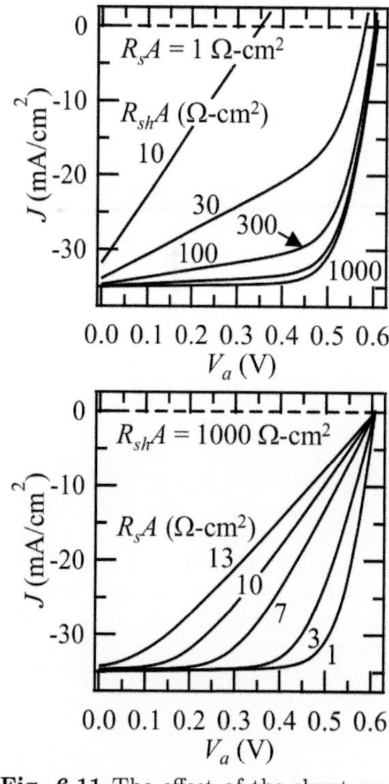

Fig. 6.11 The effect of the shunt and series resistances in a photovoltaic device on the current density - voltage graph. The upper graph has R_s = 1 Ω-cm^2 with R_{sh} varying and the lower graph has R_{sh} = 1000 Ω-cm^2 with R_s varying. See text for details of other parameter values.

Example 6.5

Determine the accuracy of eqns (6.38) and (6.39) for estimating the shunt and series resistances, respectively. Assume $J_0 = 5 \times 10^{-7}$ mA/cm^2 according to Table 6.4.

The best technique to do this is to calculate *ideal* data with eqn (6.34) and then apply the above equations to determine the resistances. The result of the calculation to find the J - V_a curve is given in Fig. 6.11,

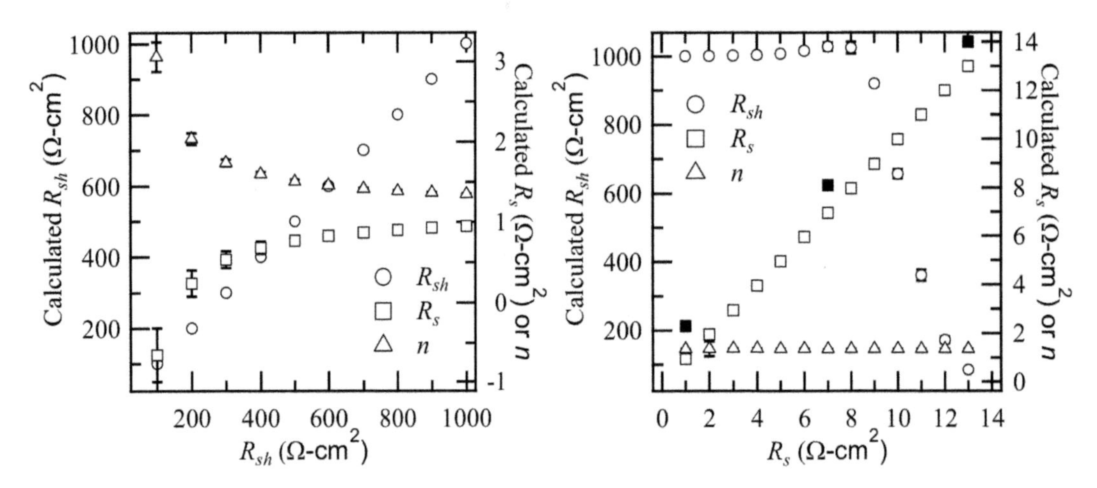

Fig. 6.12 Equations (6.38) and (6.39) were applied to the data in Fig. 6.11 with the results given in the graphs. The graph on the left has R_{sh} changing (data are from the upper graph in Fig. 6.11) while that on the right has R_s changing (data are from the lower graph in Fig. 6.11). The filled squares on the right are the result of taking the derivative of voltage with current density near V_{oc} and letting this represent R_s.

where the values for the order of magnitude for other parameters are given in Table 6.4. Specifically the values are: J_L = 35 mA/cm^2, J_0 = 5×10^{-7} mA/cm^2, n = 1.3 and V_{th} = 25.86 mV. It is clear that if R_{sh} becomes too small and R_s too high then the photovoltaic device acts more like a resistor than a properly operating device.

Now eqns (6.38) and (6.39) are applied to this data and are presented in Fig. 6.12. The techniques to estimate the resistances work well as long as R_{sh} does not become too small and R_s does not become too large. Yet, this may be too simple a conclusion. For example, letting R_{sh} = $200\,\Omega/\text{cm}^2$ and R_s = $10\,\Omega/\text{cm}^2$, two values that individually produced poor estimates of the other variable in Fig. 6.12, yields values of R_{sh} = $211 \pm 3\,\Omega/\text{cm}^2$ and R_s = $9.87 \pm 0.00_4\,\Omega/\text{cm}^2$ after applying the analysis techniques which are much better estimates than expected.[11] However, the estimate of the non-ideality factor n appears to suffer somewhat and takes the value of $1.58 \pm 0.00_4$ instead of the assumed value of 1.3, a 20% deviation. Thus, when R_{sh} and R_s approach each other in value, rather than being at their extreme values, the technique performs better.

Finally, it was mentioned above that a frequently used technique to determine R_s is to merely take the derivative of V_a with J near V_{oc}, so, the accuracy of this technique was also tested. The results are shown in Fig. 6.12. For the range of conditions chosen, eqn (6.39) is more accurate and the simple derivative technique yields a value that is almost $1\Omega/\text{cm}^2$ larger for all values of R_s. So, the more involved technique is recommended to determine more accurate values of the resistance.

[11] An error written like this: R_s = $9.87 \pm 0.00_4\,\Omega/\text{cm}^2$, the 4 as a subscript, means that the number is only accurate to the second decimal, yet, the calculated standard deviation suggests accuracy to three decimal places. So, the value may be precise, but, not necessarily accurate.

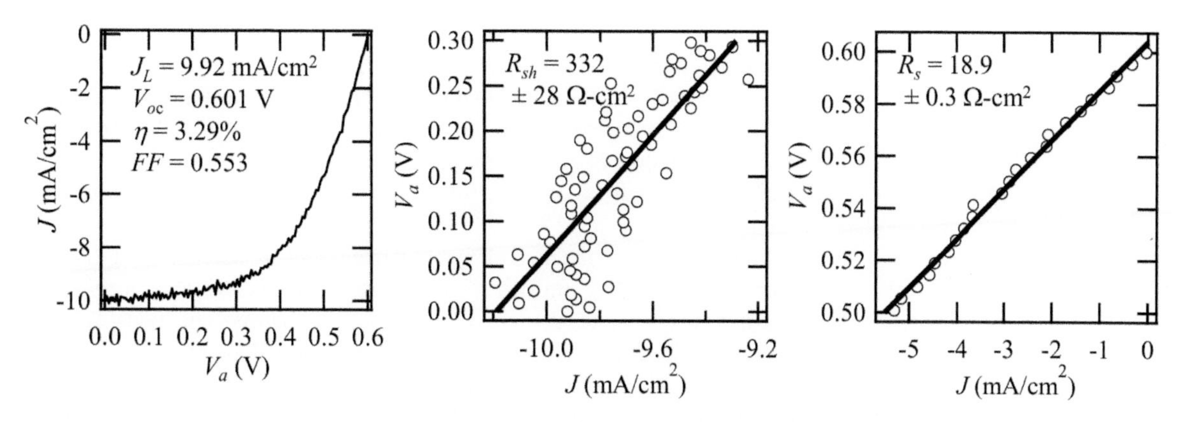

Fig. 6.13 (left) The current - voltage graph for an experimental polymer-based photovoltaic device together with various operational parameters. (center) Voltage versus current near Jsc to determine the shunt resistance and (right) near V_{oc} to find the series resistance. Data courtesy of Wenluan Zhang.

Equations (6.38) and (6.39) work well with perfect data, how do they perform with poorer quality data that has noise, a large series resistance and a low shunt resistance? This is precisely the type of data gathered from research type devices and where accurate series and shunt resistances are required so that progress and optimization can be made with the photovoltaic device.

The following example is presented to show the effect of real, noisy data on analysis. Furthermore, if the resistance is to be determined, the slope of voltage with current must be obtained, remember $V = JR$, voltage equals the product of current density and resistance times area. If the slope of current density with voltage is calculated using linear regression, then the inverse of the slope is taken as the resistance; it is usual that noisy data do not return the same resistance value for the two techniques as will be described in the example.

Example 6.6

Determine the accuracy of eqns (6.38) *and* (6.39) *data for a poor quality photovoltaic device.*

The data used for this example will be for a prototype, polymer-based photovoltaic device made from 1:1 weight ratio of poly(3-hexylthiophene) or P3HT and [6,6]-phenyl-C_{61}-butyric acid methyl ester or PCBM. This system will have a power conversion efficiency η of at most 3.5% and works by different physics than a Silicon photovoltaic device. However, for the purposes of this example it presents a great test of determining the series and shunt resistances for a system under development whose further development demands knowledge of these parameters.

The current density–voltage graph is shown in Fig. 6.13, together with light generated current density, open circuit voltage, power conversion efficiency and fill factor. The analysis according to eqn (6.39) was tried *without* success. In fact, the slope was negative implying that the thermal voltage or the non-ideality factor were negative. The reason why the analysis failed was because the shunt resistance is found to be low and the series high, concomitant with a low fill factor, which invalidates some of the assumptions in deriving the equation, such as neglecting the effect of the shunt (and series) resistances near the open circuit voltage.

The derivative of the voltage with current density was determined near J_{sc} to find R_{sh} and near V_{oc} to determine R_s, which is a reasonable method (see Fig. 6.12). The graphs to determine the resistances are shown in Fig. 6.13 together with the regression lines and the associated resistances. The shunt resistance is very low suggesting many defects in the photovoltaic device, while the series resistance is quite high most likely indicating the poor conductivity of the photovoltaic device and electrodes as well as poor charge transfer to the electrodes. Certainly much improvement is required for this photovoltaic device to be acceptable and it will be left to discussion below as to which of the resistances make the fill factor and efficiency so low.

The slope (derivative) must be obtained with voltage plotted against current density since an inverted plot of current density versus voltage will yield different values for the resistances. Only for perfect data will they be equal, this is data that follows the regression line perfectly. The reason is that linear regression minimizes the difference between the y-axis (ordinate) and the fit $a + bx$ to find a and b by using the sum of square errors (SSE). For N data points this is given by

$$SSE = \sum_{i=1}^{N} [y_i - \{a + bx_i\}]^2$$

where the data set points are (x_i, y_i). This is not the same as minimizing

$$SSE = \sum_{i=1}^{N} [x_i - \{a' + b'y_i\}]^2$$

unless, as mentioned above, perfect data are available. The intercept and slope, a' and b', respectively, for the fit of current density with applied voltage yield different resistances than those given in Fig. 6.13. One finds a similar value for the series resistance 19.1 ± 0.3 Ω-cm^2 to that when the slope of voltage with current is determined: 18.9 ± 0.3 Ω-cm^2, since the data deviate very little from the linear regression line. However, the shunt resistance is very different since the data quality is not as good: 500 ± 43 Ω-cm^2 versus 332 ± 28 Ω-cm^2. Thus, care should be taken when evaluating these resistances, particularly when analyzing poor quality data.

These resistances will be used for subsequent analysis of this photovoltaic device's performance, below. In particular, whether the shunt or the series resistance is the biggest culprit in poor performance which will be determined and this can be used to focus research and development towards an improved device.

Now that the series and shunt resistances can be determined, their effect on photovoltaic device performance is considered. The load on the device should have a resistance of $|V_m/J_m|$ (remembering that the area of the device should also be considered) and this is called the characteristic resistance or R_{ch}. Operation under this load ensures that the photovoltaic device is working at its peak power. The characteristic resistance is frequently approximated by V_{oc}/J_{sc} and can be rationalized by using eqns (6.28b) and (6.29) to arrive at

$$\hat{R}_{ch} = \left| \frac{\hat{V}_{oc} - \ln(1 + \hat{V}_{oc})}{\dfrac{e^{\hat{V}_{oc}}}{\hat{V}_{oc} + 1} - 1 - \hat{J}_L} \right| \approx \left| \frac{\hat{V}_{oc}}{-\hat{J}_L} \right| = \left| \frac{\hat{V}_{oc}}{\hat{J}_{sc}} \right| \tag{6.40}$$

where R_{ch} is normalized with V_{th} and J_0 to yield $\hat{R}_{ch}$ and J_{sc} with J_0 to give $\hat{J}_{sc}$. (The absolute value signs will be ignored from now on realizing that the resistance should always be non-negative.) The approximation comes after reviewing the order of magnitude for the parameters from Table 6.4.[12] The dimensional characteristic resistance is an important variable in photovoltaic device operation and can be estimated as V_{oc}/J_{sc}.

[12] For a crystalline Silicon photovoltaic device the dark current J_0 must be below order 10^{-8} mA/cm^2 for this approximation to hold. This is not too restrictive, however, it is important to note.

The series resistance has no effect on V_{oc} and only a moderate effect on Jsc when it very large. It does impact the fill factor though as well as the point of maximum power. The point of maximum power will be given by $V_m J_m$ and without the effect of the series or shunt resistance is denoted as P_{m0}, and V_{m0} and J_{m0} will also be used to represent the voltage and current density, respectively, under the same condition (see eqns (6.28b) and (6.29)).

A simple way to account for the effect of series resistance effect on the point of maximum power P_m is to realize that the series resistance will introduce a loss of power equal to $J_m^2 R_s$ allowing one to write

$$\begin{aligned} P_m &\approx V_{m0} J_{m0} - J_{m0}^2 R_s = V_{m0} J_{m0} \left[1 - \frac{J_{m0}}{V_{m0}} R_s \right] \\ &\approx P_{m0} \left[1 - \frac{J_{sc}}{V_{oc}} R_s \right] \equiv P_{m0} \left[1 - r_s \right] \end{aligned} \tag{6.41}$$

where r_s is the normalized series resistance and is equal to R_s/R_{ch}. By the definition of the fill factor given in eqn (6.26) one can write

$$FF_s = FF_0[1 - r_s] \tag{6.42}$$

where FF_s and FF_0 are the fill factor considering the effect of the series resistance and under no series (or shunt) resistance, respectively.

The shunt resistance should be as high as possible to prohibit current leakage. It is usually caused by defects in the device from poor manufacturing and should be avoided if possible. Its influence on the maximum power can be estimated similarly to the manner used for the effect of the series resistance. In this case the current *lost* due to the shunt resistance is estimated by $J_m = V_m/R_{sh}$ and the power lost is V_m^2/R_{sh}, so

$$\begin{aligned} P_m &\approx V_{m0}J_{m0} - \frac{V_m^2}{R_{sh}} \approx V_{m0}J_{m0}\left[1 - \frac{V_{m0}}{J_{m0}}\frac{1}{R_{sh}}\right] \\ &\approx P_{m0}\left[1 - \frac{V_{oc}}{J_{sc}}\frac{1}{R_{sh}}\right] \equiv P_{m0}\left[1 - \frac{1}{r_{sh}}\right] \end{aligned} \tag{6.43}$$

where r_{sh} is given by R_{sh}/R_{ch}. As before we can write the effect of shunt resistance on the fill factor as

$$FF_{sh} = FF_0\left[1 - \frac{1}{r_{sh}}\right] \tag{6.44}$$

where FF_{sh} is the fill factor when the shunt resistance is considered.

When both the series and shunt resistances are present the fill factor can be approximated by

$$FF = FF_0\left[1 - r_s\right] \times \left[1 - \frac{1}{r_{sh}}\right] \tag{6.45}$$

The next example will be used to demonstrate the effect of the resistances on a poorly operating photovoltaic device.

Example 6.7

Determine which resistance most affects the operation of the polymer-based photovoltaic device in Example 6.6.

The parameter values required for calculation are given in Table 6.5. The first calculation is to determine FF_0 from eqn (6.30),

Table 6.5 Parameter values for the poorly operating photovoltaic device in Example 6.6.

Parameter	Value
V_{oc}	0.601 V
J_L	9.92 mA/cm^2
η	3.29%
FF	0.553
R_sA	18.9 ± 0.3 Ω-cm^2
$R_{sh}A$	332 ± 28 Ω-cm^2

$$FF_0 = \frac{23.2 - \ln(23.2 + 1)}{23.2 + 1} = 0.827$$

which is quite high compared to the experimental value of 0.553.

Equation (6.45) is now used to determine which of the resistances most affects the cell operation. The characteristic resistance is

$$R_{ch} = \frac{0.601\text{V}}{9.92 \times 10^{-3}\,\text{A/cm}^2} = 60.6\,\Omega\text{-cm}^2$$

The normalized resistances are $r_s = 18.9\,\Omega\text{-cm}^2/60.6\,\Omega\text{-cm}^2 = 0.311$ and $r_{sh} = 332\,\Omega\text{-cm}^2/60.6\,\Omega\text{-cm}^2 = 5.48$.

The effect of the two resistances can now be determined with eqn (6.45),

$$FF = 0.827 \times [1 - 0.311] \times \left[1 - \frac{1}{5.48}\right] = 0.827 \times 0.689 \times 0.818 = 0.466$$

The calculated fill factor is approximately 15% lower than the experimentally measured value, which is acceptable considering the approximations used to derive eqn (6.45). The most telling outcome of this analysis is that the series resistance produces the greatest degradation to the photovoltaic device operation and further research should be undertaken to understand what produces this deleterious phenomenon.

6.5 Conclusion

Discussion of how a solar cell or photovoltaic device works was given in this chapter. A descriptive explanation was given at first, followed by a more detailed consideration of what happens in such a device. In this more detailed look semiconductor physics was used to determine the active region thickness that dictates how thick a solar cell should be to manufacture the most efficient device. This derivation resulted in the famous *diode equation*, central to all solar cells.

The diode is what drives the solar cell to work. An internal potential exists inside the p–n junction that sweeps light generated minority carriers to their part of the junction so they become majority carriers. Interestingly, the applied forward bias potential forces current flow in the opposite direction to the light generated current and they appear to work at odds. However, the large negative light generated current allows work to be extracted from the device and a solar cell can only do this under forward bias.

Consideration of the current–voltage graph, measured to characterize the device, was then performed. Non-idealities, such as the non-ideality factor, series resistance and shunt resistance, were placed in a more accurate diode equation that operated in parallel with the photogenerated current. This equation is useful to ascertain what affects the photovoltaic device operation, yet, is a transcendental equation, with an exponential term, making a regression fit for data challenging. So, derivatives of current with voltage at the short circuit current and open circuit voltage were used, and confirmed, to give fairly accurate values of the resistances. Then only a single parameter, the non-ideality index, must be used to fit the data resulting in a more accurate representation. This information can then be used to determine why a solar cell's operation is not optimal.

For example, if the shunt resistance is low then manufacture of the active region must be faulty and should be corrected. A high series resistance could be indicative of resistance between the active region and the electrodes, or resistance within the electrodes themselves, and a correction here may have to be made. Thus, the current–voltage graph is an important tool that can be used to optimize a given device.

The current - voltage graph can also be used to find the point of maximum power. This is the current–voltage point where the photovoltaic device should be operated since this is where one extracts the most power. The series and shunt resistances affect this point and always deleteriously. So, if research and engineering is focused on minimizing the effect of the resistances one can ultimately manufacture a more powerful device, all stemming from a good understanding of the current–voltage graph.

6.6 General references

D.M. Caughey and R.E. Thomas, 'Carrier mobilities in silicon empirically related to doping and field,' Proc. IEEE **55** (1967) 2192.

M.A. Green, 'Solar cells: Operating principles, technology and system applications,' Prentice-Hall Inc. (1982.)

M.A. Green, 'Accuracy of analytical expressions for solar cell fill factors,' Solar Cells **7** (1982–1983) 337.

A.S. Grove, 'Physics and technology of semiconductor devices,' John Wiley & Sons (1967) (This is an excellent text to learn semiconductor physics and was written by the eventual Chief Executive Officer and President of Intel Corporation. He did his PhD in chemical engineering at University of California - Berkeley, with Professor Andreas Acrivos, in the area of fluid dynamics! So, please read the Preface of this book again to realize that you really can't know the type of job you will have in the future!).

D.B.M. Klaassen, 'A unified mobility model for device simulation – II. Temperature dependence of the carrier mobility and lifetime,' Sol. State Elec. **35** (1992) 961.

A. Luque and S. Hegedus (eds), 'Handbook of photovoltaic science and engineering,' J. Wiley & Sons (2003).

A.B Sproul and M.A. Green, 'Improved value for the silicon intrinsic carrier concentration from 275 to 375 K,' J. Appl. Phys. **70** (1991) 846.

Exercises

(6.1) The effective mass for an electron in the conduction band is $1.08m_e$ and a hole in the valence band is $0.81m_e$ for Silicon; calculate the effective number density of states in the conduction and valence bands in #/cm^3. Calculate the number density of intrinsic carriers assuming the band gap energy is 1.12 eV at 300 K.

(6.2) The effective mass for an electron in the conduction band is $0.067m_e$ and a hole in the valence band is $0.47m_e$ for Gallium Arsenide; calculate the effective number density of states in the conduction and valence bands in #/cm^3. Calculate the number density of intrinsic carriers assuming the band gap energy is 1.42 eV at 300 K. Compare these values to those for Silicon in Table 6.1.

(6.3) Determine the effect of doping level on the depletion region width by independently changing the doping level in the n-type and p-type regions from 10^{14} to 10^{20} #/cm^3 while holding the other at the standard value given in Table 6.1 for Silicon. Assume there is no applied potential.

(6.4) Estimate the average distance between electrons in the n-type semiconductor and holes in the p-type semiconductors using the doping concentrations given in Table 6.1. Compare this to the minority carrier diffusion length and comment. Which is larger?

(6.5) Graph the depletion length in the p-type and n-type part of a p-n junction, and their total, assuming it is made from doped Silicon as a function of applied potential. Use the typical parameter values given in Table 6.1.

(6.6) Use eqn (6.7), developed by Sproul and Green, to calculate the number density of intrinsic carriers in #/cm^3 as a function of temperature in °C for temperature ranges typical of where a photovoltaic device may be placed. Display your results in graphical form.

(6.7) Determine the effect of temperature T on the dark current J_O determined with eqn (6.18b) for a temperature range typical of where a photovoltaic device may be placed. To do this one must know that the mobility of electrons is found to follow a $T^{-2.4}$ dependence and holes, $T^{-2.2}$, so, the diffusivity will follow $T^{-1.4}$ and $T^{-1.2}$, respectively. The temperature dependence of all other parameters may be found in the text and use typical values for a Silicon solar cell given in Table 6.1. Display your results in graphical form.

(6.8) Use eqn (6.23) and consider the results in Example 5.2, especially eqn (5.11), where the absorption coefficient is assumed constant over various energy regions, as shown in Table 5.1, and determine the light generated current for a Silicon-based solar cell. Use typical doping values given in Table 6.1; neglect the thickness of the depletion region in calculating the active region thickness and assume the active region starts right at the surface of the cell (*i.e.* $L = 0$ in eqn (6.23))

(6.9) Determine the effect of temperature on the open circuit voltage given in eqn (6.25) over the temperature range 0 to 10 °C. See Exercise 6.7 to obtain the temperature dependence of the diffusivity and use typical parameter values for doped Silicon given in Table 6.1. Remember the band gap energy will be temperature dependent too and review Table 5.3. The rule-of-thumb is that V_{oc} decreases at a rate of 2.3 mV/°C, how close is your calculation to the rule-of-thumb?

(6.10) Determine the fill factor where non-idealities are not considered, FF_O in eqns (6.30), and (6.31) for photovoltaic devices that have open circuit voltages of 0.3 to 0.7 V and present your results graphically. Do the two equations agree and how much does the fill factor change over this voltage range?

(6.11) The dark current determined when there is recombination in the depletion region J_{O2}, see eqn (6.32), is given as

$$J_{O2} = q\frac{Wn_i}{\tau_e + \tau_h}$$

by Gray (see A. Luque and S. Hegedus (2003)) and as

$$J_{O2} = \frac{\pi n_i}{2\sqrt{\tau_e \tau_h}} \frac{k_B T}{-E_{max}}$$

by Green (1982), where E_{max} is the maximum electric field strength in the depletion region, given by

$$-E_{max} = \left\{ \frac{\frac{2q}{\epsilon\epsilon_O}[\psi_O - V_a]}{\frac{1}{N_A} + \frac{1}{N_D}} \right\}$$

Are these two equations the same? Calculate J_{O2} when $V_a = 0$ and compare them; use the typical doping levels for Silicon given in Table 6.1. Different assumptions were used to arrive at these two relations accounting for their difference.

(6.12) Assume the equation for J_{O2} determined by Gray in Exercise 6.11 is the correct one for a Silicon-based solar cell and typical doping levels in Table 6.1 to determine the dark current as a function of applied voltage in eqn (6.32). Calculate J_{O1} with eqn (6.18b). Let the voltage change from 0^+ to a high number. Graph the natural logarithm of J_O as a function of V_a and determine if it is possible to separate a region where $n = 1$ and where $n = 2$. This technique is sometimes used to find J_{O1} and J_{O2}, yet, is very hard to do in practice because of other factors dictating solar cell performance. Also, fit all the data to a linear regression in a plot of $\ln(J_O)$ versus V_a and find an apparent value of n.

(6.13) Derive eqn (6.30).

(6.14) Compare eqns (6.28b) and (6.29) to the actual values of V_m and J_m determined with eqn (6.21) for V_{oc} ranging from 0.1 to 1 V at 300 K for Silicon.

(6.15) Compare eqns (6.30) and (6.31) to the actual values of FF determined with eqn (6.21) for V_{oc} ranging from 0.1 to 1 V at 300 K. Which is more accurate?

(6.16) Determine equations similar to eqns (6.28b), (6.29) and (6.30) with eqn (6.33) and demonstrate the effect of n on their values for V_{oc} ranging from 0.1 to 1 V at 300 K. Let n range from 1 to 2. Do you want a low or high value of n for a solar cell?

(6.17) Develop an approximation scheme for eqn (6.36) similar to that given by eqn (6.28b) and determine the effect of R_{sh} on its value. Use typical values for the parameters given in Table 6.4.

(6.18) Derive eqn (6.39).

(6.19) Justify the approximation made in eqn (6.40).

(6.20) Determine the accuracy of eqn (6.40) by finding the actual value of $R_{ch} \equiv V_m/J_m$ using eqn (6.35) and the typical parameter values in Table 6.4. Assume J_O is 3×10^{-9} mA/cm^2, n is 1.3 and the temperature is 300 K.

(6.21) Use eqn (6.35) to make a graph of current density versus applied voltage using typical parameter values in Table 6.4. Now append other graphs to the original by letting $R_s = 0$ in one graph where all the other parameters remain at their original value, then let $R_{sh} \to \infty$ and finally let $n = 1$. Assume J_O is 3×10^{-9} mA/cm^2 and the temperature is 300 K. Which parameter, in your estimation, has the greatest effect on solar cell operation?

(6.22) The worst possible photovoltaic device one could make is the current–voltage graph to have a straight line from the short circuit current to the open circuit voltage. Determine the fill factor for such a cell. (It is possible to argue there are more poorly operating devices, however, we would be arguing about devices that are not worth further consideration!)

The solar chimney and tower

7

The Sun can supply radiation to a photovoltaic device, as described in the previous chapter, to provide electricity for residential or industrial use. This technology is quite effective and useful. Here, heating of air with solar radiation in a chimney is initially considered to draw it through a building for a cooling effect. This is a true *passive* solar operation and has been effectively used for centuries, indeed since the time of the Roman empire if not longer. Of course, this requires proper building design, perhaps before construction begins. However, retrofitting of buildings can be effectively performed and, particularly in hot climates, natural air conditioning can be utilized to promote air circulation.

After this technology is discussed an *active* use of heated air will be considered, the so-called solar tower. Although still in the early development stage, it has potential to generate electricity in desolate areas. In fact these are big operations and a lot of flat, desolate land is required. This is a drawback due to physical isolation, however, this power plant has a unique advantage over other electricity generation technologies, no water is required! Air is heated by the Sun and it rises through a chimney where it turns a turbine that generates electricity. It's incredibly simple and potentially very useful. Even if this technology does not come to market, the analysis of how it works is useful and can be generalized to other potential energy technologies.

Dimensionless numbers and correlations between dimensionless numbers will be used to design the devices in this chapter; if the reader is not familiar with these concepts reading of Appendix C may help in understanding what they are and why they are used. The following nomenclature is used in this chapter: a solar chimney is a device that is passive in nature and promotes the flow of air while a solar tower, sometimes called a solar updraft tower, is active and generates electricity. Before either of these can be discussed though, how a chimney works is addressed, since this is the physics central to the technology.

7.1 The chimney

A chimney relies on an energy source to heat air which rises via a buoyancy effect. The analysis is relatively simple and requires use of the Bernoulli equation which comes from the fluid mechanics literature. The Bernoulli equation is valid along the path that a fluid (in our case

Solar Energy, An Introduction, First Edition, Michael E. Mackay

air) will move. This is called a streamline in the fluid mechanics literature and eqn (7.1) applies along a streamline from stations i to j. The basic equation applied from stations 1 to 2 does not consider any frictional losses and is written

$$P_j - P_i + \frac{1}{2}\rho_j v_j^2 - \frac{1}{2}\rho_i v_i^2 + \rho_j g z_j - \rho_i g z_i = 0[=]\text{Pa} \tag{7.1}$$

Here P is pressure, ρ, density, v, velocity, g, gravitational acceleration, z, the height above some arbitrary datum and [=] means 'has dimensions of.'[1] If frictional losses are considered then the right-hand side of the above equation would include *loss* terms and the equation is denoted as the *engineering Bernoulli equation.*[2] Here, for the flow of low viscosity air, neglect of frictional losses is reasonably acceptable.

[1]The Bernoulli equation relates the change in pressure to the change in kinetic and potential energies. Consider the simple example of a vertical pipe filled with water; station 1 is at the bottom and 2 at the top. The water is stationary. Using eqn (7.1) one finds $P_1 - P_2 = \rho g[z_2 - z_1]$ showing that the pressure at station 1 is dictated by the hydrostatic head of water and is greater than the pressure at station 2.

[2]The Bernoulli equation is frequently touted as being related to an energy balance. This is not strictly true, as discussed by Astarita and Mackay in the manuscript referenced at the end of this chapter. This note is sure to generate a lot of discussion...

Now consider the chimney shown in Fig. 7.1, which has a height H and an exit area of A. We apply the Bernoulli equation between stations 1 and 2 which will eventually be used to determine the flow rate of air into the chimney

$$P_2 - P_1 = -\frac{1}{2}\rho_2 v_2^2 \equiv -\frac{1}{2}\rho_3 v_3^2 \tag{7.2}$$

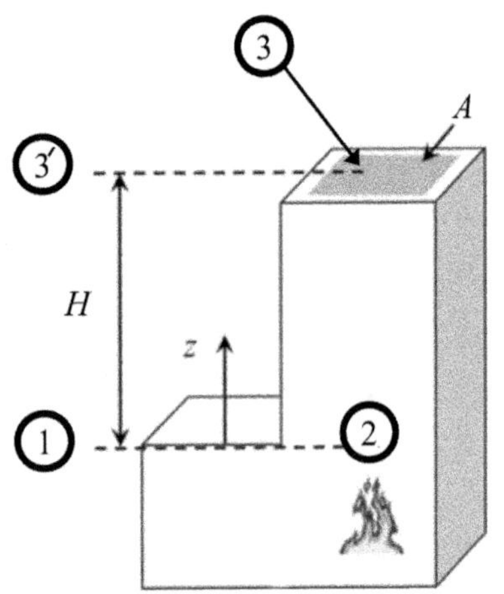

Fig. 7.1 Schematic of a chimney with a fire that heats the air and creates a draft of air from outside through the fireplace and up the chimney.

The velocity at station 1 was assumed to be zero since it is far from the fireplace entrance and the stations are assumed to be at the same height to arrive at this equation. The density and flow area at station 2 is assumed to be equal to that at 3 allowing the final equation to be written. In addition, the temperature is assumed to be uniform through the chimney which will be required below to further the analysis. An estimate of the pressure drop between station 1 and 2 is now needed.

The pressure P_1 is merely the pressure P_3' plus the hydrostatic pressure of air above it or

$$P_1 = P_{3'} + \rho_1 g H \tag{7.3}$$

A similar estimate of the pressure at station 2 yields

$$P_2 = P_3 + \rho_3 g H \tag{7.4}$$

Station $3'$ is assumed to be at the same height as station 3 which allows one to assume $P_{3'} = P_3$ and the pressure difference $P_2 - P_1$ can be written

$$P_2 - P_1 = [\rho_3 - \rho_1] g H < 0 \tag{7.5}$$

Now eqn (7.5) can be inserted into eqn (7.2) to arrive at

$$v_2 \equiv v_3 = \sqrt{2\frac{\rho_1 - \rho_3}{\rho_3} g H} \tag{7.6}$$

Rather than using density it is more convenient to use the ideal gas law and express the result in terms of temperature through $\rho = P/RT$, where R is an appropriate gas constant and T is temperature to yield

$$v_3 = \sqrt{2\frac{T_3 - T_1}{T_1} g H} \equiv \sqrt{2\frac{\Delta T}{T_1} g H} \tag{7.7}$$

where ΔT is the temperature rise. The mass flow rate $\dot{m}$ is velocity multiplied by density and flow area allowing one to write

$$\dot{m} = \rho_3 \sqrt{2\frac{\Delta T}{T_1} gH} \; A \, [=] \, \frac{\text{kg}}{\text{s}} \tag{7.8}$$

This is the working relation to demonstrate how a chimney works. The temperature rise encountered by the air generates a buoyancy force that pulls the air into the fireplace and up the chimney. A greater temperature rise and higher chimney promotes more air to be driven up the chimney.

Later in this chapter we will obtain equations for how different types of solar chimneys depend on variables such as their height. Eventually, numerical prefactors will be discarded and we will write how the mass flow rate, for example, scales with the design variables such as chimney height. The symbol ~ will be used to mean *scales as*, which is not to be confused with the ≈ symbol meaning *approximately*. This later symbol is used when numerical prefactors are included and some sort of approximation is made. Determining how something scales with variables is useful as one can rapidly ascertain which is the most important in affecting the overall design.[3] For the case of a chimney one finds

$$\dot{m} \sim \Delta T^{1/2} H^{1/2} A \tag{7.9}$$

From this result one can see the variable that affects the mass flow rate the most is the area at the chimney exit and not the height! The temperature T_1 is not included as it is measured in Kelvins and its value will not change greatly at normal terrestrial temperatures.

[3] This is sometimes called *scaling theory* which was used by the famous French Nobel laureate Pierre Giles de Gennes. In reality scaling theory has a much deeper meaning that we are using here, however, determining how a quantity is affected by various design variables is extremely useful and so we borrow this aspect from the theory.

Example 7.1

Find the flow rate of air through a large fireplace that is 10 m high.

To do this one must make an assumption as to how much the air heats up in the fireplace and we will assume that it is about 200°C. The chimney will be assumed to be round and with a 100 mm diameter. Now we can write eqn (7.8) as

$$\dot{m} = 1.3 \, \frac{\text{kg}}{\text{m}^3} \sqrt{2\frac{200 \text{ K}}{273 \text{ K}} \times 9.81 \, \frac{\text{m}}{\text{s}^2} \times 10 \text{ m}} \; \frac{\pi}{4}[0.1 \text{ m}]^2 = 0.122 \, \frac{\text{kg}}{\text{s}} \tag{7.10}$$

by making suitable assumptions for the ambient air temperature and air density (see Appendix B). We had to make an assumption about the air temperature of the flue gas (*i.e.* ΔT), however, scaling theory demonstrates that if we assume it to be twice as hot, this only increases the mass flow rate by 40%. So, the magnitude of the mass flow rate is not that sensitive to this variable.

Regardless, this is a reasonably large flow rate and opponents of the use of *open* fireplaces in houses frequently cite this as a reason to not use a fireplace in the winter. Cold air must come from outside the house with an overall effect of increasing the heat load on the house.

A way to gauge the magnitude of this mass flow rate is to calculate the *air change rate per hour*, meaning how many hours it will take to changeover the entire air mass in a room, house or building. This is usually abbreviated as ACH or *air changes per hour*. Assume this is a 200 m^2 house with rooms that are 3 m high on average, making a house volume V of 600 m^3. From the above we find that there are approximately 0.56 ACH (ACH = $\dot{m}/\rho V$, where the mass flow rate should be in units of mass per hour). The ACH recommended for living areas is 0.35 ACH, so, the fireplace essentially doubles the ACH. If the flue gas temperature is the same and the diameter of the chimney doubled then the ACH increases to about 2 ACH, which becomes a very large increase in the ACH and will substantially increase the heat load for the central furnace.

It is understood that a fire will produce radiant energy which will heat the house, yet, many studies have shown that the effect of burning a fire in an old-style fireplace will result in an overall cooling effect on the house. This is for two reasons: firstly, the fire draws cold, outside air into the house at a rate far exceeding the stoichiometric need of the chemical reaction to produce fire, resulting in a cooling effect, and secondly, most of the heat exits through the chimney in the form of hot gases. It has been estimated that approximately 80% of the energy is lost through the chimney, so, it is fairly easy to determine that the net effect is to cool the house rather than heat it when using an open fireplace.

Wind can affect the operation of a chimney by increasing or decreasing the flow rate. Chimney caps have been engineered to reduce adverse wind effects, such as allowing the wind to go down into the chimney, and subsequently the house, as well as to eliminate rain or snow from entering.

Wind can also cause the draft within a chimney to increase through the same principle that allows a perfume atomizer to operate.[4] With reference to Fig. 7.1, and applying the Bernoulli equation between stations 3′ and 3, one arrives at

$$P_3 - P_{3'} = -\frac{1}{2}\rho_3 v_w^2 \tag{7.11}$$

where v_w is the wind velocity. It is assumed that the wind velocity is negligible at station 3′, and its magnitude increases as it moves around the chimney. One can write eqn (7.5) as

$$P_2 - P_1 = P_3 - P_{3'} + [\rho_3 - \rho_1]gH = -\frac{1}{2}\rho_3 v_w^2 + [\rho_3 - \rho_1]gH \tag{7.12}$$

Using this equation one can arrive at

[4] An atomizer works by flowing air past a tube which reduces the pressure at station 2, as shown in the figure.

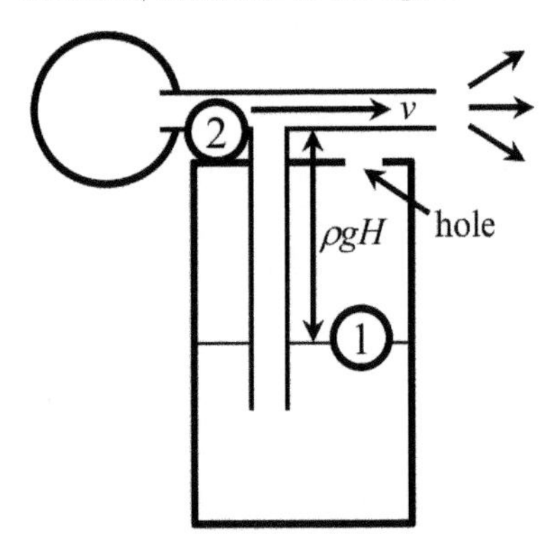

The bulb at the left is squeezed pushing air past the top of the suction tube. Applying Bernoulli's equation between stations 1 and 2 yields $P_2 - P_1 = -\frac{1}{2}\rho_{\text{air}} v^2$, where ρ_{air} is the density of air and v is the velocity in the tube. A hole in the bottle prevents a vacuum occurring at station 1. The hydrostatic head, or pressure, for the perfume in the tube is $\rho g H$, where ρ is the perfume density. As long as $P_1 - P_2$ is greater than $\rho g H$ then perfume will flow through the tube and will be atomized onto your skin. The design equation is $v \geq \sqrt{2\rho g H/\rho_{\text{air}}}$. A restriction is frequently placed just above the suction tube to increase the velocity allowing more perfume to be atomized each time the bulb is squeezed.

$$\dot{m} = \rho_3 \sqrt{2 \frac{\Delta T}{T_1} gH + v_w^2} \, A \tag{7.13}$$

This relation overestimates the influence of wind on the chimney draught and one can write

$$\dot{m} = \rho_3 \sqrt{2 \frac{\Delta T}{T_1} gH + C_w v_w^2} \, A \tag{7.14}$$

where C_w is a correction factor and is of order 0.1 through simulations of a solar tower, discussed below, by Ming *et al.* Theoretical estimation of C_w requires comprehensive solution to fluid mechanics differential equations, which is possible. Strict application of the Bernoulli equation as we have done does not yield a good analytical result, however, coupling the Bernoulli results in eqn (7.14) with numerical simulation provides a good compromise and a simple equation that can be used.

Another correction should also be applied to eqn (7.8) to account for frictional and other effects, which yields

$$\dot{m} = \rho_3 C_c \sqrt{2 \frac{\Delta T}{T_1} gH + C_w v_w^2} \, A \tag{7.15}$$

where C_c is a constant of order 0.67.[5] This too is an empirical correction factor and an exact model of the chimney at hand could be accurately studied using one of the many fluid mechanics software packages available. For our purposes though, a simple relation like that given in eqn (7.15) is good enough to understand the parameters that affect the operation of a chimney or the solar chimney and tower discussed below.

[5]Corrections to eqn (7.8) yield eqn (7.15) with values for the constants of: $C_w = 0.1$ and $C_c = 0.67$. These parameters depend on a variety of factors and are only approximate corrections. Equation (7.15) should be used if flow non-idealities are considered, and not eqn (7.14).

7.2 The solar chimney

Now that we know how a chimney works we can use the above analysis to determine a way to *cool* a house with a chimney! This is similar to the cooling effect mentioned in the above example where cooler, outside air is directly brought into the house. Air circulation is promoted through the house or building by heating air in a chimney with the Sun to draw air into the building. These types of air conditioners have been used for centuries, if not millennia, to draw air through a house and promote cooling by convection and/or a temperature decrease. The solar chimney has the added advantage that the mass flow rate of air though the building increases with power input from the Sun, which is what is wanted; an increased cooling effect when it is required.

Moreover, if the outside air is too hot during the day a damper is used to stop air flow. When the outside air cools down the damper is opened and air is then drawn through the building to provide cooling. This is accomplished by having a large thermal mass in the solar chimney that can store energy to provide it to the air, even when the Sun has set.

A schematic of a solar chimney is given in Fig. 7.2 demonstrating the design variables including the azimuthal angle, θ_Z, and the angle

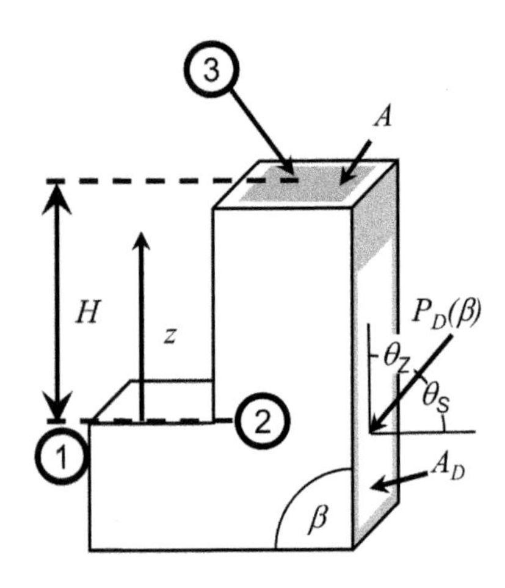

Fig. 7.2 Schematic of a solar chimney where the Sun heats air within the chimney through a transparent window causing a natural draft to occur upwards. The power to the device supplied by the Sun, $P_D(\beta)$, occurs at the angle β, which in the figure is 90°.

perpendicular to the device area, θ_S, relative to the radiation received by the device, $P_D(\beta)$, where β is the angle the device makes with the horizontal. The chimney is assumed to be perpendicular to the ground and so β is 90°, which is not always the case.

A key component of the solar chimney is the transparent window of area A_D that allows the radiation to heat the air (or thermal mass) within the chimney. We will assume that any radiation transmitted through the window will be absorbed by the air and that it has the value $P_D(\beta)$, which will be discussed in greater detail below. The analysis is begun by applying the First Law of Thermodynamics (FLOT) between stations 1 and 3 to arrive at what will be called the operating line

$$\dot{m}\left[C_P\left[T_3 - T_1\right] + gH\right] = P_D(\beta)A_D \qquad \text{operating line} \qquad (7.16)$$

where kinetic energy effects have been neglected, a good assumption, which is why applying the FLOT between stations 1 and 2 and 2 and 3 was not considered. The heat capacity of air is given the symbol C_P and the temperatures of the two stations are T_1 and T_3 as discussed above. The heat capacity was assumed constant over that temperature range and $C_P[T_3 - T_1]$ approximated the change in enthalpy. In many solar chimneys there is a black absorber plate on the wall opposite to the transparent material, to absorb the solar radiation, only to re-emit it to the surrounding air.

Unfortunately, there is only one equation and two unknowns: $\dot{m}$ and $\Delta T \equiv T_3 - T_1$. In this case the FLOT is called the *operating line*, now another equation is required.

A *design line* is needed which could be developed to include detailed calculations of heat transfer to the air within the solar chimney and how that affects $\dot{m}$. Here we will assume that *all* the radiation reaching the air within the chimney will be absorbed by the air to give a maximum possible heating effect. This is not a poor assumption, merely a necessary one. In general the solar chimney will not be thermally isolated from the environment, making detailed heat transfer calculations difficult and full of many assumptions.

So, how can one determine the solar chimney performance? Fortunately we have eqns (7.8) or (7.15) to provide the design line that is based on the Bernoulli equation and fluid mechanics, the result of which is shown below, where wind effects have not been included,

$$\dot{m} = \rho_3\sqrt{2\frac{\Delta T}{T_1}\,gH}\;A \qquad \text{design line} \qquad (7.8)$$

The solar chimney performance is now determined by the intersection of the operating line and the design line to find the *operating point*. In other words, the two simultaneous equations are solved to find $\dot{m}$ and ΔT, which define the operating point. The following example demonstrates this principle.

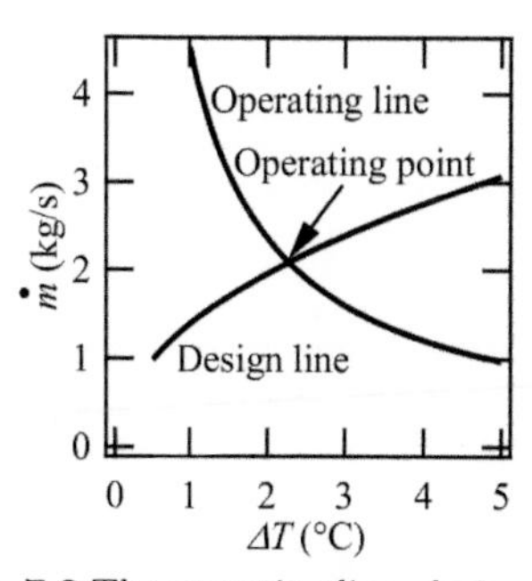

Fig. 7.3 The operating line, design line and operating point for the solar chimney discussed in Example 7.2.

Example 7.2

Find the operating point of a solar chimney which receives 500 W/m² of radiation and has a device area (window) of area 10 m². The exit area from the 10 m high chimney is 1.5 m² while the ambient air temperature is 35 °C. Neglect the effect of wind and the empirical constant C_c is assumed to be equal to one.

First the operating line needs to be calculated. The heat capacity of air can be found in Appendix B and will be assumed to be constant with a value of 1007 J/kg-K. Various values of ΔT (*i.e.* $T_3 - T_1$) will be assumed and the mass flow rate determined with eqn (7.16). The result of this calculation is given in Fig. 7.3.

Now the design line can be calculated from eqn (7.8) by assuming various values of ΔT and calculating $\dot{m}$. This is also shown in the figure and the intersection of the two lines defines the operating point. For the conditions here, the mass flow rate is about 2 kg/s and the temperature rise is 2 °C.

The density of air at 35°C is 1.17 kg/m³ using the data in Appendix B, so, the volumetric flow rate of air into the house is 1.7 m³/s. Assuming there is a central hallway in the house, through which the air is drawn, with dimensions of 2 m by 3 m, this corresponds to a velocity of about 300 mm/s, which is substantial. Of course, if the air is forced to be drawn from an underground chamber the cooling effect can be even greater (see Exercise 7.3). This will certainly reduce the mass flow rate though, as the frictional drag of the air with the walls in the chamber will produce an additional pressure drop that will have to be considered.

The operating point in the above example was found by the intersection of the operating and design lines, however, it is possible to combine eqns (7.8) and (7.16) to find

$$\dot{m}\left[C_P \frac{T_1}{2gH}\left[\frac{\dot{m}}{A\rho_3}\right]^2 + gH\right] = P_D(\beta)A_D \tag{7.17}$$

which can be written

$$\dot{m}^3 + 3p\dot{m} + 2q = 0 \tag{7.18}$$

where

$$p \equiv \frac{2[gH\rho_3 A]^2}{3C_P T_1}$$
$$q \equiv -\frac{P_D(\beta)A_D gH[\rho_3 A]^2}{C_P T_1}$$

This third-order equation can be solved and the only real solution is

$$\left.\begin{aligned} \dot{m} &= -2r\sinh\left(\frac{\phi}{3}\right) \\ \sinh(\phi) &= \frac{q}{r^3}\ ,\ r = -\sqrt{p} \end{aligned}\right\} \quad \text{operating point} \tag{7.19}$$

For the conditions encountered in most solar chimneys one can approximate $\sinh(\phi)$ with $\frac{1}{2}\exp(\phi)$ according to its definition (as well as $\sinh(\phi/3)$ with the appropriate relation) to find

$$\dot{m} \approx \left[2\frac{P_D(\beta)A_D gH[\rho_3 A]^2}{C_P T_1}\right]^{1/3} \quad \text{approximate operating point} \tag{7.20}$$

This equation could have been determined by noting the relative order of magnitude for each term in eqn (7.18) and eliminating the $3p\dot{m}$ term.

Both eqns (7.19) and (7.20) have ρ_3 as a variable that will be a function of the temperature rise within the chimney and is a function of the mass flow rate. Thus, a circular solution of these equations with eqn (7.8) or (7.15) (assuming v_w is zero) will have to be done. Yet, the temperature rise in solar chimneys is small and one can use ρ_1 to find the mass flow rate with little error, especially for a solar chimney. The mass flow rate at the operating point in Example 7.2 is found to be 2.09 kg/s using eqn (7.19) and 2.12 kg/s from eqn (7.20). The difference between the two is a mere 1-2%, which is good enough for design calculations!

The equation for the approximate operating point has further use since one can determine how the solar chimney operating point scales with design variables as given here,

$$\dot{m} \sim P_D(\beta)^{1/3} A_D^{1/3} H^{1/3} A^{2/3} \tag{7.21}$$

This has a different scaling to the design variables in eqn (7.9) for the chimney. It is amazing how little the mass flow rate, and hence the cooling effect, is affected by the power input $P_D(\beta)A_D$ as well as the chimney height H. Again, as with the normal chimney, the variable to which the mass flow rate is most sensitive is the chimney area A.

Previous experimental studies have shown that

$$\dot{m} \sim P_D(\beta)^{0.46-0.57} H^{0.60} A^{0.71-0.76} \tag{7.22}$$

While the scaling with the chimney area A is close to that predicted by the simple model discussed here, the scaling with $P_D(\beta)$ and H is not close. The reason is certainly due to the simplicity of the model presented where heat transfer effects have not been considered. This can be seen in how the efficiency of an experimental device η scales with power input

$$\eta \equiv \frac{\dot{m}C_P\Delta T}{P_D(\beta)A_D} \sim P_D(\beta)^{0.19-0.30} \tag{7.23}$$

which is the rate of enthalpy rise for the air divided by the rate of energy supplied to the device. The experiments show that efficiency is slightly

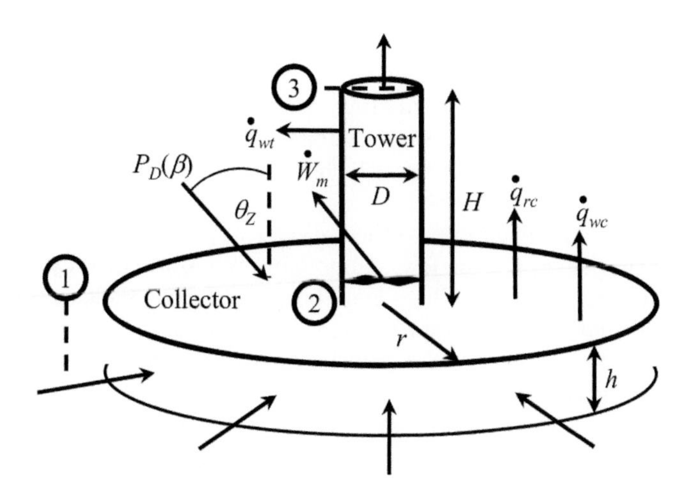

Fig. 7.5 Schematic of a solar tower used to produce electricity by turning a turbine located at the bottom of the tower (chimney). The solar collector, of radius r, is used to heat the air via the greenhouse effect that rises through the tower, of diameter D and height H.

affected by the power with a fairly small power law exponent. Yet, it was assumed here that the efficiency was 100% and that all the power from the Sun was absorbed by the air, making the efficiency independent of $P_D(\beta)$ or to scale as $P_D(\beta)^0$.

7.3 The solar tower

The passive solar cooling effect promoted by the solar chimney is a powerful element of building design especially in warm climates. However, the flow of air is not harnessed to generate useful work. Remarkably, Leonardo da Vinci (1452–1519) envisioned a spit that was turned by the rising air in a fireplace as shown in Fig. 7.4, that can transform the air temperature rise into work.

Fig. 7.4 A spit turned by natural convection within a fireplace as envisioned by Leonardo da Vinci. Used with permission from TopFoto.

The idea of a solar tower is to put a flat, circular *cover* or *collector* a distance h above the ground, typically 1-2 m, that will heat the air underneath it, as shown in Fig. 7.5. The cover radius r can be quite large and of order hundreds of meters to produce utility scale electricity. This necessitates that the land where the tower is built should be very flat and devoid of obstructions.

A turbine is placed at the base of the tower (chimney) that is turned by the rising air within the tower of diameter D and height H. Some proposed solar towers are so massive that there are several turbines arranged circumferentially around the entrance to the tower. Furthermore, the tower height would make it one of the largest structures in the world!

In terms of an actual solar tower a pilot–scale plant was built in Manzanares, Spain and used between 1982 and 1989 to produce 50 kW of electrical power. The collector radius was 122 m and the tower diameter and height were approximately 10 m and 195 m, respectively. So, even a small, pilot–scale solar tower is big.

Why are there no utility-scale solar towers used today? Part of the answer may center around the size and scale of these power plants. They will be huge, yet, several have been proposed around the world. Also, because of their size, they can only be placed in desolate areas far away from where the electricity is needed. Finally, the power generated will be somewhat intermittent with little or none produced at night. A solution to this is to put material, like bags of water, under the cover to absorb radiation during the day, which is released at night to continue operation.

There are some advantages to these electrical power plants. An obvious one is that there will be few maintenance issues as the only moving parts will be the turbines. There is no burning of fossil fuels to power a steam cycle that turns a steam turbine. Another, perhaps less obvious advantage, yet, one that will surely become more important over the next century, is that water is not required (apart from the water used to store heat). Coal-fired and nuclear fission power plants use a lot of water and thermal (as well as air) pollution is a challenge to their ecological-based operation. Water is expected to become a rarer commodity in the future and a solar tower could alleviate some of the environmental strain imposed on the Earth. Imagine a solar tower farm in northern Africa to produce electricity, transmitted to Europe; this was proposed although it never came to fruition.[6]

[6] Electrical losses are minimized by transmitting at high voltage, making a solar farm in northern Africa for Europe feasible. Consider that transmission losses occur at a resistance R and the European Community requires a certain amount of power, $P_{\text{need}} \equiv IV$, where I and V are the current and voltage, respectively. The power losses create heat and are given by, $P_{\text{heat}} = IV = I^2R$; remember $V = IR$. The transmission current is $I = P_{\text{need}}/V$ making the heat power loss, $P_{\text{heat}} = P_{\text{need}}^2 R/V^2$. So, the power loss scales with inverse voltage squared and transmitting electricity at very high voltage is dangerous, yet, it is extremely efficient.

The analysis presented below is simplified and aimed at understanding the variables that affect production of electricity from the solar tower. Engineering details of the building structure and mass flow rates and pressures required to turn the turbine are not considered and are certainly needed for a true design.

The First Law of Thermodynamics is applied between station 1, far from the collector, and station 3, just at the exit of the tower, to determine the operating line for the solar tower resulting in

$$\dot{m}\left[C_P\left[T_3 - T_1\right] + \frac{1}{2}v_3^2 + gH\right] = \left[P_D(0) - \dot{q}_{rc} - \dot{q}_{wc}\right] A_C - \dot{q}_{wt} A_T - \dot{W}_m$$

Here $\dot{m}$ is the air mass flow rate, C_P, the air heat capacity which is assumed to be constant over the temperature range considered, T_1 at station 1 and T_3 at station 3, v_3, the air velocity at station 3, H, the tower height, $P_D(0)$, the insolation absorbed within the device, $\dot{q}_{rc}$, the rate of heat transfer due to radiation from the cover to the surroundings, $\dot{q}_{wc}$, the rate of heat transfer due to wind convection from the cover to the surroundings, $\dot{q}_{wt}$, the rate of heat transfer due to wind convection from the tower to the surroundings, A_C, the cover area, A_T, the tower external area and $\dot{W}_m$, the amount of electrical power generated by the turbine. The radiative heat transfer losses from the tower are ignored, as will be justified after reading the discussion below. All signs for the rate of heat transfer and work have been explicitly written.

Before considering the magnitude of the heat transfer terms it is useful, in order to determine the *operating line*, to look at the magnitude of the terms on the left-hand side of the above equation. The heat capacity of air is approximately 1 kJ/kg-K and the temperature rise will be of order 10 K, making this term, order 10 kJ/kg. The potential energy term gH will be of order 3 kJ/kg assuming the tower height to be 300 m, a good assumption.[7] Now the kinetic energy term is much smaller as the velocity in these devices is of order 10 m/s, making it of order 0.03 kJ/kg, so it can be neglected. This allows us to write, with some generality, the *operating line* for a solar tower

[7] We will use half order of magnitudes and the logarithmically spaced *half* is 3.

$$\dot{m}\left[C_P\left[T_3 - T_1\right] + gH\right] = \left[P_D(0) - \dot{q}_{rc} - \dot{q}_{wc}\right] A_C - \dot{q}_{wt} A_T - \dot{W}_m \; [=] \; \mathrm{W} \qquad \text{operating line} \quad (7.24)$$

Some of the heat transfer terms can be neglected since they are very small. Consider radiant heat transfer from the cover to the sky

$$\dot{q}_{rc} = \sigma_S e_C \left[T_C^4 - T_{\mathrm{sky}}^4\right] \qquad (7.25)$$

with σ_S being the Stefan-Boltzmann constant, e_C, the emissivity of the cover, T_C, the cover temperature and T_{sky}, the effective sky temperature for radiative heat transfer (an approximate relation is $T_{\mathrm{sky}} = 0.0552\, T_{\mathrm{amb}}^{1.5}$ where T_{amb} is the ambient air temperature). Assume the ambient air temperature is 35 °C and that the air flowing under the cover rises by 20 °C making the cover temperature 55°C. If the emissivity of the cover is 0.05, one can determine $\dot{q}_{rc} = 10.3\,\mathrm{W/m^2}$ which, as it turns out, is quite small and negligible.

The next heat transfer term is due to wind blowing over the cover, which can be written

$$\dot{q}_{wc} = h_w\left[T_C - T_{\mathrm{amb}}\right] \qquad (7.26)$$

where h_w is the heat transfer coefficient from a flat surface. There are many correlations for this and after an extensive survey of the literature the following relation best represents all data (see the article written by Palyvos, referenced at the end of this chapter)

$$h_w(\mathrm{W/m^2\text{-}K}) = 7.4 + 4.0 v_w(\mathrm{m/s}) \qquad \text{for} \qquad 0 \le v_w \le 4.5 \text{ m/s} \quad (7.27)$$

where v_w is the *free* wind velocity 10 m above the surface. Assuming wind velocities of 0 and 3 m/s, as well as the cover temperature being 20°C above ambient, one finds $\dot{q}_{wc}$ equal to 74.0 W/m^2 and 194 W/m^2, respectively. Clearly, this rate of heat transfer is much greater than radiative and is not in fact negligible even when the wind is not blowing and heat transfer is solely from natural convection.

The final heat transfer term to consider is heat transfer from the tower to the surroundings,

$$\dot{q}_{wt} = h_{wt}\left[T_T - T_{\mathrm{amb}}\right] \qquad (7.28)$$

where h_{wt} is the heat transfer coefficient for wind blowing on the outside of the cylindrical tower and T_T is the tower external surface temperature. We use a standard Nusselt number (Nu)–Reynolds number (Re) dimensionless correlation for air blowing past a cylinder to estimate h_{wt}, see Appendix C where dimensionless correlations are explained,

$$Nu \equiv \frac{h_{wt}D}{k_{\text{air}}} = 0.35 + 0.56Re^{0.52}$$
$$Re \equiv \frac{\rho_{\text{air}} v_w D}{\mu_{\text{air}}} \tag{7.29}$$

where k_{air}, ρ_{air} and μ_{air} are the thermal conductivity, density and viscosity of air. Again, for a 20 °C temperature rise of the air in the tower, assuming the tower diameter to be 10 m and using values for the physical parameters of air from Appendix B, one finds h_{wt} changes from 9.23×10^{-4} W/m^2-K to 2.71 W/m^2-K as the wind velocity increases from 0 to 3 m/s. This is an over three order of magnitude change in the heat transfer coefficient, a vastly different change than occurs for heat transfer from a flat surface like the cover. The rate of heat transfer increases by the same amount from 0.0185 W/m^2 to 54.2 W/m^2. So, when there is no wind this heat transfer rate is truly negligible, while in a moderate wind it is significant, albeit much less than that from the cover. Furthermore, the tower area is much smaller than the cover so the overall magnitude is even less.

Some solar towers have been proposed, making the chimney almost the highest structure in the world, perhaps 800 m high. Will the wind conditions change for such a tall structure? As one might guess the velocity of wind changes with height above ground level and is usually represented by

$$v_w = v_{w0}\left[\frac{z}{z_0}\right]^n \tag{7.30}$$

where z is height above ground level and v_{w0} is the velocity at the reference height z_0, usually taken as 10 m. The power law parameter n has values of 0.10–0.15 for smooth, uninhabited places on Earth, which is how the terrain would be at a solar tower installation. The Manzanares facility in Spain had a tower that was almost 200 m high. So, if the wind velocity was 3 m/s at a height of 10 m it would be 4–5 m/s at the top of the tower making the heat transfer coefficient increase by about 25%. An 800 m tower would experience winds of order 5–6 m/s and a 60% increase in h_{wt}. Even with this increase the rate of heat transfer is not that significant when compared to the rate from the much larger cover.

Example 7.3

Find the operating point for the solar tower pilot plant in Manzanares, Spain, assuming wind speeds of 0 and 3 m/s. The insolation is 840 W/m^2 and the surrounding air temperature is 35 °C. Use eqn (7.8) rather than (7.15) for the design line to determine how much wind affects the heat transfer rate.

Table 7.1 Physical dimensions and typical operating conditions for the solar tower at Manzanares, Spain.

tower height (H)	194.6 m
tower diameter (D)	10.16 m
collector radius (r)	122.0 m
typical air temperature rise (ΔT)	20 °C
typical power output ($\dot{W}_m$)	50 kW

This facility operated under the conditions given in Table 7.1, with reference to Fig. 7.5. The design line for this power plant is easily calculated by using eqn (7.8) together with the data in the table and is presented in Fig. 7.6. Obviously, as the rate of the temperature rise increases the greater will be the mass flow rate.

The operating line is calculated from eqn (7.24) using the equations for the various heat losses given in eqns (7.25) to (7.29). Two wind velocities were assumed, 0 and 3 m/s with the results given in Fig. 7.6.

The effect of heat transfer is first discussed assuming a constant wind velocity of 3 m/s. The upper graph is used to show how the various heat transfer rates affect the calculation. First, if $\dot{q}_{rc}$ and $\dot{q}_{wt}$ are neglected the design line and operating point move a small amount. For example, the operating point moves from a temperature rise of 19.6 °C and 1380 kg/s to 19.9 °C and 1390 kg/s, a relatively small change. In fact, the curve labeled *only* $\dot{q}_{wc}$ is two curves: one including $\dot{q}_{wc}$ and $\dot{q}_{rc}$ and another including only $\dot{q}_{wc}$. So, $\dot{q}_{rc}$ has an extremely small contribution compared to the other terms, as expected. Finally, if all heat transfer terms are neglected the operating point changes, a much larger amount to a temperature rise of 24.1 °C and mass flow rate of 1510 kg/s. Remarkably, for all the cases studied, the temperature rise is calculated to be almost exactly that given in Table 7.1 regardless of the assumptions made.

Next consider how wind will affect the solar tower performance that is shown in the lower graph. Even though the wind is not blowing there will be contributions from $\dot{q}_{wc}$ and $\dot{q}_{wt}$ due to natural convection instead of forced convection. In this case the operating point changes to 22.3 °C and 1460 kg/s, demonstrating how important heat transfer is to operation. However, when there is little wind, neglecting all heat transfer considerations is not that poor an assumption since the mass flow rate is within 3–4% between the two scenarios.

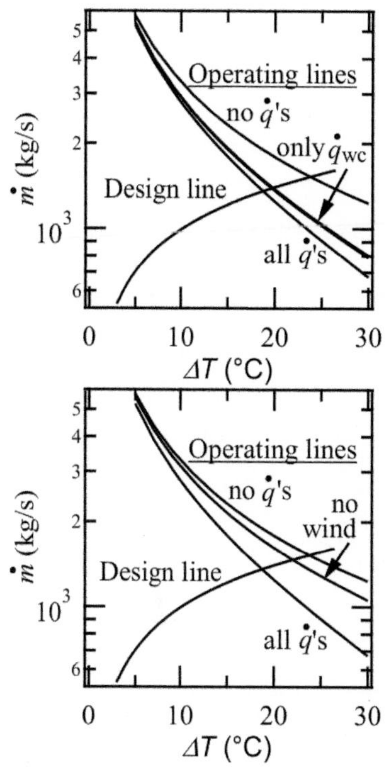

Fig. 7.6 The operating line and design line for the solar tower discussed in Example 7.3. The upper graph is used to demonstrate the effect of heat transfer on the operating line, the design line is unaffected by this effect. The operating line labeled *all $\dot{q}$'s* includes $\dot{q}_{wc}$, $\dot{q}_{rc}$ and $\dot{q}_{wt}$; that labeled *only* $\dot{q}_{wc}$ includes $\dot{q}_{wc}$ solely; and that labeled *no $\dot{q}$'s* does not consider heat transfer at all. The wind velocity was 3 m/s in all cases. The lower graph has calculations that consider the effect of wind on heat transfer where the curve labeled *no wind* has a wind velocity of zero, while in previous calculations a velocity of 3 m/s was assumed.

It was shown in the above example that the operating line is given by eqn (7.24) while the design line can be determined from either (7.8) or (7.15). The above example shows that heat transfer from the solar tower to the surroundings is important under most conditions, however, if heat transfer considerations are neglected it is possible to write an equation very similar to eqn (7.18) with p and q given by

$$p = \frac{2\left[gH\rho_3 A\right]^2}{3C_P T_1} \tag{7.31a}$$

$$q = -\left[P_D(0)A_C - \dot{W}_m\right]\frac{gH\left[\rho_3 A\right]^2}{C_P T_1} \tag{7.31b}$$

This is a useful equation, despite its limitation of not considering heat transfer, since it is possible to find an analytical solution for the mass flow rate and to determine how it scales with design variables. In addition, as shown in the above example, when there is no wind, neglecting heat transfer is not that poor an assumption. From eqns (7.18) and (7.31), while using the same approximation of simplifying eqn (7.19) to (7.20), one finds find the *approximate operating point* (or approx. oper. pt.)

$$\dot{m} \approx \left[\frac{2gH\left[\rho_3 A\right]^2}{C_P T_1}\left[P_D(0)A_C - \dot{W}_m\right]\right]^{1/3} \quad \text{approx. oper. pt.} \tag{7.32}$$

Of course, it is possible to use eqn (7.19) to find a more accurate solution than eqn (7.32). Regardless, the scaling relation given below shows the same scaling as found with the solar chimney in eqn (7.21)

$$\dot{m} \sim P_D(\beta)^{1/3} A_C^{1/3} H^{1/3} A^{2/3} \tag{7.33}$$

as expected. In terms of linear dimensions the mass flow rate scales as $r^{2/3}D^{4/3}$ suggesting that the tower diameter is a very important design parameter.

The importance of this is clarified by considering the maximum possible work that can be obtained from a solar tower. To do this we assume that the turbine within the tower operates similarly to a wind turbine. This is not exactly true, however, we can estimate the maximum possible electrical work by doing so. The major assumption is that the kinetic energy of the air is completely transferred to rotating the turbine blades.[8] This is not possible for a turbine located centrally within the solar tower and the situation is slightly more complicated, for our purposes though, it is appropriate to write the FLOT at station 2 in Fig. 7.5 as

$$\dot{m}\left[\frac{1}{2}v_{2'}^2 - \frac{1}{2}v_2^2\right] \approx -\frac{1}{2}\dot{m}v_2^2 = \dot{W}_{m,\max} \tag{7.34}$$

where stations 2 and 2′ are just at the entrance to and exit from the turbine, respectively, and it was assumed that all the kinetic energy provided by the moving air was transferred to the turbine. The sign of the work term is negative, as work is done by the system on the surroundings which will be ignored for now on. This equation can be combined with eqn (7.32) to arrive at, assuming $\rho_3 = \rho_2$ (or equivalently $T_3 = T_2$),

$$\dot{W}_{m,\max} = \frac{P_D(0)A_C gH}{C_P T_1}\frac{1}{1 + \dfrac{gH}{C_P T_1}} \approx \frac{P_D(0)A_C gH}{C_P T_1} \tag{7.35}$$

[8] A wind turbine works by air coming to the turbine and transferring its kinetic energy to the turbine's rotational energy. Since the air loses velocity, due to its reduction of energy, the area over which it flows must increase according to the mass balance. This is demonstrated in the figure between stations 1 and 2 and the mass balance $\rho_1 A_1 v_1 = \rho_2 A_2 v_2$ which implies $A_2/A_1 \approx v_1/v_2 > 1$, where ρ, A and v are the air density, flow area and wind velocity, respectively.

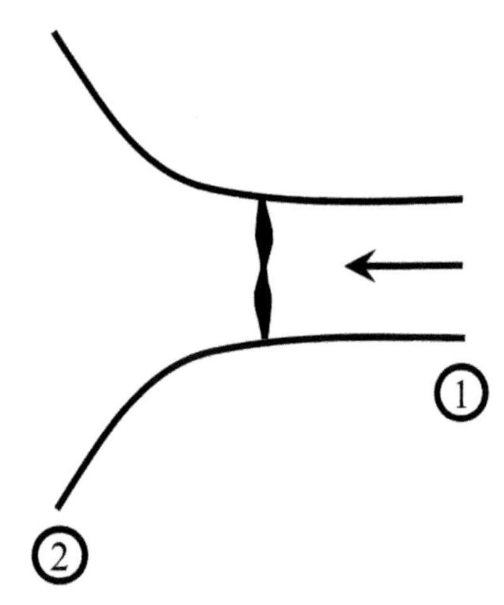

Using eqn (7.34), and the definition of the mass flow rate, the maximum work (power) from a wind turbine is $\frac{1}{2}\rho_1 A v_1^3$, where A is the area swept by the turbine's blades. The power scales as $D^2 v_1^3$, where D is the turbine diameter, demonstrating how the power is very sensitive to wind velocity as well as turbine diameter. The sensitivity of power to wind velocity is critical since a 10% velocity reduction corresponds to a 30% power loss and this is why power companies are extremely cautious as to where a wind farm is located.

where the approximation is valid for most solar towers based on their typical dimensions.

Mass flow rate is the most important variable in the solar chimney to provide cooling, however, in the solar tower it is the maximum work which is most important. The maximum power scales as

$$\dot{W}_{m,\max} \sim P_D(0) A_C H \tag{7.36}$$

Now it is apparent that the maximum power generated by a solar tower depends explicitly on the amount of power absorbed by the working fluid (air) $P_D(0)A_C$ and the tower height H, as might be expected. More precisely, it is obvious that the maximum power will scale directly with $P_D(0)A_C$ since this is the amount of energy available to it over a given time period. The direct scaling with height is strictly due to the buoyancy effect. The potential energy inside the tower is $\rho_3 gH$ and that outside the tower is $\rho_1 gH$; their difference is the driving force for the *chimney effect* which scales directly with H. Since air has such a low density the height must be extreme to generate enough potential energy. Would water be a better working fluid? Do Exercises 7.7 to 7.9 to find out!

The tower diameter does not affect the maximum power generated by a solar tower, demonstrating that this variable is not as important as implied by eqn (7.33). However, the diameter affects, among other variables, the velocity of air in the tower, which is important to improving turbine efficiency so its influence is certainly not negligible.

The overall efficiency of a solar tower is actually quite small which ultimately contributes to the large size required. Taking the maximum power and dividing it by the power supplied to the air $P_D(0)A_c$ one finds the solar tower efficiency η

$$\eta = \frac{gH}{C_P T_1} \tag{7.37}$$

A large solar tower would only have a 1% efficiency!

Example 7.4

Find the operating point for the solar tower pilot plant in Manzanares, Spain using the approximate operating point given in eqn (7.32). *Data for the power plant are given in Example 7.3. Also determine the maximum power that can be generated with this power plant using eqn* (7.35) *and compare it to the typical power output. Neglect the effect of wind and use eqn* (7.8) *as the design line.*

The approximate operating point is found when heat losses are not considered and the temperature rise will be greater than that encountered in the real power plant. As such, the mass flow rate will also be overestimated since the buoyancy effect offered to the air is more than the true situation. Regardless, the equation to be used is

$$\dot{m} \approx \left[\frac{2gH\,[\rho_3 A]^2}{C_P T_1} \left[P_D(0) A_C - \dot{W}_m \right] \right]^{1/3} \tag{7.32}$$

together with eqn (7.8) written as

$$\Delta T = \left[\frac{\dot{m}}{\rho_3 A} \right]^2 \frac{T_1}{2gH} \tag{7.38}$$

The second equation is needed because ρ_3 is a variable in the first equation, which is the air density at T_3, and is unknown, and an iterative procedure must be performed. Air density as a function of temperature is given in Appendix B as $\rho_3(\mathrm{kg/m^3}) = 1.297 - 3.757 \times 10^{-3}\ T_3(°\mathrm{C})$. The solution procedure is to assume a value for T_3 and calculate ρ_3 then determine $\dot{m}$ from eqn (7.32) and finally calculate the temperature rise from eqn (7.38). Continue this iteration until the assumed T_3 and the calculated $T_3 = T_1 + \Delta T$ are equal. When this is done a mass flow rate of 1540 kg/s and a temperature rise of 25.3 °C are found. If eqns (7.24) and (7.8) (or (7.38)) are solved simultaneously, neglecting all heat transfer considerations, then one finds 1510 kg/s and 24.1 °C, which is fairly close to the approximate operating point.

The maximum power that can possibly be generated by the solar tower is given by eqn (7.35). Inserting numbers and performing the calculation results in a maximum power of 242 kW. As shown in Table 7.1, the typical power output is 50 kW, so, the efficiency is approximately 20% based on the possible power that could be developed. Of course, the overall efficiency is only 0.6% based on the insolation falling on the cover using eqn (7.37).

7.4 Conclusion

A novel solar tower was discussed, building on the concepts that make a chimney and solar chimney work. Although it would be massive in size to generate industrial-scale electricity, and there are no known power plants in operation, the physics and engineering involved in its concept are so simple there would seem to be potential for their use in the future. Furthermore, as water becomes a more valuable and rarer commodity, the fact that water is not used at all is a great advantage. The poorest land can also be used, freeing more valuable land and other natural resources for other uses.

The solar chimney has been used for many years to help to cool a house in a hot climate. Solar energy is stored in the chimney during the day and when the outside temperature falls at night a damper in the chimney is opened to allow air circulation through the house. This simple design is extremely effective, especially when there is no power available to cool the house.

Massive solar towers take this principle to another level and convert heat into electricity. Although conversion of rising hot air within a chimney to useful work was envisioned by Leonardo da Vinci many years ago, no significant, commercial solar tower has been erected. However, the incentive to build one may change, driven by energy security concerns, among other reasons.

These solar powered devices are only 1% efficient necessitating that they must be huge in scale. For example, in the USA, electrical consumption is approximately 4×10^{12} kW-h/year which is an instantaneous power of 4.6×10^{11} W. Assuming that there is an average of 250 W/m^2 of insolation available to the tower then the collector area must be 425 km × 425 km! Doubling the insolation reduces this to approximately 300 km × 300 km, so, this technology will certainly not replace coal-fired power plants.

Yet, there are targeted places that could truly warrant use of this technology. Furthermore, water is not required and all the engineering principles are present, the technological risk is not as large as rolling out a completely new technology and building a new infrastructure. It could be useful for isolated, desolate places that do not require much power, and in the evening could be used for shelter. However, the economics of building this compared to installing a solar photovoltaic array is not clear.

7.5 General references

G. Astarita and M.E. Mackay, 'The generalized engineering Bernoulli equation (GEBE) and the first and second laws of thermodynamics for viscoelastic fluids,' J. Rheol. **40** (1996) 335.

S.A.M. Burek and A. Habeb, 'Air flow and thermal efficiency characteristics in solar chimneys and Trombe Walls,' Energy and Buildings **39** (2007) 128.

J.A. Duffie and W.A. Beckman, 'Solar energy of thermal processes,' John Wiley and Sons, 3rd Edition (2006).

T. Ming, X. Wang, R. Kiesgen de Richter, W. Liu, T. Wua, Y. Pan, 'Numerical analysis on the influence of ambient crosswind on the performance of solar updraft power plant system,' Renew. Sustain. Energy Rev. **16** (2012) 5567.

J.A. Palyvos, 'A survey of wind convection coefficient correlations for building envelope energy systems' modeling,' App. Thermal Eng. **28** (2008) 801.

D. Ryan and S.A.M. Burek, 'Experimental study of the influence of collector height on the steady state performance of a passive solar air heater,' Solar Energy **84** (2010) 1676.

Exercises

(7.1) The draft in a chimney is important to ensure smoke goes up the chimney rather than back into the building. Draft is measured in terms of the pressure drop into a fireplace ($P_1 - P_2 \equiv \Delta P$, see Fig. 7.1) and a draft of 20 Pa is considered good for a fireplace in a household residence. The equation for draft is written

$$\Delta P = CP_1 H\left[\frac{1}{T_1} - \frac{1}{T_3}\right]$$

where C is a group of variables to make a constant factor. Determine C and calculate T_3 if the chimney is 6 m high and ΔP is 20 Pa.

(7.2) It is proposed to make a solar chimney that has a rock wall, rather than a transparent wall as in Fig. 7.2, which absorbs the solar radiation during the day to release it only at night when required. This is accomplished by closing a damper during the day and when the outside air temperature becomes low enough, at night, it is opened. Assume the rock heats to 40 °C while the outside air temperature is 20 °C at night. The rock gradually (linearly) cools to the surroundings temperature over 5 h. Determine the mass flow rate of air over the 5 h period, in kg/s, and the ACH as a function of time. Assume the house has a volume of 400 m^3 and the rock wall is 10 m high by 2 m wide by 1 m thick. Also assume the flow area for the air through the chimney to be 1.5 m^2 and the exit air temperature, the same as the rock. Rock has a heat capacity of 1 kJ/kg-K and a density of 2500 kg/m^3.

(7.3) A solar chimney can be used during daylight hours to draw hot air underground, where it is cooled, and then circulated through the house, as shown in the figure below. Using the assumptions given below, determine the temperature of the house, T_2. Assume the cool air enters the house at 13 °C call this T_1, enters the chimney at temperature T_2 and exits the chimney at temperature T_3. Also, assume the house air temperature T_2 is the same throughout the entire house (this is sometimes called the well-mixed assumption). Use the First Law of Thermodynamics to determine T_2 (see eqn (3.12))

$$\dot{m}C_p\left[T_2 - T_1\right] = \dot{Q}_{in}$$

assuming the house is the system, the solar chimney is thermally isolated from the house and there are no heat losses from the house. Here $\dot{Q}_{in}$ is the rate of solar energy input into the house. To find this, assume the insolation is 500 W/m^2 and the (effective) area of the house that the insolation can be transmitted through is 30 m^2 (multiplying these two together yields $\dot{Q}_{in}$). Of course $\dot{m}$ is the mass flow rate of air through the house and C_p is the heat capacity of air.

The solar chimney is 4 m high (this is short since the air exits from the second floor after travelling through first floor), the device area A_D to gather the insolation is 13 m^2 and the chimney exit (flow) area is 2 m^2.

Finding the temperature T_2 will require you to use the design and operating lines for the solar chimney together with the FLOT discussed above.

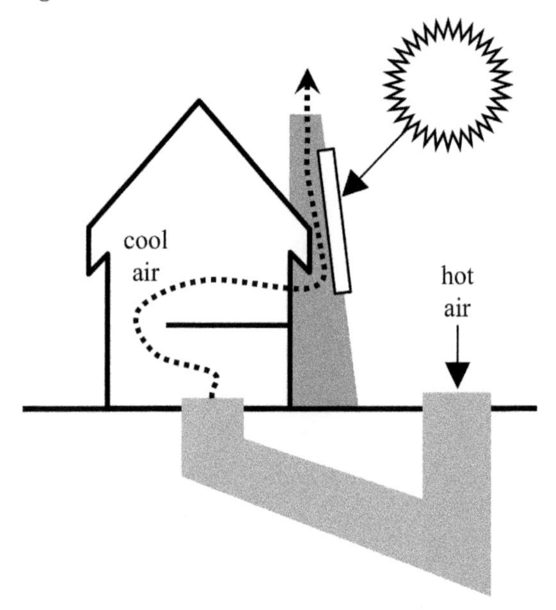

(7.4) Determine if Leonardo da Vinci was right and a spit could be turned by rising air within a chimney, as shown in Fig. 7.4. Assume the temperature rise of the air is 200 °C, the diameter and height of the cylindrical chimney are 1 m and 10 m, respectively, and the ambient air temperature is 300 K. To do this one must realize that the rate of kinetic energy imparted by flowing air to the fan (turbine) is $\dot{KE} = \frac{1}{2}\dot{m}v^2$, where $\dot{m}$ is the air mass flow rate and v the air velocity. See the discussion which follows

Example 7.3. Assume the fan or turbine sweeps a circular area almost equal to the chimney area. The minimum power to turn the spit is 0.25 hp.

(7.5) Repeat Exercise 7.4 for a solar chimney. In this case assume the chimney is also circular in cross-section, has the same dimensions as above and the half facing the Sun is transparent and the other half is the absorber which absorbs all the radiation. Use eqn (7.20) to find the rate of kinetic energy generation assuming $P_D(\beta) = 500$ W/m^2. Comment on the amount of power generated in light of the size and scale of proposed solar tower facilities.

(7.6) Pretorius and Kröger (Trans. ASME, **128** (2006) 302) have given the following relation for the heat transfer coefficient from the cover of a solar tower, which is more complicated than eqn (7.27) and is written as

$$h(\mathrm{W/m^2\text{-}K}) = \frac{0.2106 + 0.0026 v_w \left[\dfrac{\rho_{\mathrm{air}} T_m}{\mu_{\mathrm{air}} g \Delta T}\right]^{1/3}}{\left[\dfrac{\mu_{\mathrm{air}} T_m}{g \Delta T C_{P,\mathrm{air}} k_{\mathrm{air}}^2 \rho_{\mathrm{air}}^2}\right]^{1/3}}$$

where T_m is the mean temperature of the cover and air and all other variables are defined in the chapter. The properties of air are evaluated at T_m. All variables are in MKS units and the temperature is in Kelvins. Compare this relation to eqn (7.27) for various wind velocities (from $0 \rightarrow 4.5$ m/s) using a range of temperature differences (from $5 \rightarrow 45$ K) and assume the outside air temperature is 300 K. Comment on your results.

(7.7) Assume water is the working fluid for a solar tower. Use eqn (7.35) to determine what the maximum power can be for a *water* tower compared to one where air is the working fluid.

(7.8) Repeat Example 7.3 where the working fluid is water and not air.

(7.9) Repeat Example 7.4 where the working fluid is water and not air.

(7.10) Determine the power generated in the Manzanares, Spain solar tower when the wind is blowing and do not include the effect of insolation. In other words, assume values of the wind velocity from 0 to 50 m/s and determine how much wind would contribute to the power output from the solar tower.

(7.11) It has been proposed to make a *solar tower* on the side of a mountain (S.V. Panse *et al.* Energy Conv. Man. **52** (2011) 3096), which is an interesting concept since the cover and tower are integrated into one. The cover is placed on the side of the mountain and the air channeled up to the top where turbines are located. Assume the mountain is 500 m high and has a slope of 25° from the horizontal with a circular base, the cover encircles one-quarter of the base and follows the conical contour to the apex (*i.e.* the cover, 'covers' one-quarter of the conical, mountain surface area) and is 2 m above the mountain surface. The useful insolation is 500 W/m^2, this is the rate of energy transfer that is absorbed by the air, and the inlet air temperature is 300 K while the outlet area at the top of the mountain is 100 m^2; again the one-quarter of the mountain circumference will be covered by the cover at the top of the cover. Determine the mass flow rate, temperature rise (*i.e.* the operating point) and maximum work that could be generated in the inclined solar tower. Assume the actual work output is 20% of the maximum to report the amount of work that could be generated; is this a lot? Since the cover area is so large, heat transfer must be considered, let the wind velocity be 4 m/s. Also, you must consider the assumptions that go into the derivation of eqn (7.8). The temperature just inside the cover will not be vastly different to just outside the cover making it difficult to estimate the draft into the chimney. In addition, it was assumed the air had a constant temperature within the chimney which will certainly not be true. So, assume the area in eqn (7.8) is $\sqrt{A_2 A_3}$ while the air density in the chimney is $\sqrt{\rho_2 \rho_3}$, making the equation

$$\dot{m} = \sqrt{\rho_2 \rho_3} \sqrt{2 \frac{\Delta T}{T_1} g H} \sqrt{A_2 A_3}$$

You may also have to make other engineering assumptions!

(7.12) Consider Exercise 7.11 from an optical point of view. Assume the solar zenith angle θ_Z is 25°; find the projected area of the cover at solar noon to estimate the actual area that will be exposed to the Sun. Of course, the curved surface will reflect light differently than a flat surface, however, the point of this exercise is to ascertain how much of the covered area is useful in terms of heating air. Assume the mountain is in the northern hemisphere and the conical segment faces to the south.

(7.13) Air conditioning used to cool buildings in the USA accounts for 5% of electricity use. Suppose you live in the desert and want to power your air conditioning unit during the day with a solar tower and require 3500 W to do so. Assume $P_D(0) = 800$

W/m^2 and T_1 = 40 °C; design the solar tower assuming local building codes only allow a 50 m high chimney.

(7.14) It has been proposed by the balloon manufacturer and adventurer, Per Lindstrand, to make an inflatable tower that is held up by balloons. The design calls for a 1 km high tower with a 7 km radius collector, which would create 130 MW of power and over a year would produce 281 GWh of electricity. Consider these numbers and indicate if they are correct. Incidentally, it is stated that the inflatable tower would only cost $US 20-million while a conventional tower of similar dimensions would cost $US 750-million, a considerable saving.

(7.15) Explain why you would use water rather than any other common material to absorb insolation under the cover of a solar tower.

The flat plate solar energy collector

A flat plate solar energy collector is a passive device that harnesses solar radiation to heat a working fluid. The working fluid can be water that is directly used or another liquid that then transfers the absorbed energy to water. This later system would be used when there is a chance for the water to freeze if it were exposed to external temperatures.

As the reader will find, these systems are very low technology, however, they are very efficient. Efficiencies, in terms of how much insolation is transferred to water, are over 50% which is much greater than photovoltaic devices and are even more efficient than coal-fired power plants which are around 35% efficient. Thus, although these devices are not technologically appealing they are well worth installing at your residence or place of work due to their efficiency.

Perhaps they should be the first solar powered device one should purchase since their price is fairly low and they will have a long service life. In addition, they require low cost maintenance, unlike solar panels. Maintenance can be high for photovoltaic devices because the power inverter, which converts DC to AC current, must be replaced at least once during the life of a photovoltaic system, adding a cost of between $2,000 and $20,000 (depending on the system size) over the device's entire usable life.

So, solar thermal hot water heating systems that directly convert energy from the Sun to a useable form seem more cost effective. They should save the average household on the order of $200–$300 per year and take about 10–15 years to pay off the system, which will last for approximately 20 years.

Although the above argument points to having a solar thermal hot water system, there are some that argue it is cheaper and more effective to heat water with photovoltaically generated electricity and an air source heat pump (see the reference for Holladay at the end of this chapter). This argument does not follow the general rule-of-thumb that one should never use high quality energy to make low quality, if at all possible. In other words, use electricity to power electrical devices rather than to heat water. Of course, the decision as to whether solar thermal or solar photovoltaic is the right choice depends on your location, water demand, system cost, etc. and careful decision-making should be performed. By the end of this chapter the reader should be able to make that informed decision.

A significant deviation from other textbooks and analysis techniques is made in this chapter. The *heat removal factor* (F_R) and other parameters are often introduced and used to estimate the useful heat ($\dot{Q}_u$) transferred to the working fluid as it flows through the pipes. The useful heat is defined below and will be estimated *via* basic engineering heat transfer calculations. The heat removal factor technique finds $\dot{Q}_u$ with charting techniques and/or correlations that have been developed over a period of decades. Here the reader is introduced to the basic principles that can be used to design a solar hot water heating system from the beginning.

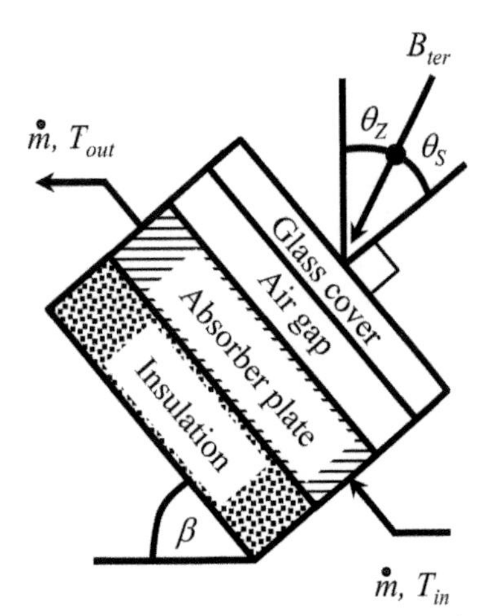

Fig. 8.1 Schematic of a flat plate solar energy absorber device that makes hot water, showing the various components and angles. The mass flow rate, $\dot{m}$, and water temperature in and out, T_{in} and T_{out}, respectively, are also given in the figure.

This decision was made for two reasons. Firstly, it is believed that the correlative approach was introduced because radiative heat transfer has a difference of T^4 between surfaces, which complicates hand calculation of design equations and which almost always involves iteration. The reader now has powerful calculation tools that can easily solve coupled algebraic equations involving non-linear terms. Thus, the base equations are given rather than charts and correlations. Secondly, new materials are being developed and this could change how these systems are designed. For example, an effort exists to make flat plate solar collectors out of plastic which would most likely prohibit the use of older correlations. Thus, the reader would be faced with a challenge that might seem insurmountable in designing such a system. It is hoped that by the end of this chapter the reader will have a good grasp of how to design a contemporary system, and that knowledge can be generalized to more innovative devices. Dimensionless numbers and correlations between dimensionless numbers will be used to design the devices in this chapter; if the reader is not familiar with these concepts, reading Appendix C may help one to understand what they are and why they are used.

8.1 The basic system

Fig. 8.2 Picture of a tube in good thermal contact with the absorber plate (fin). In this design the tube is centrally located within the absorber plate rather than being soldered or welded on the back. Photo courtesy of Sun Ray Solar.

The basic system is shown in Fig. 8.1, which consists of a glass cover over an absorber plate which is in thermal contact with the flowing water that enters and exits the device. Insulation at the back of the device, as well as around the edges (not shown), reduces thermal losses. The glass cover is required to reduce thermal losses on the device front, just as a greenhouse would experience.

Water flows into the absorber plate system at a mass flow rate $\dot{m}$ and typically does so through pipes that are soldered or welded to the back of the plate to yield good thermal contact. An example of another design is shown in Fig. 8.2, which is a system that has the riser pipe incorporated into the absorber plate. This system allows the fins to interlock together and the risers are soldered into a header tube to make a compact system. The water goes in and out of the system through the header tubes, which uniformly forces water into the riser tubes and collects the hot water at the top of the device.

To simplify analysis the glass cover which is present in many contemporary systems will initially be omitted. Inclusion of a cover adds difficulty to the analysis which does not allow the reader to first focus on the heart of the system, the absorber plate and transfer of absorbed energy to flowing water. Yet, insulation will be included in the basic design to minimize heat transfer from the back.

The amount of radiation striking the southward facing device (in the Northern hemisphere) at an angle of β from the horizontal to maximize the input of solar radiation ($P_D(\beta)$) is given by

$$P_D(\beta) = B(\beta) + D(\beta) + A(\beta)[=]\frac{\text{W}}{\text{m}^2} \tag{8.1}$$

as discussed in Chapter 2. The symbol [=] means 'has dimensions of.' Each term in eqn (8.1) is just at the external surface of the absorber plate (or at the external surface of the cover should it be used) for the beam component ($B(\beta)$), diffuse component ($D(\beta)$) and the reflected or albedo component ($A(\beta)$). These terms can be determined with the principles given in Chapter 2, and for reference $B(\beta) = B_{ter}\cos(\theta_S)$, where B_{ter} is the direct beam insolation at the surface of the device and θ_S is the solar angle. The radiation has not been absorbed by internal processes at this point and when absorbed it will be written as $aB(\beta)$, where the absorbance a is assumed to be a constant over the solar spectrum and not a function of incident angle.

A good absorber will be a selective surface, meaning that the absorbance for high energy solar radiation will be near one (large), yet, its emittance for low energy radiation will be near zero (small). This is because the Sun produces high energy radiation and this must be absorbed as efficiently as possible. Since the absorber will be at a much lower temperature than the Sun, say on the order of 100 °C, the emittance, which is equal to the absorbance, for low energy radiation must be low to minimize thermal losses.[1] This forms the basis of a selective surface, a large absorbance for high energy radiation and a low emittance (absorbance) for low energy radiation.

[1] Remember Kirchoff's law where it was proved that the emittance equals the absorbance, in Chapter 3.

The flux of radiation, $aP_D(\beta)$, represents the amount of energy entering the device which must now be converted to useful heat ($\dot{Q}_u$), determined by application of the First Law of Thermodynamics (FLOT) to the flowing water,

$$\dot{Q}_u = \dot{m}C_{pw}\Delta T\,[=]\,\text{W} \tag{8.2}$$

where C_{pw}, is the water heat capacity and ΔT, is the difference $T_{out} - T_{in}$, with T_{in} and T_{out} being the inlet and outlet temperature, respectively, to the flat plate solar energy absorber. Applying the FLOT to the device operating at steady state (see Fig. 8.3) yields

$$\dot{m}C_{pw}\Delta T = [aP_D(\beta) - \dot{q}_{rad} - \dot{q}_{conv} - \dot{q}_{bot}]A_D \tag{8.3}$$

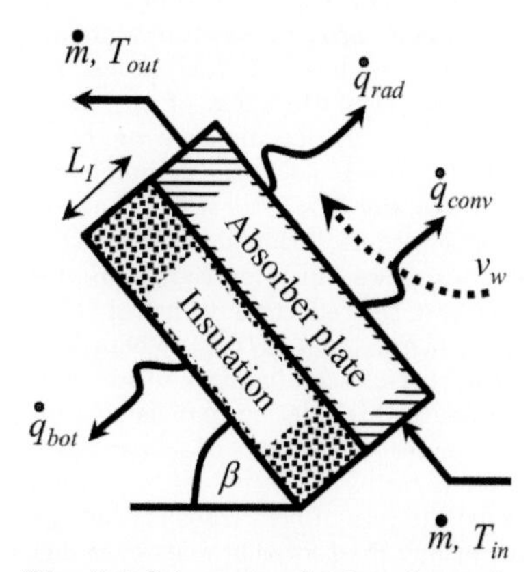

Fig. 8.3 Schematic of a flat plate solar energy absorber without a cover.

It is obvious that $\dot{Q}_u$ is merely the rate of sensible heat rise for the water entering into the flat plate absorber of area A_D. There are three heat loss terms $\dot{q}_{rad}$, which are that due to radiation, $\dot{q}_{conv}$, due to convective heat transfer and $\dot{q}_{bot}$ is the rate of heat transfer from the bottom of the device, as shown in Fig. 8.3. Note a heat transfer rate that is capitalized like $\dot{Q}$ is the total heat transfer rate in Watts, while a lower case like $\dot{q}$ is the heat transfer rate in Watts/m^2. Each heat transfer term is now considered in turn.

Radiative heat transfer from the absorber plate is given by

$$\dot{q}_{rad} = e_p \sigma_S \left[T_p^4 - T_{sky}^4\right] [=] \frac{\text{W}}{\text{m}^2} \tag{8.4}$$

where e_p is the emissivity of the absorber plate, σ_S, the Stefan-Boltzmann constant, T_p, the absorber plate temperature and T_{sky}, the sky temperature for radiative heat transfer.[2] Obviously one wants e_p to be as small as possible, as discussed above.

[2] A simple relation for the temperature of the sky is $T_{sky} = 0.0552 T_a^{1.5}$ where T_a is the surrounding air temperature and all temperatures are in Kelvins.

The convective heat transfer rate is a function of the wind velocity v_w. If it is zero then only natural convection is present, while if the wind is blowing then forced convection occurs.[3] The heat transfer rate is given by

[3] Heat transfer is much more efficient for forced convection. You have felt this effect before if you have placed your finger or toe in hot water. Putting it in the hot water at first feels comfortable, if you move your finger or toe it will suddenly feel much hotter due to a greater rate of heat transfer under convective (moving) conditions. Of course, you know this too when it is cold outside, wind makes it feel much colder and forced convection leads to the *wind chill factor*.

$$\dot{q}_{conv} = h_{conv} \left[T_p - T_a\right] \tag{8.5}$$

where h_{conv} is the convective heat transfer coefficient and T_a is the air (ambient) temperature. This is *Newton's law of cooling* where the rate of heat transfer is proportional to the temperature difference. There are many correlations for the heat transfer coefficient, Duffie and Beckman recommend

$$h_{conv}(\text{W/m}^2\text{-K}) = \text{MAX}\left(5, \frac{8.6\, v_w(\text{m/s})^{0.6}}{L_H(\text{m})^{0.4}}\right) [=] \frac{\text{W}}{\text{m}^2 \text{ - K}} \tag{8.6}$$

where the function MAX($\cdot$) means maximum and L_H is the cube root of the house volume upon which the flat plate absorber is placed. If the device is on the ground or the designer chooses not to use this correlation then another for air flowing over a flat plate can be used, as given by Palyvos,

$$h_{conv}(\text{W/m}^2\text{-K}) = 7.4 + 4.0\, v_w(\text{m/s}) \tag{8.7}$$

Now consider the final heat transfer rate through the bottom (and sides) of the device. This is approximated as due to heat transfer through insulation placed on the bottom, which is written as

$$\dot{q}_{bot} = \frac{k_I}{L_I} \left[T_p - T_a\right] \tag{8.8}$$

where k_I is the thermal conductivity of the insulation with thickness L_I. This is *Fourier's law of heat conduction*.[4] The assumption was made that the insulation represents the greatest resistance to heat transfer through the bottom of the device and is a good assumption except under extreme conditions. Furthermore, the insulation temperature next to the absorber plate was assumed to be at the temperature of the absorber

[4] Fourier's law for conduction of heat can be compared to Newton's law for convection of heat. The thickness L_I is the physical thickness of the insulator, however, for air blowing by a surface there is no easily definable air thickness, say L_{air}, over which the heat transfer occurs. Boundary layer theory addresses this, for right now we calculate the effective value of L_{air} through $h_{conv} \equiv k_{air}/L_{air}$, where k_{air} is the thermal conductivity of air. Rearranging this relation and using typical values one finds: $L_{air} \approx 2.5 \times 10^{-2}$ W/m-K/[5 W/m^2-K] = 5 mm. Thus, the rate of heat transfer through air, or any fluid for that matter, is dictated by the effective heat transfer layer thickness adjacent to a solid substrate.

plate T_p which is again a reasonable assumption as *spray-on* insulation, such as polyurethane, is now being used which has good thermal contact with the plate. Also, the insulation was assumed to be at the external air temperature at the bottom of the device which is a good assumption for good insulation. Finally, radiative heat transfer from the bottom was ignored which a simple calculation can confirm to be correct. Note if heat transfer from the edge is to be considered then the effective device area at the bottom can be increased, or a new heat transfer term can be added similar to $\dot{q}_{bot}$.

The design of the flat plate solar hot water heater is performed by calculating the temperature rise ΔT of water flowing through the device under a given insolation $P_D(\beta)$. If all the terms are put into eqn (8.3) one realizes that all of them can be assumed, are known or are based on ambient conditions, except two: ΔT and T_p. We have one equation and two unknowns; another equation is required that will be addressed below. Before doing this eqn (8.3) is re-written to emphasize that it is the *operating line*,

$$\dot{m}C_p\Delta T = [aP_D(\beta) - \dot{q}_{rad} - \dot{q}_{conv} - \dot{q}_{bot}]A_D \quad \text{operating line} \qquad (8.3)$$

The FLOT is taken as the operating line meaning that it indicates the values of T_p and ΔT that are possible. The FLOT is merely a balance of energies, or in our case the rate of energy transfer. If one term is smaller then another must be larger so the equation is balanced.

To understand this, consider the operating line shown in Fig. 8.4 generated with eqn (8.3) using the standard conditions for a solar thermal flat plate absorber given in Table 8.1. The graph may seem confusing at first; as the absorber plate temperature is increased the temperature rise of the water is decreased. This can be understood by remembering the FLOT (*operating line*) is an energy balance and if energy goes into heating the plate then it cannot heat the water. Likewise if energy is used to increase the temperature of the flowing water then less is available to increase the absorber plate temperature, so, it falls.

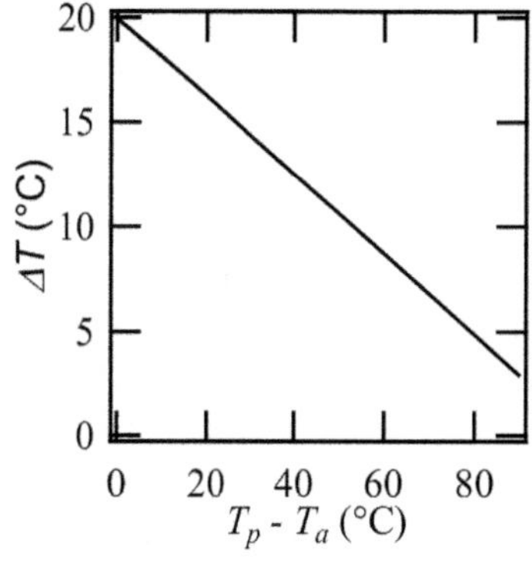

Fig. 8.4 Operating line for the water temperature rise ΔT in a solar hot water heater as a function of $T_p - T_a$. The standard conditions were used as given in Table 8.1.

The abscissa of the graph in Fig. 8.4 is written as $T_p - T_a$ as this is one way to present the data; of course there are other variables used and not just $T_p - T_a$. One should realize that for a flat plate device without a cover the minimum temperature that the plate can have is T_a since it cannot fall below that unless the inlet water temperature is very low.[5] A calculation where part of the system can go below the ambient temperature is considered in Example 8.4, however, for right now, the minimum plate temperature is assumed to be equal to the air temperature. At this condition the maximum temperature rise is approximately 20 °C. Of course, as will be determined below, the conditions required to achieve this temperature rise are extreme and so a more moderate temperature rise will occur. Since the temperature rise will be less than this, the question arises: The temperature of domestic hot water is typically 50 °C, how is the water temperature increased? This is answered in Section 8.4.

[5] This is not exactly true. Since the sky temperature is lower than the air, under natural conditions, it is possible for water to freeze at night even when the air temperature is above the freezing point. This occurs by water radiating energy to the low temperature sky until it freezes. But, one should realize that the Earth is radiating energy too which would be absorbed by the water and this needs to be taken into account. So, whether the water freezes or not would be the result of a complicated calculation requiring detailed information about the surrounding terrain as well as the temperature of the sky.

Table 8.1 Standard conditions used in designing a solar thermal flat plate absorber device in this chapter. Each device is 2.5 m long and 1.2 m wide and two are connected in parallel. The water flows along the length direction and it is assumed that there is no heat loss as the water exits the device. These conditions are for a solar hot water heater used for a typical residence. The insolation was assumed to be 1000 W/m^2, similar to the AM1.5G spectrum, allowing ready comparison to solar photovoltaic devices, and unless otherwise stated, this is direct beam insolation.

Insolation (B_{ter})	1000 W/m^2
Device tilt (β)	50°
Solar angle (θ_S)	20°
Device area (A_D)	6 m^2
House length scale (L_H)	10 m
Air temperature (T_a)	10 °C
Wind velocity (v_w)	3 m/s
Water mass flow rate ($\dot{m}$)	0.06 kg/s
Inlet water temperature (T_{in})	15 °C
Absorber plate short wavelength absorbance (a)	0.9
Absorber plate long wavelength emissivity (e)	0.1
Insulation thermal conductivity (k_I)	0.026 W/m-K
Insulation thickness (L_I)	50 mm
Cover thickness (L_c)	3.2 mm
Absorber plate–cover air gap (L_{pc})	40 mm
Cover refractive index (n_c)	1.53
Cover absorptivity (α_c)	0.04 cm^{-1}
Cover emissivity (e_c)	0.88
Number of riser pipes (N_p)	16 (8 in each)
Riser pipe length (L_p)	2.5 m
Riser pipe diameter (D_p)	12.5 mm

The results in Fig. 8.4 show what is possible, how do we find the absorber plate temperature–water temperature rise for our system? This will be the *operating point* and it will be found at the point where the *operating line* intersects the *design line*. The design line is found by considering the operation of the solar thermal device through detailed heat transfer calculations, which will be discussed immediately after the system configuration and an introduction to heat transfer of flowing water is considered.

The system in Table 8.1 has two modules connected in parallel and is similar to many systems installed at residences, see Fig. 8.5. The eight riser pipes in each module that carry the water are aligned to the long direction and assumed to be equal to the module length, which is 2.5 m, thus, the riser pipe length is 2.5 m. The pipes are connected to a larger diameter pipe on each end by a hole cut in its side to make a manifold to

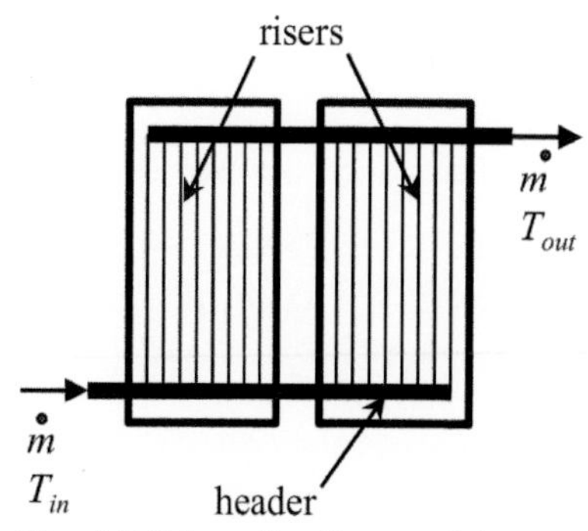

Fig. 8.5 Schematic of two modules connected in parallel with the header and riser pipes indicated.

distribute the water to and then collect the water from the risers which are called headers. The water will exit the top header and will go back to the house and into a storage tank where it is recirculated. When this system is mounted on a house roof the top header is closest to the roof apex (ridge) and the device tilt angle β is typically $|L| + 10°$, where L is the latitude, as discussed in Chapter 2.

Another piping arrangement is the *serpentine* which takes a single pipe and bends it back and forth so it covers the entire device area, where one end enters the bottom and the other exits the top. This could change the heat transfer from that occurring under laminar conditions to turbulent, and how it affects the operating point is considered in Exercise 8.5. Laminar and turbulent flow are described below should the reader not be familiar with this concept.

The water is heated in all designs by the absorber plate transferring heat (energy) to the flowing water in the pipes, since heat transfer is driven by a temperature difference between two systems in thermal contact. The flat plate absorber device has water entering a pipe at T_{in} and exiting at T_{out} which is in thermal contact with an absorber plate at temperature T_p which is assumed to be constant over the device area. What temperature difference do we use that drives heat transfer from the plate to the flowing water? Do we use $T_p - T_{in}$ or $T_p - T_{out}$ or ...?

The answer is the log mean temperature difference or ΔT_{lm} whose derivation can be found in an undergraduate heat transfer textbook. The textbook by Middleman, referenced at the end of this chapter, is particularly good and clearly written. The log mean temperature difference comes about by considering a differential energy balance between the water flowing in the pipe in thermal contact with the absorber plate, then integrating from the inlet to exit to arrive at

$$\Delta T_{lm} = \frac{[T_p - T_{out}] - [T_p - T_{in}]}{\ln\left(\dfrac{T_p - T_{out}}{T_p - T_{in}}\right)} = \frac{\Delta T}{\ln\left(\dfrac{T_p - T_{in}}{T_p - T_{in} - \Delta T}\right)} \tag{8.9}$$

One can think of ΔT_{lm} as an average temperature difference driving heat transfer and the natural logarithm comes about through an integral of the form $d[T_p - T]/[T_p - T]$.

Now the rate of heat transfer from the absorber plate to the water $\dot{Q}_w$ is

$$\dot{Q}_w = h_w A_p \Delta T_{lm} \tag{8.10}$$

where h_w is the heat transfer coefficient for the water in thermal contact with the pipe wall and A_p is the pipe area. If there are N_p pipes with diameter D_p and length L_p then $A_p = N_p \pi D_p L_p$ and is the heat transfer area for the water flowing within the device. The key factor now is the heat transfer coefficient, which must be calculated. Negligible thermal resistance for the pipe wall itself is assumed since most thermal resistance occurs between the water and pipe wall.

There are many correlations in the literature to find h_w and all involve three dimensionless numbers: the Nusselt number (Nu), Reynolds number (Re) and Prandtl number (Pr).[6] It is important to note that the dimensionless numbers are determined for water at the mean temperature T_m given by

[6] See Appendix C for the importance and definitions of dimensionless numbers.

$$T_m = \frac{1}{2}\left[T_{out} + T_{in}\right] \tag{8.11}$$

Since the water properties are a function of temperature and the temperature rise is not known *a priori*, one ultimately arrives at an iterative solution, as is the nature of most heat transfer calculations involving flow. This will be discussed in more detail below.

The heat transfer coefficient depends strongly on whether the flow in the pipe is laminar or turbulent. Laminar flow means the fluid travels at the same position relative to the pipe center-line along the pipe length, while in turbulent flow the fluid will deviate strongly from 'straight line' flow. Turbulent flow is much more efficient to induce heat transfer, and so is desirable, although due to engineering constraints, solar thermal flat plate hot water heaters typically operate under laminar flow conditions, at least in residential applications. Regardless, for completeness, Nusselt number correlations will be given for both laminar and turbulent flow.

Laminar flow has $Re < 2100$ and Nu is found from the following correlations:

$$\begin{array}{lll} \text{if} \quad RePrD_p/L_p < 12 & \text{then} & Nu = 3.66 \\ \text{if} \quad RePrD_p/L_p > 12 & \text{then} & Nu = 1.6[RePrD_p/L_p]^{1/3} \end{array} \tag{8.12}$$

Since k_w will be known Nu allows calculation of h_w ($Nu = h_w D_p/k_w$, see Appendix C). Now it is clear why iterative calculations will be required. One needs h_w to calculate the rate of heat transfer to the water which will cause its temperature to rise. However, k_w is a function of temperature and it is needed to determine h_w. Thus, one assumes an exit temperature and calculates T_m, then k_w is evaluated at this temperature, as are all the other physical properties to determine Re and Pr. Then Nu is evaluated, h_w is calculated and the temperature rise found. If it does not agree with the assumed temperature rise the process is repeated until convergence is calculated (see below).

Turbulent flow has $Re > 2100$ and follows the relation (although there are many correlations, this is a fairly accurate one)

$$Nu = \frac{\frac{1}{2}f[Re - 1000]Pr}{1.07 + 12.7\sqrt{\frac{1}{2}f}\left[Pr^{2/3} - 1\right]}\left[\frac{\mu(T_m)}{\mu(T_p)}\right]^{0.11} \tag{8.13}$$

where the viscosity ratio is evaluated at the mean and pipe wall temperatures and f is the *Fanning* friction factor which is a dimensionless pressure drop over the pipe length,

$$f \equiv \frac{\Delta P}{2\rho_w V_w^2} \times \frac{D_p}{L_p} \tag{8.14}$$

where ΔP is the actual pressure drop, ρ_w, the density of water and V_w, the average water velocity in the pipe (see Appendix C and eqn (C.1)). The friction factor for both laminar and turbulent flow is a function of the Reynolds number and is given by

$$f = \frac{16}{Re} \quad \text{laminar flow} \tag{8.15a}$$

$$f = \frac{0.079}{Re^{1/4}} \quad \text{turbulent flow} \tag{8.15b}$$

although f is not required for the Nusselt number correlation for laminar flow and is included for completeness. The reader should realize there are many f–Re correlations that take factors such as pipe roughness into account, which can noticeably affect the pressure drop in turbulent flow. One may notice that if calculations are performed near the laminar–turbulent transition there may be curious changes in the heat transfer coefficient; this is to be expected since the correlations presented above do not take into account the *transition region*, which occurs over a fairly large range of Reynolds numbers near 2100 until the flow is fully turbulent. If operating within the transition region the reader should search for heat transfer correlations that consider this and ensure that the correlation is for heat transfer for a constant temperature pipe wall that is hotter than the fluid (it actually makes a difference if the water is hotter than the pipe).

Now it is possible to write an equation for the *design line*. The sensible heat taken from the pipe (absorber plate) to the water is equal to that which causes the water temperature to rise, which can be written mathematically as

$$\begin{aligned} \dot{m}C_p\Delta T &= h_w A_p \Delta T_{lm} \\ &= h_w A_p \frac{\Delta T}{\ln\left(\dfrac{T_p - T_{in}}{T_p - T_{in} - \Delta T}\right)} \quad \text{design line} \end{aligned} \tag{8.16}$$

The *operating point* is found by simultaneously solving eqns (8.3) and (8.16) since one now has two equations and two unknowns, T_p and ΔT.

Example 8.1

Find the operating point of a single flat plate solar collector using the standard conditions given in Table 8.1. Assume there is no cover.

First the *operating line* is calculated using the standard conditions given in the table and using eqn (8.3) which was previously plotted in Fig. 8.4. This is graphed again in Fig. 8.6 where the convective heat transfer coefficient of Duffie and Beckman (eqn (8.6)) was used to describe the heat transfer from the absorber plate to air, h_{conv}. Also, only the direct beam component is considered in the calculation of $P_D(\beta)$.

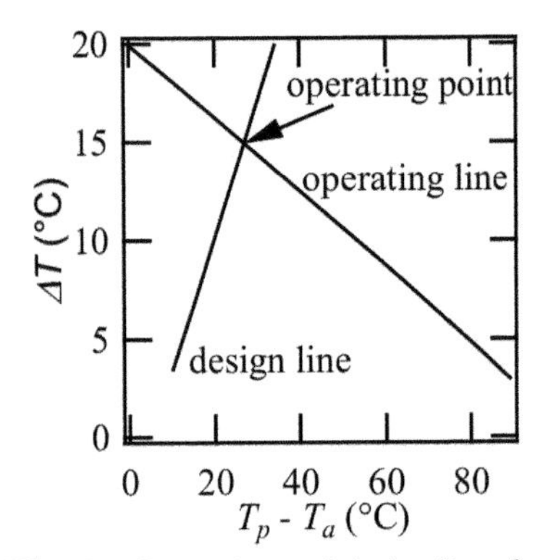

Fig. 8.6 Operating and design lines for the water temperature rise ΔT in a solar hot water heater as a function of $T_p - T_a$. The operating point is also shown. The standard conditions were used as given in Table 8.1 and no cover was used.

Now the *design line* is determined by assuming a value for T_p and calculating h_w for the conditions in Table 8.1. The temperature rise ΔT is not known so it is assumed, then the physical properties of water are determined with the correlations given in Appendix B and used to calculate Re and Pr. The Reynolds number is of order 700–800, so the flow is laminar, and for laminar flow one needs to calculate $RePrL_p/D_p$ with eqn (8.12). This term is almost constant since the most temperature dependent property, the viscosity, is not present in the product $Re \times Pr$ and one finds it is approximately equal to 13 for the conditions in Fig. 8.6. This in turn makes h_w also approximately constant at a value of 182 W/m^2-K. The design line can be calculated with eqn (8.16) and is shown in the figure.

The *operating point* is found by simultaneously solving eqns (8.3) and (8.16) and is represented by the intersection of the two lines in Fig. 8.6. Thus, the operating point is found to be; T_p = 37.0 °C and ΔT = 14.9 °C, which is a reasonably large temperature rise.

The reader may note that the equation for the design line can be simplified and rewritten as

$$\Delta T = [T_p - T_{in}] \times \left[1 - \exp\left(-\frac{h_w A_p}{\dot{m} C_{pw}}\right)\right] \tag{8.17}$$

by using the definition of ΔT_{lm}.[7] This is a somewhat simpler form to use, yet, does not eliminate iteration in the calculation since ΔT, or really the local temperature of the water flowing through the pipes, influences h_w. Regardless it is an easier version to use and is valid for both laminar and turbulent flow.

[7] In reality eqn (8.17) is the solution to the heat transfer problem for flow in a pipe with a constant wall temperature. It can be rearranged to the form of eqn (8.16) to define the log mean temperature difference.

Assuming laminar flow, which is what occurs for the standard conditions, the important combination of parameters is $RePrD_p/L_p$, which can be written as

$$\frac{RePrD_p}{L_p} \equiv \frac{4}{\pi} \frac{\dot{m} C_{pw}}{N_p L_p k_w} \tag{8.18}$$

The ratio of C_{pw}/k_w is fairly constant for liquid water and equal to 6620 m-s/kg and will vary by ±10% over the temperature range 0–100 °C. Thus, if the reader wants to sacrifice some accuracy, an iterative calculation does not have to be performed to determine the design line.

Using this approximation, the design line is found by first calculating Re to determine whether the flow is laminar or turbulent. Assuming it is laminar, which is usually the case, at least for systems installed on houses, one can directly calculate $RePrD_p/L_p$ from eqn (8.18). Then eqn (8.12) can be substituted into eqn (8.17) to yield

$$\frac{\Delta T}{T_p - T_{in}} = 1 - \exp\left(-11.5 N_p \frac{k_w L_p}{\dot{m} C_{pw}}\right) \quad (RePrD_p/L_p < 12) \tag{8.19a}$$

$$\frac{\Delta T}{T_p - T_{in}} = 1 - \exp\left(-5.45\left[N_p \frac{k_w L_p}{\dot{m} C_{pw}}\right]^{2/3}\right) \quad (RePrD_p/L_p > 12) \qquad (8.19b)$$

These equations are reasonably accurate for determining ΔT as a function of T_p by assuming C_{pw}/k_w = 6620 m-s/kg. In Example 8.1, calculation of the design line would deviate from the iterative solution by less than 1.5%, which is acceptable. So the above equations can be used directly with eqn (8.3) to determine the operating point with reasonable accuracy. Furthermore, if an iterative solution is required it is merely coupled with the solution of a fourth-order algebraic equation.

The minimum temperature the absorber plate can have is near the surrounding air temperature for the standard conditions given in Table 8.1. At this condition all possible heat from the Sun has been garnered and transferred to the flowing water. Strictly speaking this is not exactly true since the absorber plate is radiating energy to the sky, which is at a lower temperature than the air, however, for our purposes it is assumed that the minimum plate temperature is equal to the air temperature (see note 5).

This means that all the heat loss terms; $\dot{q}_{rad}$, $\dot{q}_{conv}$ and $\dot{q}_{bot}$ are zero (or near zero) and all the absorbed radiation is given to the flowing water, which can be seen upon inspection of eqn (8.3). At this point one would have

$$\Delta T_{max} = \frac{a P_D(\beta) A_D}{\dot{m} C_{pw}} \qquad (8.20)$$

which represents the maximum temperature rise the flowing water can have, ΔT_{max}. In Example 8.1 the maximum temperature rise is 21.5 °C and so the operating point is 6–7 °C lower than this. In this calculation it was assumed that the radiative heat transfer term is zero, keeping this in the calculation, and assuming the minimum absorber plate temperature is T_a finds ΔT_{max} = 19.9 °C.

8.2 The effect of a cover

Many flat plate solar absorbers have a cover to reduce heat loss via the *greenhouse effect*; a schematic is shown in Fig. 8.7. This complicates the analysis for two reasons. Firstly, the amount of radiation that reaches and is absorbed by the absorber plate must be determined. This is written $\|a\tau\| P_D(\beta)$, where the symbol $\| \bullet \|$ represents the result of a thorough calculation and is not merely the product of the absorber plate absorbance and glass cover transmittance. It is normally called the absorbance–transmittance product though. The full calculation results in an infinite series that accounts for light transmission through the cover, then absorption by the absorber plate and then reflection by the absorber plate. The reflected radiation can then be transmitted back through the cover, absorbed or reflected back to the absorber plate. This process occurs multiple times and the result in summing the terms

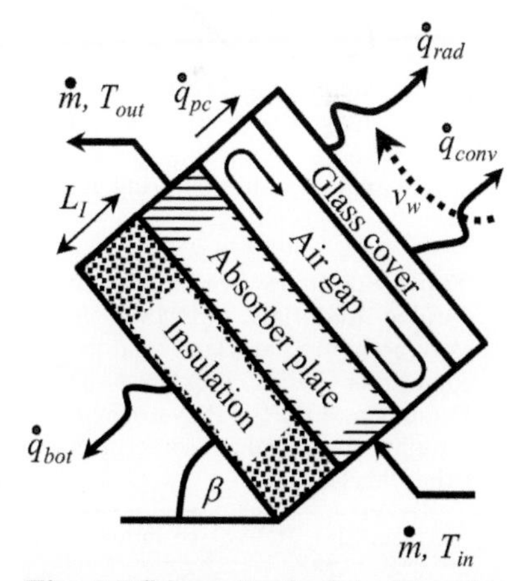

Fig. 8.7 Schematic of a flat plate solar energy absorber without a cover. The action arrows in the air gap represent the air trapped between the glass cover and absorber plate that recirculates due to natural convection.

is that $\|a\tau\| \approx a\tau$, where a is the absorbance of the absorber plate and τ is the transmittance of the cover; this approximate relation will be used here after reviewing the more detailed calculation.

The second reason that the analysis is complicated is that there are two heat transfer rates which occur in series: those from the absorber plate to the cover ($\dot{q}_{pc}$) and from the cover to the air ($\dot{q}_{ca}$). Each, though, has two heat transfer rates that occur in parallel: radiative ($\dot{q}_{rad}$) and convective ($\dot{q}_{conv}$). This certainly makes the calculation complicated since an iterative solution must be pursued, yet, not impossible. This is not a challenge for present day equation solvers.

We will first determine how much radiation reaches the absorber plate and consider transmission of the beam component; different procedures must be used for the diffuse and albedo radiation components that are considered later. Reference to Fig. 8.8 shows that the beam component B_{ter} makes an angle θ_S with the normal to the glass cover surface. The radiation is refracted at an angle θ_T which is given by *Snell's Law* to be

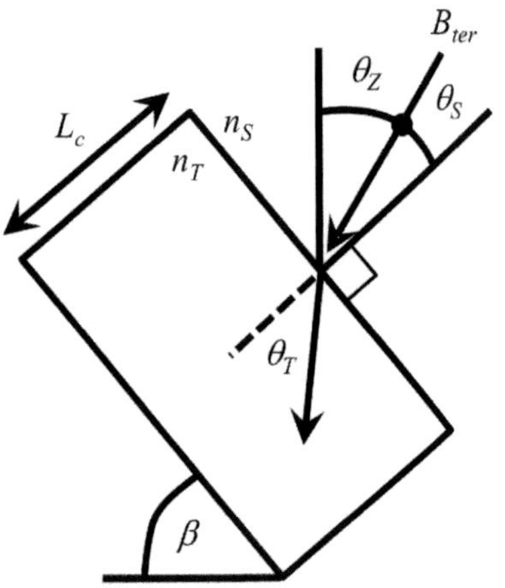

$$\frac{n_T}{n_S} = \frac{\sin(\theta_S)}{\sin(\theta_T)} \qquad \text{Snell's Law} \qquad (8.21)$$

where n_S is the refractive index of the surrounding medium (air which has $n_S \approx 1.0$) and n_T is the refractive index of the cover material. Many if not most covers are made of low-iron glass to reduce the absorption of radiation; there are some that use other materials and their refractive indices are listed in Table 8.2.

Fig. 8.8 Schematic of a cover used in a flat plate solar energy absorber. The light is refracted according to Snell's law to an angle θ_T after entering at θ_S.

Table 8.2 Refractive indices for various materials. PMMA is Polymethymethacrylate and is sometimes called Plexiglass or Perspex and PET is Polyethylene terephthalate and is sometimes called Mylar.

Material	Refractive index
Air	1.00
Water	1.33
PMMA	1.49
Glass	1.53
PET	1.64

The beam component is unpolarized radiation (light) and must be considered to have two polarization components. Before this is discussed a brief description of waves is presented. Electromagnetic (light) waves are transverse waves whose electric field oscillates transversely, or perpendicular to, their direction of propagation (x-axis), as seen in Fig. 8.9. The amplitude of the wave is given by the magnitude of the electric vector (E). Unpolarized light has the direction of the electric vector in any arbitrary direction perpendicular to the x-axis (*i.e.* in the y–z plane) for any given wave. Polarized light has the all the waves' electric vectors oscillating in only one direction, say the y-direction. A polarizer is used to filter out all other directions. There are other types of waves, sound waves for example are longitudinal waves, as shown in Fig. 8.9. A longitudinal wave has its oscillation in the direction of propagation and for sound waves the amplitude is given by the pressure. High pressure at the apex of the wave is called *condensation*, while low pressure is called *rarefaction* and is at the wave nadir. Because of this sound waves cannot be polarized. You may have noted that sound waves are longitudinal waves upon viewing a speaker as it oscillates to produce regions of high and low pressure, thereby making *sound*.

Now consider the beam component of solar radiation and the two polarization directions required to analyze light propagation through the cover. One polarization direction is perpendicular to the plane made by the incident beam and the surface normal while the other will be parallel to this plane. The symbols $\perp$ and $\|$ will be used as superscripts to the variables below for the perpendicular and parallel components, respectively. If the variable has the superscript $\square$, such as $\rho^{\square}$, the same equation can be used for either the $\perp$ or $\|$ component and the $\square$'s within the equation are changed to either $\perp$'s or $\|$'s accordingly.

When the radiation strikes the glass cover an amount is initially reflected from the surface and dictated by ρ_0, see Fig. 8.10, that can be determined by

$$\rho_0^{\perp} = \frac{\sin(\theta_T - \theta_S)^2}{\sin(\theta_T + \theta_S)^2} \tag{8.22a}$$

and

$$\rho_0^{\|} = \frac{\tan(\theta_T - \theta_S)^2}{\tan(\theta_T + \theta_S)^2} \tag{8.22b}$$

Determining the total amount of radiation reflected and transmitted through the cover is the result of considering multiple reflections from the bottom and top of the cover as shown in Fig. 8.10. The result of the analysis is an infinite series which yields for the reflectance ρ, transmittance τ and absorbance a

$$\rho^{\square} = \rho_0^{\square}\left[1 + \frac{\tau_d^2\,[1-\rho_0^{\square}]^2}{1-[\rho_0^{\square}\tau_d]^2}\right] \tag{8.23a}$$

$$\tau^{\square} = \tau_d\frac{[1-\rho_0^{\square}]^2}{1-[\rho_0^{\square}\tau_d]^2} \tag{8.23b}$$

$$a^{\square} \equiv 1 - \rho^{\square} - \tau^{\square} = [1-\tau_d]\,\frac{1-\rho_0^{\square}}{1-\rho_0^{\square}\tau_d} \tag{8.23c}$$

where d is the distance the radiation must propagate through the cover which equals $L_c/\cos(\theta_T)$, where L_c is the cover thickness. Also, the quantity τ_d, which is the single pass transmittance, is given by

$$\tau_d = \exp(-\alpha \times d) \tag{8.24}$$

where α is the absorption coefficient; values for various materials are given in Table 8.3. One finds the total amount of light reflected or transmitted by taking the arithmetic average of the perpendicular and parallel components *via*

$$\rho = \frac{1}{2}\left[\rho^{\perp} + \rho^{\|}\right] \tag{8.25a}$$

$$\tau = \frac{1}{2}\left[\tau^{\perp} + \tau^{\|}\right] \tag{8.25b}$$

In most cases the cover will absorb very little light. For example, using the absorption coefficient for low-iron glass given in Table 8.3 and the cover thickness given in Table 8.1 the amount of radiation transmitted for a single pass through the cover is given by eqn (8.24) to equal 0.9987, assuming the beam component is normal to the cover. Since $\tau_d \approx 1$ one finds $a^{\square} \approx 0$ in eqn (8.23c) allowing eqn (8.23b) to be simplified to

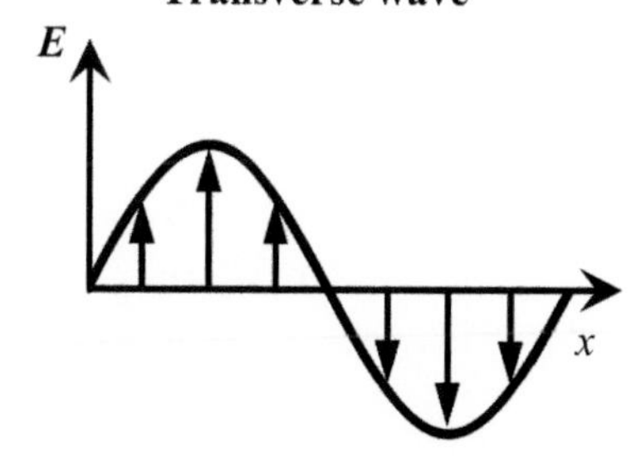

Fig. 8.9 Transverse and longitudinal waves.

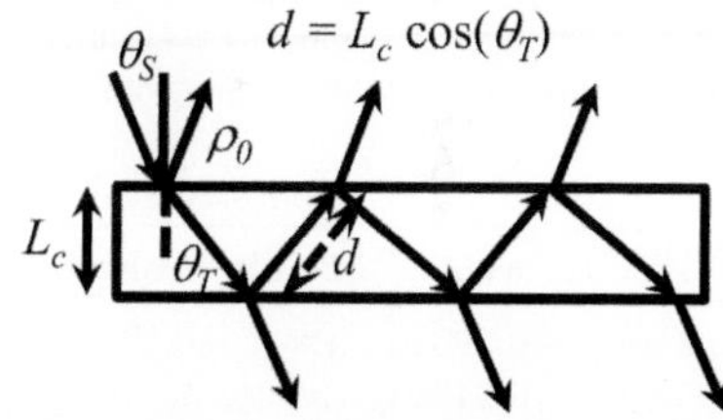

Fig. 8.10 Multiple reflection occurs when radiation propagates through a transparent material.

Table 8.3 Absorption coefficient for various materials. PMMA is Polymethymethacrylate and is sometimes called Plexiglass or Perspex and PET is Polyethylene terephthalate and is sometimes called Mylar.

Material	α (cm^{-1})
Air	$\approx 10^{-7}$
Water	0.07
PMMA	2.0
Glass (low iron content)	0.04
PET	2.0

$$\tau^{\Box} \approx \tau_d \frac{1-\rho_o^{\Box}}{1+\rho_o^{\Box}} \equiv \tau_d \times \tau_r^{\Box} \tag{8.26}$$

where the definition of $\tau_r^{\Box}$ is given by comparing the two equations. The quantity $\tau_r^{\Box}$ is frequently introduced to emphasize that this is the part of the radiation that is available after the initial reflection, while τ_d is that initially transmitted through the glass. Regardless, this approximate relation is fairly accurate and can be used with little worry for applications involving covers for solar applications.

Using the above, either the exact or approximate relation, one can find the amount of beam radiation that is transmitted through the cover (τ_B) from

$$\tau_B = \frac{1}{2}[\tau^{\perp} + \tau^{\parallel}] \tag{8.27}$$

Detailed discussion of the above and that which follows, concerning transmission and absorption of radiation in a solar energy collector, can be found in the textbook of Duffie and Beckman referenced at the end of this chapter.

Example 8.2

Determine the amount of light transmitted through the cover for the standard conditions listed in Table 8.1. Compare the exact and approximate techniques.

The first quantity to determine is θ_T using Snell's law given by eqn (8.21) which is found to be 12.94°. Now $\rho_0^{\perp}$ and $\rho_0^{\parallel}$ need to be calculated with eqns (8.22a) and (8.22b), respectively, and are 0.05148 and 0.03684. Finally, the single pass transmittance τ_d can be calculated from eqn (8.24) and is 0.9870.

Now, the perpendicular and parallel components of the transmittance (eqns (8.23b) and (8.26)), as well as the total transmittance (eqn (8.27)), can be calculated and are shown in Table 8.4. As can be seen there is very little difference in the values and one must go to four significant figures to see it. Thus, the approximate technique is more than adequate to determine the transmittance through the glass cover.

Table 8.4 Comparison of the exact and approximate techniques to calculate the transmittance of radiation through the glass cover under the standard conditions given in Table 8.1.

Exact		Approximate	
variable	value	variable	value
$\tau^{\perp}$	0.8902	$\tau^{\perp}$	0.8903
$\tau^{\parallel}$	0.9168	$\tau^{\parallel}$	0.9168
τ_B	0.9035	τ_B	0.9036

The transmittance for the diffuse and albedo components, τ_D and τ_A, respectively, are quite different since they are scattered radiation and come from all directions (isotropic) unlike the beam component. Simulations and modeling have been performed where it is found one should use an apparent angle for the total amount of radiation transmitted through the cover which can be written

$$\tau_D \approx \exp\left(-\frac{\alpha L_c}{\cos(\theta_D)}\right) \quad \text{where} \quad \theta_D \equiv 60° \tag{8.28a}$$

and

$$\tau_A \approx \exp\left(-\frac{\alpha L_c}{\cos(\theta_A)}\right) \quad \text{where} \quad \theta_A \equiv 90° - 0.5778\beta + 0.002693\beta^2 \quad (8.28b)$$

Now it is possible to find $\|\alpha\tau\|$ and hence the amount of solar energy transferred to the device with one more approximation. Detailed calculations are required to find $\|\alpha\tau\|_B$, $\|\alpha\tau\|_D$ and $\|\alpha\tau\|_A$ for each component of the radiation where we now have to consider reflection from the absorber surface to the cover plate and back. Instead, it is found that each is given approximately by the product of the transmittance through the cover multiplied by the absorbance of the absorber plate as discussed above. This may be expected since the materials used in the device would be optimized. For example, the glass cover material would be made to transmit as much radiation as possible and the absorber material would absorb almost all of the radiation. Certainly, this would not apply in general, yet, using this approximation one can write

$$\|\alpha\tau\| P_D(\beta) \approx \alpha\tau_B B(\beta) + \alpha\tau_D D(\beta) + \alpha\tau_A A(\beta) \quad (8.29)$$

Example 8.3

Compare the transmittance for the beam, diffuse and albedo components for the standard conditions listed in Table 8.1.

The beam component has already been calculated in Example 8.2; now we calculate the transmittance for the diffuse and albedo components with eqns (8.28a) and (8.28b) for the standard conditions given in Table 8.1. The results of the calculation are given in Table 8.5 and, remarkably, the direct beam's transmittance is the lowest! However, the beam component is usually the largest component of the total insolation so is most likely the largest contributor.

Table 8.5 Values of the transmittance for the direct beam, diffuse and albedo components through the glass cover under the standard conditions given in Table 8.1.

variable	value
τ_B	0.9035
τ_D	0.9747
τ_A	0.9666

We are now in a position to model the flat plate solar energy collector since the amount of radiation that can reach the absorber plate and heat the flowing water can be determined. However, this is the amount of energy coming into the flat plate solar collector and there will be heat loss terms that must be considered.

Reference to Fig. 8.7 shows that there are two rates of heat transfer through the top of the system: that from the plate to the cover, $\dot{q}_{pc}$, and that from the cover to the surrounding atmosphere, $\dot{q}_{ca}$. These two rates *must* be equal, at steady state operation, since they operate in series, allowing us to write

$$\dot{q}_{top} = \dot{q}_{pc} = \dot{q}_{ca} \quad (8.30)$$

where $\dot{q}_{top}$ is the rate of heat transfer from the top of the device. This is an important constraint to consider when designing the system and allows one to determine the heat flow from the top. There are also two heat transfer rates that operate in parallel within $\dot{q}_{pc}$ and $\dot{q}_{ca}$, which are convective and radiative heat transfer, given by

$$\dot{q}_{pc} = h_{pc}[T_p - T_c] + \frac{\sigma_S[T_p^4 - T_c^4]}{\dfrac{1}{e_p} + \dfrac{1}{e_c} - 1} \tag{8.31a}$$

and

$$\dot{q}_{ca} = h_{ca}[T_c - T_a] + \sigma_S e_c[T_c^4 - T_{sky}^4] \tag{8.31b}$$

where h_{pc} is the convective heat transfer coefficient for the air between the absorber plate and cover, h_{ca}, the convective heat transfer coefficient between the cover and air, T_p, the absorber plate temperature, T_c, the cover temperature, T_a, the air or ambient temperature, σ_S, the Stefan-Boltzmann constant, e_p, the (low energy or long wavelength) emissivity of the absorber plate and e_c, the emissivity of the cover. The curious form for radiative heat transfer in eqn (8.31a) comes from an analysis involving radiative energy exchange between the absorber plate and cover whose derivation can be found in the references at the end of this chapter.[8]

[8] Radiative heat transfer between bodies is dictated by *view factors* that indicate the relative amount of radiation intercepted by the two surfaces. The result of the calculation for two parallel plates results in the radiative heat transfer part in eqn (8.31a) and for a differential area surrounded by a hemisphere in the equivalent term in eqn (8.31b).

The same assumption will be made for heat transfer from the bottom of the system as was made in the previous section: the major resistance will be heat transfer through the insulation. Again, since heat transfer for processes in series is limited to the lowest value, here we assume that heat flow through the insulation rather than from the device to the air is limiting, so, this will represent the heat flow from the bottom ($\dot{q}_{bot}$) written as

$$\dot{q}_{bot} = \frac{k_I}{L_I}[T_p - T_a] \tag{8.32}$$

where it is assumed that the top of the insulation is at the plate temperature and the outside of the device is at the ambient temperature, both reasonable assumptions, as discussed in the previous section.

The First Law of Thermodynamics or FLOT can now be written

$$\dot{m}C_{pw}[\Delta T] = [\|a\tau\| P_D(\beta) - \dot{q}_{top} - \dot{q}_{bot}]A_D \quad \text{operating line} \tag{8.33}$$

and is the operating line for the flat plate solar energy collector. The temperature difference ΔT is $T_{out} - T_{in}$.

Further details are required to determine the operating line. The heat transfer coefficient for the air between the absorber plate and cover (h_{pc}) will be discussed first and is dictated by the Rayleigh number that is used to describe *natural convection* under the influence of a temperature difference. If air is confined between two flat plates at different temperatures, that are placed at an angle from the horizontal, then the air will move up along the hot plate, turn around and fall back along the cool plate. The air circulating motion increases the heat transfer rate and is influenced by the temperature difference, the gap between the two plates and the material's physical properties. The Rayleigh number Ra is used to gauge the effect of recirculation and written as

$$Ra = \frac{g\chi\rho^2 C_p}{k\mu}\Delta T_{pc} L_{pc}^3 \equiv C_{Ra}\Delta T_{pc} L_{pc}^3 \tag{8.34}$$

where g, is the gravitational constant, χ, the thermal expansion coefficient, ρ, the density, C_p, the heat capacity, k, the thermal conductivity, μ, the viscosity, ΔT_{pc}, the temperature difference between the two plates (in our case this will be $T_p - T_c$), L_{pc}, the gap between the absorber plate and cover and C_{Ra}, the group of physical constants that is given in Appendix B. The smaller the Rayleigh number the less heat transfer occurs.[9] Note that the air properties are determined at $T_m \equiv \frac{1}{2}[T_p + T_c]$, which is typical of heat transfer calculations.

[9] Have you ever wondered how fiberglass insulation works? The fibers prevent air movement to reduce natural circulation of air between a hot and cold wall on the side of your house, for example. The thermal conductivity of glass is about 40 times greater than air and so this increase in conduction is sacrificed by the great reduction in overall heat transfer through elimination of air movement. Note that the amount of glass in the insulation is actually quite small and is of order 0.5 mass% which greatly reduces its influence in heat transfer.

The Nusselt number used to find h_{pc} can be determined with this correlation,

$$Nu \equiv \frac{h_{pc}L_{pc}}{k} = 1 + 1.44\left[1 - \frac{1708}{Ra\cos(\beta)}\right]^{+} \times \left[1 - \frac{1708\sin(1.8\beta)^{1.6}}{Ra\cos(\beta)}\right] + \left[\left\{\frac{Ra\cos(\beta)}{5830}\right\}^{1/3} - 1\right]^{+} \quad \text{for } 0 \le \beta \le 60^\circ \text{ and } Ra \le 10^5 \tag{8.35}$$

with the superscript + meaning that if the quantity in the brackets is negative then the entire term is zero. As before, the heat transfer coefficient between the cover and air h_{ca} can be determined with eqn (8.6) or (8.7).

The *method of solution* to find the operating point is an iterative one. This is because the temperatures within the device, such as the absorber plate temperature, T_p, and cover temperature, T_c, are not know *a priori*, and this influences the physical properties of air. A suggested method is:

Operating line

(1) Know: All the parameters in Table 8.1
(2) Determine: $\|a_T\|P_D(\beta)$ using eqn (8.29)
(3) Assume: T_p value
(4) Iterate: Change T_c until $\dot{q}_{pc} = \dot{q}_{ca}$ (eqns (8.31a) and (8.31b)) which will involve calculating the heat transfer coefficients that will depend on temperature. Use the equations in Appendix B to find the effect of temperature on the properties of air, which should be evaluated at the mean temperature at each position (*i.e.* $T_m = \frac{1}{2}[T_p + T_c]$ for heat transfer from the plate to cover and $T_m = \frac{1}{2}[T_c + T_a]$ for heat transfer from the cover to air)
(5) Calculate: $\dot{q}_{bot}$ from eqn (8.32)
(6) Calculate: ΔT from eqn (8.33)
(7) Repeat: Go to item 3 and assume another T_p value to calculate the operating line for various values of T_p

The following example demonstrates how to perform the calculation.

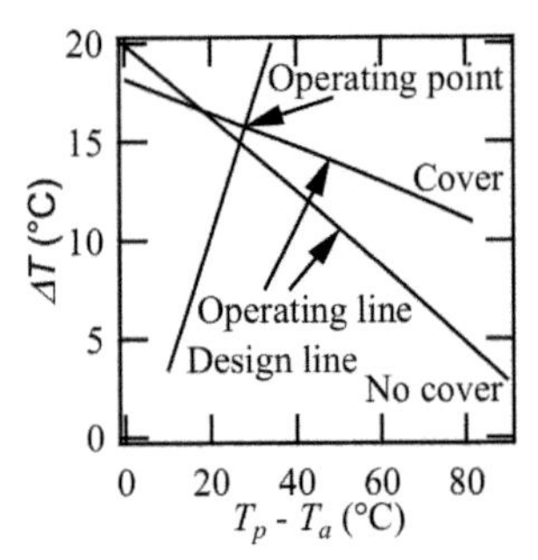

Fig. 8.11 Operating and design lines for the water temperature rise ΔT in a solar hot water heater as a function of $T_p - T_a$. The standard conditions were used as given in Table 8.1 and no cover and one cover were used.

Example 8.4

Find the operating point of a flat plate solar collector using the standard conditions given in Table 8.1. Assume there is a cover and compare the result to that found with no cover in Example 8.1.

The *design line* is the same as that found in Example 8.1 and it is reproduced in Fig. 8.11. The *operating line* is calculated using the standard conditions given in Table 8.1 and the transmittance values found in Example 8.3. However, the standard conditions have only a direct beam component and so only τ_B is required, which was initially calculated in Example 8.2. This allows one to determine $\|a\tau\|P_D(\beta)$ from eqn (8.29) quite easily using τ_B and the absorber plate absorbance.

Rather than stating the various equations that have to be considered in the above text, outlined on the previous page, the results of the calculation are given in Table 8.6 for the reader to compare to their own calculation. The operating line for the standard conditions, using eqn (8.6) for the heat transfer coefficient h_{ca}, is plotted in Fig. 8.11.

Some comments should be made about the results presented in Table 8.6. The very top row of data has both T_p and T_c arbitrarily set to T_a, the ambient air temperature, and represents the condition where the (almost) maximum value of the useful heat $\dot{q}_u$ can be obtained. The next four rows of results, between the double horizontal lines, allow T_c to vary for a given T_p so that $\dot{q}_{pc} = \dot{q}_{ca}$, the result that follows our procedure discussed above. Since $T_c < T_a$ there will be convective heat transfer from the air to the cover while there is radiative heat transfer from the cover to the *sky*, which has an effective temperature lower than ambient. This is why T_c can be smaller than T_a. The net result is that $\dot{q}_{ca} \equiv 0$ for the data in row 2 and is the lowest temperature that T_c can have. Thus, $\dot{q}_{pc}$ is also identically zero since the condition of $\dot{q}_{pc} = \dot{q}_{ca}$ must be followed and so $T_p \equiv T_c$. Amazingly, since $T_p < T_a$ there is heat transfer from the surroundings through the bottom of the device and $\dot{q}_{bot} < 0$! The end result is that $\dot{q}_u$ is slightly higher than that in row 1!

In reality, achieving the conditions given between the double horizontal lines in Table 8.6 will be difficult and a very large mass flow rate of water would have to be used, as will be discussed in the next section. Furthermore, whether these conditions can be achieved at all are questionable. The Earth, and the house upon which the device may sit, will be radiating heat too and since the rates of heat transfer are so small under these conditions these perturbations will most likely overcome the very isolated conditions upon which the calculation is made. Regardless, one can start calculations at the very smallest value that T_c can have, $T_{c,min}$, given by

$$h_{ca}[T_{c,min} - T_a] = \sigma_S e_c[T_{c,min}^4 - T_{sky}^4] \tag{8.36}$$

Of course, this value for T_c necessitates that $\dot{q}_{top} = 0$ and so one finds $T_p = T_{c,min}$ which can be the temperature where the calculation be-

Table 8.6 Temperatures and heat flows for the flat plate solar collector with a cover using the standard conditions given in Table 8.1.

T_p (°C)	ΔT (°C)	T_c (°C)	$\dot{q}_{top}$ (W/m^2)	$\dot{q}_{bot}$ (W/m^2)	$\dot{q}_u$ (W/m^2)
10.0	18.2	10.0	0.0	0.0	764.2
2.516	18.3	2.516	0.0	−3.9	768.0
10.0	17.9	3.8	14.4	0.0	749.8
20.0	17.2	6.1	38.5	5.2	720.4
30.0	16.4	8.5	65.5	10.4	688.3
35.8	15.9	10.0	82.1	13.4	668.7
38.2	15.8	10.6	89.1	14.7	662.8
40.0	15.6	11.1	94.5	15.6	654.1
50.0	14.7	13.8	125.2	20.8	618.1
60.0	13.8	16.7	157.7	26.0	580.4
70.0	12.9	19.6	191.6	31.2	541.2
80.0	11.9	22.7	227.1	36.4	500.7
90.0	10.9	25.8	263.9	41.6	458.6

gins. This will also make $T_p < T_a$ adding heat to the system through the bottom!

The above discussion seems pedantic, however, when the reader begins designing these devices using basic heat transfer calculations and then reads the literature, it becomes very confusing. Doing the above detailed calculations and only considering the information for $T_c \geq T_a$ as relevant to the design is a good way of reconciling differences in design procedures.

Now, the data beginning in row 6, just below the second double line, represents that where $T_c \geq T_a$. The first of these rows has $T_c = T_a$ and one can see at this condition T_p = 35.8 °C, which is quite high. So, if the cover temperature is forced to equal the air then the absorber plate is actually quite warm. The next row, which is shaded, is the operating point and the following rows are the result of further calculation.

The intersection of the design line and operating line is the *operating point*, which is found to be at T_p = 38.2 °C and ΔT = 15.8 °C. The operating point is close to that for the system with no cover which was T_p = 36.9 °C and ΔT = 14.9 °C.

The cover does not seem to improve operation much. However, the amount of useful heat $\dot{Q}_u$ calculated using eqn (8.2) does increase with the cover from 3759 W with no cover to 3977 W with a cover, a 6% increase. Thus, 6% more energy will be delivered to the water during any given time period and is substantial.

The efficiency of the flat plate solar collector can be found from

$$\eta = \frac{\dot{Q}_u}{P_D(\beta)A_D} = \frac{\dot{m}C_{pw}\Delta T}{B_{ter}\cos(\theta_S)A_D} \tag{8.37}$$

With the efficiency defined this way, the device that has a cover is more efficient at 70.5% compared to 66.7% without a cover. Yet, the cover allows less radiation to reach the absorber plate since $\|a\tau\| = 0.8132$, compared to the system operating with no cover where one simply has absorbance at $a = 0.9$. However, the absorbed radiation is treated more efficiently when a cover is present since the heat losses are reduced. The heat loss through the top of the device having no cover at its operating point is 204 W/m^2 compared to 82.9 W/m^2 when a cover is used. One can now appreciate how the cover reduces heat transfer to promote the *greenhouse effect.* One should also appreciate that using a cover that minimizes reflective losses and has low emissivity as well as low absorbance is a must to ensure efficient operation.

Finally, to conclude the discussion, we consider the efficiency of these devices, which is quite high and of order 70%. The challenge though is that the quality of the energy it produces, hot water, is not that great. If one could produce high pressure steam or electricity at this efficiency a major technological breakthrough would be had. Nevertheless, the ability to make hot water for consumer and industrial use cannot be denied and these devices should be seriously considered for use.

8.3 The effect of mass flow rate and device area

In order to understand the effect of mass flow rate and device area on the operating point of a flat plate solar energy collector, eqn (8.20) is modified slightly by writing it in a more general form to include transmittance through the cover,

$$\Delta T_{max} = \frac{\|a\tau\| P_D(\beta) A_D}{\dot{m} C_{pw}} \tag{8.38}$$

This equation, which neglects heat losses, will be used to understand how $\dot{m}$ and A_D affect the operating point. Since heat losses are not included, the design equation is not required and ΔT_{max} represents the operating temperature rise in the device.

Now one can guess that as the mass flow rate increases, the temperature rise will decrease while if the area is increased the temperature rise will increase. This is certainly what will be observed, however, direct proportionality to either variable is not found. Interestingly, eqn (8.38) is a reasonable approximation to the temperature rise found in a device having a cover since heat losses are minimized in the operating device.

First consider the effect of mass flow rate on device performance, particularly for the devices having a cover. Consider the data given in Table 8.7 and the effect of increasing the mass flow rate from 0.06 kg/s (approximately 1 gal/min) to 0.12 kg/s, given in the upper half of the table (A_D = 6 m^2). As expected ΔT decreases when $\dot{m}$ is increased, however, it does not halve when $\dot{m}$ is doubled. Heat transfer considerations prohibit this. This is due to an increase in h_w from more efficient heat transfer and from the absorber plate temperature being lower, which reduces radiant energy losses when $\dot{m}$ is larger. Glancing at other trends in the data for the larger area systems reveals similar observations. Also, there is a slight increase in $\dot{Q}_u$ due to favorable heat transfer conditions at the larger water flow rate. For reference, the design and operating lines are shown in Fig. 8.12 when $\dot{m}$ is doubled from the standard condition.

Values of ΔT_{max} are also given in the table. As long as the system has a cover and the operating temperature is not too high then ΔT_{max} is a good first approximation to the temperature rise one may expect in a solar hot water heater. This is because the low operating temperature, that approaches the air temperature and cover reduces heat losses, which forces the operating point to move closer to ΔT_{max}. So, one can estimate the temperature rise and useful heat to within 20% by knowing a minimal number of parameters for the system, allowing a good first design to be made.

Does an increase in the mass flow rate raise the operating cost for the system since the pump used to force water through riser tubes must consume more energy? The mechanical work or rate of energy required to drive the pump $\dot{W}_m$ can be found from the FLOT in an intermediate form, as given in eqn (3.11), by assuming that the riser pipes are horizontal, there is no velocity change and there is no temperature change (so the internal energy change for a liquid is essentially zero as they are basically incompressible) to be $\dot{W}_m = \dot{m}\hat{V}_w \Delta P = [\dot{m}/\rho_w]\Delta P$, $\hat{V}_w$ is the specific volume of water. Assuming laminar flow, which is a good assumption for a residential system, one can use the definition of the friction factor (eqn (8.14)) and the friction factor–Reynolds number correlation for laminar flow (eqn (8.15a)) to find the pressure drop and ultimately $\dot{W}_m$,

$$\dot{W}_m = \frac{128}{\pi} \frac{\mu_w}{\rho_w^2} \frac{L_p}{D_p^4} \dot{m}^2 \approx 4 \frac{\text{m}^2}{\text{kg-s}} \dot{m}^2 \tag{8.39}$$

if all other flow loss terms, other than those in the riser pipes, are neglected. The approximation represents an equation for the standard conditions where physical parameter and dimensional values have been inserted in the equation. Since $\dot{m}$ is of order 0.1 kg/s the power is extremely small, less than 1 W. Even accounting for other pressure losses in the system, like flow through valves and the header, and using a heat transfer fluid other than water, which can have a viscosity 10–100 times that of water, the power required to drive the pump is minimal. So, using a higher water mass flow rate is not economically or energetically limiting. However, if there is an electrical power outage the system will not work!

Table 8.7 The effect of mass flow rate and device area on the temperature rise in a flat plate solar energy collector with and without a cover. The numbers in parentheses after the temperature rise are the useful heat $\dot{Q}_u$ in kW. The standard condition for the mass flow rate and area are 0.06 kg/s and 6 m^2, respectively, and are given in these units in the table. All other conditions are given in Table 8.1.

Variable	No cover	Cover
$\dot{m}/A_D$	0.06/6	0.06/6
ΔT(°C)	14.9 (3.75)	15.8 (3.98)
ΔT_{max}(°C)	20.2 (5.08)	18.2 (4.58)
$\dot{m}/A_D$	0.12/6	0.12/6
ΔT(°C)	8.0 (4.02)	8.2 (4.13)
ΔT_{max}(°C)	10.1 (5.08)	9.1 (4.58)
$\dot{m}/A_D$	0.06/12	0.06/12
ΔT(°C)	26.8 (6.74)	29.7 (7.48)
ΔT_{max}(°C)	40.4 (10.1)	36.4 (9.16)
$\dot{m}/A_D$	0.12/12	0.12/12
ΔT(°C)	14.9 (7.50)	15.8 (7.95)
ΔT_{max}(°C)	20.2 (10.2)	18.2 (9.16)

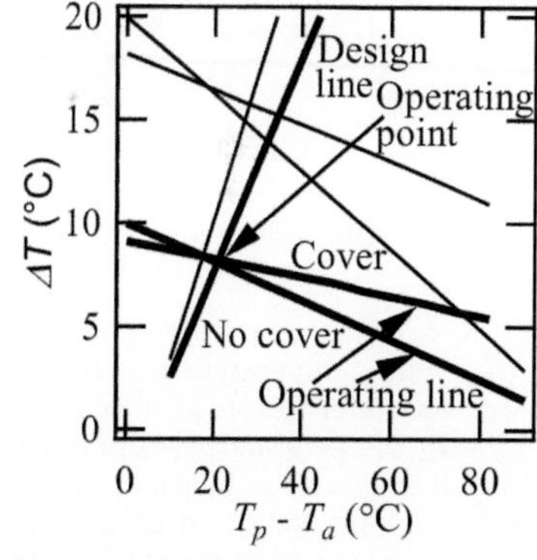

Fig. 8.12 The effect of increasing the mass flow rate from the standard condition of 0.06 kg/s (see Table 8.1) to 0.12 kg/s. The thick lines are for a mass flow rate of 0.12 kg/s while the fine lines are for a mass flow rate of 0.06 kg/s.

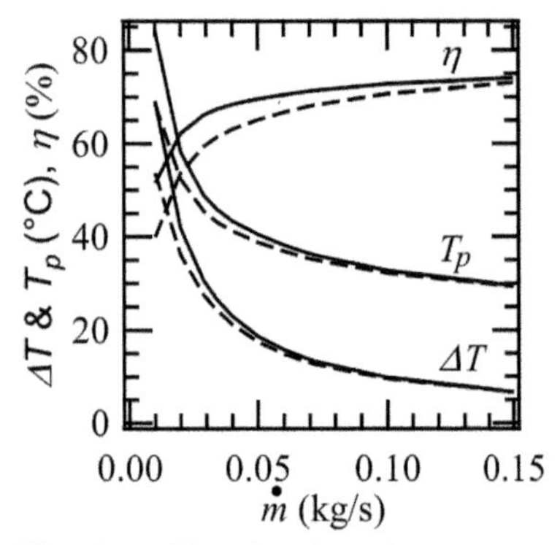

Fig. 8.13 The absorber plate temperature (T_p), flowing water temperature rise (ΔT) and efficiency (η) for a solar hot water heater as a function of water mass flow rate ($\dot{m}$). The standard conditions were used as given in Table 8.1, except $\dot{m}$ was varied and no cover (dashed lines) and one cover (full lines) were assumed.

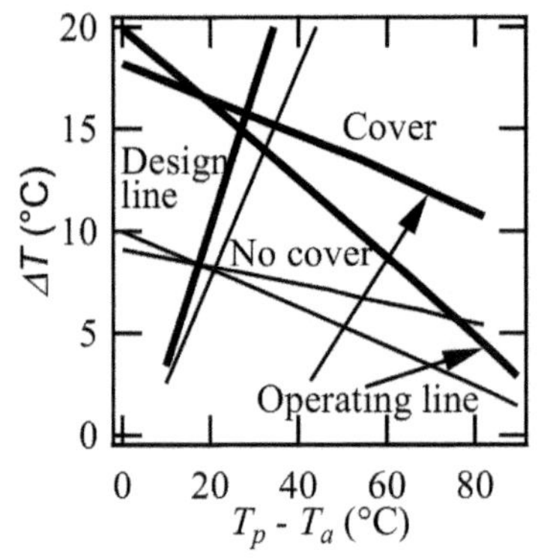

Fig. 8.14 Design and operating lines for the water temperature rise ΔT in a solar hot water heater as a function of $T_p - T_a$. The standard conditions were used as given in Table 8.1, except $\dot{m}$ was 0.12 kg/s and no cover and one cover were assumed to generate the fine lines. The results are the same as in Fig. 8.12. The thick lines are the result of a calculation that has $\dot{m}$ equal to 0.12 kg/s and with the device area A_D doubled to 12 m^2.

The effect of mass flow rate on ΔT, T_p and the efficiency η is shown in Fig. 8.13. Several observations can immediately be made. As already discussed ΔT falls in value upon an increase in $\dot{m}$, as does T_p. Since the overall system temperature decreases, the efficiency improves due to a reduction in heat losses from the system. Ultimately, at infinite mass flow rate, the efficiency would achieve the theoretical value, $\dot{Q}_u/P_D(\beta)A_D$, which is equal to $\|a\tau\|$. For the standard conditions with a cover this ultimate efficiency is 0.813.

Should the area be doubled, two things occur (see Table 8.7). Firstly, the overall area is increased to allow more insolation for the devices to use and this immediately increases ΔT. Secondly, the number of riser pipes operating in parallel increases from 16 to 32 which reduces the velocity of water in the risers. The temperature rise does not double though nor does the useful heat.

The reason is twofold. The lower velocity in the larger area system reduces the heat transfer coefficient from the riser pipe wall to the flowing water which reduces ΔT and $\dot{Q}_u$. Furthermore, the lower velocity forces the system to operate at a higher absorber plate temperature which increases radiative heat transfer. Both these effects inhibit a direct doubling of ΔT and $\dot{Q}_u$, as can be seen in the data in Table 8.7.

The result obtained from calculating the operating and design lines for the system having doubled the device area and mass flow rate from the standard conditions is given in Fig. 8.14. Firstly, the thin lines in the figure are the same as those in Fig. 8.12, where the only change from the standard conditions is that the mass flow rate of water was increased to 0.12 kg/s. Clearly, doubling the area increases ΔT as well as T_p, as discussed above. The slope of the operating line also increases, which is the result of increased radiative heat transfer for the elevated system temperature.

8.4 Recirculation of water

The temperature rise of water from a flat plate solar energy collector is small and of order 20 °C which is not hot enough for residential hot water which is delivered at approximately 50°C. This shortcoming is overcome by recirculating the water within a tank, as shown in Fig. 8.15. The useful heat is constantly delivered to the water in the system and it gradually rises in temperature with time. How is the temperature rise found?

The strict answer to this question is that coupled differential equations would have to be solved; one for the flat plate solar collector and the other for the tank. The differential equation is the Unsteady State First Law of Thermodynamics (US-FLOT) which is written

$$\frac{d(m_0\hat{H})}{dt} + \dot{m}\times\left[\hat{H}_2 + \frac{1}{2}v_2^2 + gh_2\right] - \dot{m}\times\left[\hat{H}_1 + \frac{1}{2}v_1^2 + gh_1\right] = \dot{Q} + \dot{W}_m \quad (8.40)$$

where m_0 is the total mass of water in the system, $\hat{H}$, the enthalpy, v, the velocity of fluid entering the system, g, the gravitational acceleration, h, the height, $\dot{Q}$, the sum of all heat flows to and from the system and $\dot{W}_m$, all mechanical work terms. The subscripts '1' and '2' are for conditions entering and exiting the system, respectively. The form of this equation would make the operating line a function of time, which is difficult to analyze.

This challenge is simplified by making the *pseudo-steady state approximation* or PSSA. The PSSA is frequently used for reaction sequences where one reaction is much faster than another. Consider this reaction sequence

$$A \rightleftharpoons B \qquad \text{(slow reaction)} \qquad [1]$$
$$B + C \rightleftharpoons D \qquad \text{(fast reaction)} \qquad [2]$$

If reaction [2] is much faster than [1], the reactants and product concentration in [2] can be assumed to be constant and it is at equilibrium. The kinetics for the much slower reaction [1] can then be solved.

For the system at hand, the flat plate solar energy collector and the water storage tank, the collector has much faster kinetics than the tank since the mass of water in the pipes within the collector is much less with a much faster average velocity than that in the tank. So, the operating point for the collector will not change, as a first approximation, for a given temperature change of the water stored within the tank. Note though that the water input temperature to the collector will slowly vary which could affect the operating point since the physical properties of water will change. This will be ignored here and the operating point for the collector having a water input temperature at the standard condition, 15°C as a stand-alone system will be used.

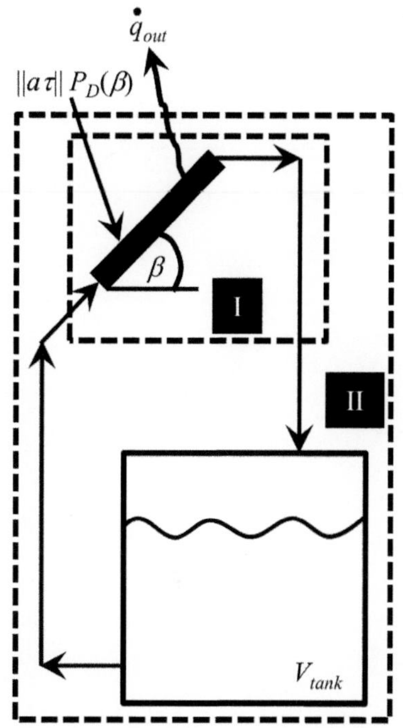

Fig. 8.15 Schematic of a flat plate solar energy collector with a recycle tank of volume, V_{tank}. System I is the flat plate device while System II is the entire system, which has $\|a\tau\|P_D(\beta)$ energy flow entering and $\dot{q}_{out}$ exiting.

The energy input $\|a\tau\|P_D(\beta)$ and output $\dot{q}_{out}$, which is the sum of $\dot{q}_{top}$ and $\dot{q}_{bot}$, are shown in the figure and it is their difference multiplied by the device area A_D which gives the useful heat, written as

$$\dot{Q}_u = \dot{m}C_{pw}\Delta T = 0.06 \text{ kg/s} \times 4195 \text{ J/kg-K} \times 15.8 \text{ °C} = 3977 \text{ W} \qquad (8.2)$$

for the standard conditions with a single cover. This is System I in the figure to give the rate of energy added to the tank.

Now one can apply the US-FLOT, eqn (8.40), to the System II, noting there is no mass flow into or out of the system nor mechanical work, to arrive at

$$\frac{d(m_0\hat{H})}{dt} = \dot{Q}_u$$

If a pump were used to recirculate the water then there would be a work term included in the above equation. The enthalpy $\hat{H}$ is written as $C_{pw}[T - T_0]$, where T is the temperature at any time and T_0 is the initial temperature taken as the reference temperature for the enthalpy (*i.e.* the enthalpy is assumed to be zero at T_0). Substituting this in the equation it can then be easily integrated to yield

$$T = T_0 + \frac{\dot{Q}_u}{m_0 C_{pw}} t \tag{8.41}$$

This equation can be used to estimate the temperature of water in the tank or the amount of time it takes to reach a certain temperature. Inherent assumptions in arriving at this final equation are that the heat capacity of water is constant with temperature, a good assumption; the insolation is constant, probably not a good assumption over a day; and the water in the tank is well mixed so the temperature throughout the tank is homogeneous.

Example 8.5

Find the amount of time it takes to heat 250 kg of water in a tank that is initially at 15 °C to 50 °C for a flat plate solar energy collector using the standard conditions given in Table 8.1 Assume there is a cover.

One can take eqn (8.41) and rearrange it to

$$t = \frac{m_0 C_{pw}}{\dot{Q}_u}[T_f - T_0]$$

where T_f is the final temperature. The time to achieve the given temperature is easily calculated by

$$t = \frac{250 \text{ kg} \times 4195 \text{ J/kg-K}}{3977 \text{ W}}[50 - 15] \text{ K} = 9230 \text{ s} = 2.6 \text{ h}$$

This is an incredibly short time! In reality the insolation will be below 1000 W/m^2 and not constant, which will increase the time. Typically it takes a day or so for the tank to achieve operating temperature.

8.5 Stagnation temperature

Equation (8.41) shows that the temperature will continue to increase with time, which can be a problem, especially when the water does not flow to put some heat load on the collector. At this condition, when water is not flowing, the device will increase in temperature and reach a steady-state temperature called the *stagnation temperature*. This can be a challenge to the materials used and could warp or otherwise damage the device, as well as either boiling the water or chemically degrading the heat transfer fluid if water is not used. The stagnation temperature, T_{stag}, is found by balancing the various heat flows which can be written

$$\|a_T\| P_D(\beta) = \dot{q}_{top} + \dot{q}_{bot} \tag{8.42}$$

Consider $\dot{q}_{top} = \dot{q}_{pc} = \dot{q}_{ca}$ first, the two components are

$$\dot{q}_{pc} = h_{pc}[T_{stag} - T_c] + \frac{\sigma_S[T_{stag}^4 - T_c^4]}{\dfrac{1}{e_p} + \dfrac{1}{e_c} - 1} \quad (8.31a)$$

and

$$\dot{q}_{ca} = h_{ca}[T_c - T_a] + \sigma_S e_c[T_c^4 - T_{sky}^4] \quad (8.31b)$$

The bottom heat transfer rate is

$$\dot{q}_{bot} = \frac{k_I}{L_I}[T_{stag} - T_a] \quad (8.32)$$

One assumes that when solving the above equations $T_p = T_{stag}$, which is a reasonable approximation since the absorber plate will be at the highest temperature and, at equilibrium, all parts of the device will achieve the same temperature. The method of solution is given below.

Stagnation temperature

(1) Know: All the parameters in Table 8.1

(2) Determine: $\|a\tau\|P_D(\beta)$ using eqn (8.29)

(3) Assume: T_{stag} value

(4) Iterate: change T_c until $\dot{q}_{pc} = \dot{q}_{ca}$ (eqns (8.31a) and (8.31b)) which will involve calculating the heat transfer coefficients that will depend on temperature. Use the equations in Appendix B to find the effect of temperature on the properties of air, which should be evaluated at the mean temperature at each position (*i.e.* $T_m = \frac{1}{2}[T_{stag} + T_c]$ for heat transfer from the plate to cover and $T_m = \frac{1}{2}[T_c + T_a]$ for heat transfer from the cover to air)[10]

(5) Calculate: $\dot{q}_{bot}$ from eqn (8.32)

(6) Validate: if eqn (8.42) is satisfied

(7) Check: if validation is not obtained then go to item 3 and assume another T_{stag} until validation occurs; otherwise the solution is found

[10] One can use the correlations in Table 9.3 for temperatures substantially above 100 °C rather than the correlations in Appendix B.

The stagnation temperature can be quite high and approach 200 °C! The next example shows this is, in fact, true.

Example 8.6

Find the stagnation temperature using the standard conditions given in Table 8.1 Assume there is a cover.

We follow the method of solution given above to find the stagnation temperature. First the insolation absorbed by the system is $\|a\tau\|P_D(\beta) =$ 764.2 W. If we define $\dot{q}_{out} = \dot{q}_{in} + \dot{q}_{bot}$ we can iterate until $\dot{q}_{out} = \|a\tau\|P_D(\beta)$, with the results shown in Table 8.8

Table 8.8 The absorber plate temperature, cover temperature and heat transfer rate out of the system to determine the stagnation temperature of a flat plate solar energy collector having one glass cover under the standard conditions given in Table 8.1.

T_{stag} (°C)	T_c (°C)	$\dot{q}_{out}$ (W/m^2)
170	53.5	700.2
190	61.4	820.8
180	57.4	758.9
181	57.8	764.9

First an absorber plate temperature of 170°C was assumed and $\dot{q}_{out}$ was too low so the plate temperature was increased to 190°C, which was found to be too high as $\dot{q}_{out}$ was now too large. An intermediate temperature was assumed which was much closer and after one more iteration a temperature of 181°C represents a fairly accurate solution. The stagnation temperature is very high and materials selection should be made to ensure that they can withstand such temperatures.

8.6 Conclusion

In this chapter the operation of a flat plate solar energy collector has been considered. A considerable deviation from the previous literature has been made to incorporate engineering heat transfer calculations, rather than correlative techniques which have been presented in the past. The reason for this was given in the introduction. The previous technique has its roots in the pioneering work of Hottel from the 1940s and 1950s which has served well for over 60 years. As new materials and designs are developed a more flexible technique may have to be used which has served as motivation for this chapter.

The efficiency of these devices is impressive: greater than 50%. If any other energy generating device could work at this efficiency level, this would be a major technological breakthrough. Of course, turbines at the bottom of dams are extremely efficient, approximately 90%, however, there is only a limited amount of hydroelectric power available. The energy developed from flat plate collectors is relatively low quality as it makes low temperature water and not high pressure steam. If it could be developed to boil water at high pressure then this would represent a major breakthrough. Of course this is the subject of the next chapter where the insolation is concentrated to do just that and a solar-thermal power plant is discussed. Regardless, a flat plate solar energy collector is a very useful device and has great utility.

8.7 General references

J.A. Duffie and W.A. Beckman, 'Solar energy of thermal processes,' John Wiley and Sons, 3rd Edition (2006)

M. Holladay, 'Solar thermal is dead,' GreenBuildingAdvisor.com, March 23, 2012

K.G.T. Hollands *et al.* 'Free convection heat transfer across inclined air layers,' J. Heat Trans. (ASME) **98** (1976) 189

S.A. Kalogirou, 'Solar energy engineering: Processes and systems,' Elsevier (2009)

S. Middleman, 'An introduction to mass and heat transfer,' John Wiley and Sons (1998)

J.A. Palyvos, 'A survey of wind convection coefficient correlations for building envelope energy systems' modeling,' App. Thermal Eng. **28** (2008) 801.

Exercises

(8.1) Apply Snell's law for a 0° and 90° incidence angle and determine simple equations.

(8.2) Since heat transfer is so important to this chapter consider these observations and explain them in terms of heat transfer. Why does a piece of metal feel cold compared to a piece of wood? In a similar vein, why would a gemstone, like diamond or sapphire, feel colder than a piece of glass (*i.e.* SiO_2)? This later observation is supposedly a way to determine if someone is trying to sell you a piece of *cut glass* rather than the real thing.

(8.3) The surface of a space station orbiting the Earth has an absorbance of 0.1 and an emissivity of 0.6; determine the equilibrium temperature of the station when in the Sun. Assume the temperature of space for radiative heat transfer purposes is 0 K.

(8.4) If you want to minimize heat transfer from the space station when it is not in direct sight of the Sun, should its geometry be a cylinder or a sphere?

(8.5) Find the operating point for a flat plate solar energy collector that has one pipe that bends back and forth behind the absorber plate (serpentine arrangement) using the standard conditions given in Table 8.1. Assume its length is 16 × 2.5 m so the overall pipe length is the same as the standard conditions and that there is one cover.

(8.6) Prove that putting black paint on an absorber plate (like Aluminum) would yield a poor device. Black paint has a high energy (short wavelength) absorbance of 0.96 and a low energy (long wavelength) emittance of 0.88. Do this by performing a design using the standard conditions in Table 8.1 and there is no cover. Compare with the results in Example 8.1. A collector that does not have a cover is frequently used to heat swimming pool water.

(8.7) Assume the device in Example 8.1 is actually used to heat swimming pool water. The pool contains 70 m^3 of water, estimate the temperature rise in a day if there is 5 h of insolation (this time was chosen since the insolation in the example is 1000 W/m^2 which is high; in reality, the insolation will be smaller and for a longer time). Neglect direct heat transfer to and from the pool itself, just determine how much the solar water heater will heat up the pool. In general, the pool temperature will rise between 3 and 8 °C over several days of good weather, according to some manufacturers and installers.

(8.8) Use the standard conditions, with a cover, in Table 8.1 except use *air* as the working fluid rather than water and determine the operating point. What is the thermal efficiency of this device? Compare the results to those where water is the operating fluid.

(8.9) Find the steady-state stagnation temperature for a solar thermal hot water heater that has a white absorber plate, with all other design features the same as in Table 8.1 (see Example 8.6). The absorber plate is selective in nature with a high energy (short wavelength) absorbance of 0.2 and a low energy emissivity (long wavelength) of 0.9.

(8.10) Determine the stagnation temperature for the system in Example 8.6 if there is no cover and compare the two stagnation temperatures.

(8.11) A standard correlation given in the literature for the efficiency η of a flat plate solar energy collector is

$$\eta = \eta_o - \frac{k_1 \Delta\theta + k_2 \Delta\theta^2}{P_D(\beta)}$$

where η_o is the efficiency due to optical limitations and $\Delta\theta = T_p - T_a$. See, for example, the equation given by Quaschning (V. Quaschning, 'Renewable energy and climate change' J. Wiley & Sons (2010)) where $k_1 = 3.97$ W/m^2-K and $k_2 = 0.01$ W/m^2-K^2. Firstly, in our terminology, we can write $\eta_o = \|a\tau\|$ and represents the ultimate limit that the device could operate (see the discussion in Example 8.4 where this condition would have $T_p = 10$ °C). Secondly, the equation actually represents the operating line and only presents the possible efficiencies of the device. Use this equation to calculate the efficiency of the device as a function of $\Delta\theta$ and compare it to the results given in Example 8.4. Assume that $T_c \geq 10.0$ °C.

(8.12) Simon and Harlamert (NASA Technical Memorandum, NASA TM X-71427, 'Flat plate collector performance evaluation: The case for a solar simulation approach,' (1973)) give the following data for efficiency η of a solar flat plate collector as a function of $x \equiv [T_p - T_a]/P_D(\beta)A_D$, find the product $\|a\tau\|$ and an overall heat transfer coefficient U_o from the data. This can be done by taking eqn (8.33) and rearranging it to the form,

$$\frac{\dot{m}C_{pw}[T_{out} - T_{in}]}{P_D(\beta)A_D} \equiv \eta = \|a\tau\| - U_o \times x$$

U_o is an overall heat transfer coefficient made of a variety of terms and one can come to this form of equation by substituting the temperature difference in radiative heat transfer with a factored term: $T_2^4 - T_1^4 \equiv [T_2 - T_1] \times \{[T_2^2 + T_1^2][T_2 + T_1]\}$. The term in the curly brackets is relatively constant with temperature (prove it if $T_2 = 50$–100 °C and $T_1 = 25$ °C) and can be incorporated into U_o. Determine U_o with the variables that make up $\dot{q}_{pc}$ and $\dot{q}_{ca}$ in eqns (8.31a) and (8.31b), respectively, and eqn (8.32) for $\dot{q}_{bot}$. Calculate U_o assuming reasonable values for the terms (note $h_{pc} \approx 3$ W/m^2-°C) and compare it to the value you find from the experimental data; are they close? One must realize that $\dot{q}_{pc}$ and $\dot{q}_{ca}$ operate in series and an overall heat transfer coefficient from the absorber plate to the cover and then to the air U_{top} can be found from $\dot{q}_{top} = U_{top}[T_p - T_a]$, where $U_{top}^{-1} \equiv h_{pc}^{-1} + h_{ca}^{-1}$ with $\dot{q}_{pc} \equiv h_{pc}[T_p - T_c]$ and $\dot{q}_{ca} \equiv h_{ca}[T_c - T_a]$. Here h_{pc} and h_{ca} are heat transfer coefficients that incorporate the factored T^4 temperature differences as well as other terms. The top and bottom heat transfer rates operate in parallel so $U_o = U_{top} + k_I/L_I$. In the table x is given in units of 10^{-2} m^2-°C/W, so, the first value of x is 2.90×10^{-2} m^2-°C/W.

x	η	x	η	x	η
2.90	0.706	5.86	0.682	10.5	0.535
3.07	0.729	6.27	0.596	10.7	0.500
3.29	0.734	6.72	0.610	10.8	0.534
3.32	0.677	7.36	0.593	11.0	0.395
4.06	0.711	7.39	0.553	12.7	0.414
4.12	0.722	8.32	0.607	12.9	0.454
4.99	0.657	8.74	0.595	15.6	0.316
5.15	0.665	8.83	0.535	16.3	0.296
5.44	0.665	9.29	0.629	20.6	0.237
5.56	0.634				

Solar thermal energy generated electricity

Solar thermal energy generated electricity (STEGE) is produced similarly to a typical, coal-fired power station used today, with the only difference being the heat source. The power cycle for a contemporary power plant occurs by first pulverizing coal into a fine powder which is then ignited to generate heat within a furnace. Water is pumped through pipes within the furnace that are in thermal contact with the inferno and high pressure steam is produced. This is passed through a turbine that turns a generator to produce electricity. The exhaust from the turbine is low pressure steam that moves through a condenser in thermal contact with a heat sink to produce liquid water. The water is then pumped back through the boiler and the cycle is closed.

Of course a key factor in this technology is the heat source. Coal is used in approximately 50% of the power plants in the United States and 42% worldwide; other sources are oil, natural gas, nuclear energy and to a much lesser extent the Sun. All these energy sources perform the same function, *boil water*. A nuclear power plant does nothing else other than use a controlled thermonuclear reaction to boil water! This is an important point since nuclear power is frequently equated to a high technology energy source. Certainly, controlling the nuclear reaction and all the periphery instrumentation and equipment surrounding the nuclear reactor is fairly sophisticated, yet, what it does is merely boil water to turn a turbine; not high technology at all.[1]

[1] If one were to ask the average person how a nuclear power plant operates they would certainly indicate something akin to the power source used in science fiction stories like the Starship Enterprise that used *dilithium crystals*. A nuclear power plant merely uses a thermonuclear reaction to boil water.

A solar energy generated power plant will operate on the same principle as a coal-fired plant; why is high pressure steam necessary? It is required to turn a turbine, similarly to a person blowing on a pinwheel. The high pressure steam is forced over the blades of the turbine which is exhausted at a lower pressure and the lost energy is used to turn the shaft. The rotating shaft has permanent magnets on it, away from the blades of course, which are located within wound wire. The rotating magnetic field produces an electric field within the wires and so electricity is produced. This is a simple and cost effective way to produce electricity and the entire technology rests on producing a rotational motion. A hydroelectric plant and wind turbines do the same thing by using flowing water and the wind, respectively, to produce electricity. Indeed it has been proposed to put turbines on the seafloor and catch tidal currents to the same effect. Basically, linear motion is translated

Solar Energy, An Introduction, First Edition, Michael E. Mackay

Table 9.1 Energy distribution in a typical 500 MW (subcritical) pulverized coal-fired power plant. *Source:* L.C. White (1991), 'Modern Power Station Practice' in Volume G: Station Operation and Maintenance (1991) 3rd edition, British Electricity International, Pergamon Press, Oxford, UK.

Component	Percentage
Condenser loss	52.5
Electrical output	39.0
Boiler losses	5.5
Turbine losses	1.5
Works auxiliaries	1.0
Radiation losses	0.5
Total (input)	100

to rotational and the fact that the magnetic and electric fields are perpendicular to each other is used to make electricity.

The efficiency of the coal-fired power plant is reasonably good since it translates about 40% of the coal energy content to electrical energy; the distribution of energy use and loss in a typical power plant is shown in Table 9.1. Surprisingly, most energy is lost to the cooling part (condenser) of the process. This is a necessary part of the operation, as explained in Chapter 3, a cycle does not work unless heat is rejected. Regardless, over 50% of the energy is lost in this component to the process which is immense in absolute magnitude. Consider the rest of the losses to understand how much energy is lost to the condenser. Even though they are small and of order 1%, if one were to consider that this is a 500 MW power plant, a 1% loss means that there is 5 MW of power which is not being used! The percentage is small, yet, the absolute number is staggering. The average residential household in the United States uses electricity at a rate of 1.3 kW on average and so a 1% loss is equivalent to 3800 households, quite a large number. Fifty times more energy is rejected to the environment *via* the condenser.

Although power systems engineering is not the focus of this chapter, due to the similarity of all these technologies; *i.e.* coal, nuclear, gas, oil and solar, in producing electricity, the basic Rankine cycle is discussed. This will give the reader a grounding in power cycles and then the way energy (heat) is produced in a STEGE plant can be fully appreciated, as discussed in subsequent sections. Dimensionless numbers and correlations between dimensionless numbers will be used to design the devices in this chapter, if the reader is not familiar with these conceptsn reading Appendix C may help in understanding what they are and why they are used.

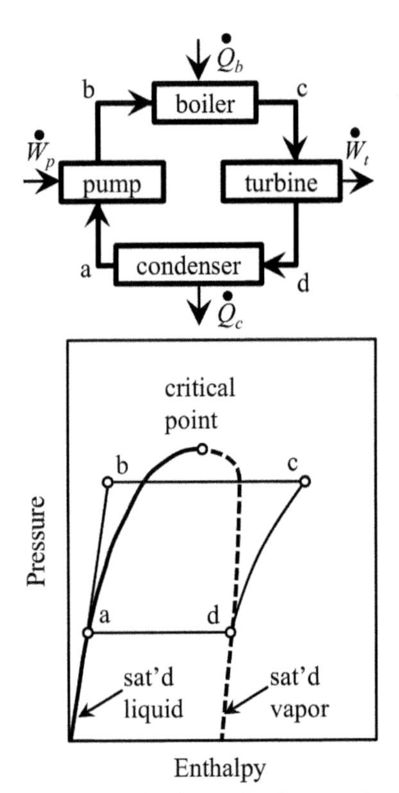

Fig. 9.1 The basic Rankine cycle. The upper figure shows the components making up the cycle while the lower figure shows the cycle on a pressure–enthalpy diagram whose shape is similar to that for water. Points a, b, c and d represent various states of water at the entrance to and exit from various components making the power cycle.

9.1 The Rankine cycle

The Rankine cycle is the cycle of choice to produce electricity and is shown in Fig. 9.1 in a pressure (P)–enthalpy (H) diagram. Water is the usual working fluid in the cycle. Although there are more details to the cycle than shown, such as a reheat component, this will be ignored here so the reader may understand the basics of the cycle and the concept behind its design. It differs from the basic Carnot cycle in important aspects which will be discussed within the text below.

The four components to the cycle are the pump, boiler, turbine and condenser. We will begin with the pump and work around the cycle explaining how the steps between each component occur and the reason why they are designed that way. The **state a** is saturated liquid, although the liquid may be cooled slightly below the saturation state in practice. A saturated liquid means it is just at its boiling temperature for the given pressure, yet, it is in the liquid state. If the temperature were increased very slightly then a bubble of vapor would be produced. If the temperature is decreased then it is technically called a compressed liquid. The normal boiling point of water is 100 °C at 1 atm. pressure so water at ambient conditions is a compressed liquid since the pressure required to boil water at 20 °C is very low (2.3 kPa ≈ 0.0023 atm). The reason one desires a liquid at **state a** is because the pump does not work well at all if it is a mixture of vapor and liquid. Furthermore, it does not pay to use energy to decrease the temperature of the water below the saturation temperature (boiling point) since this is wasted energy. Thus, the target is to have saturated liquid water at **state a**.

In the Rankine cycle the liquid is pumped *reversibly* and *adiabatically* to **state b** at a higher pressure and a required power input of $\dot{W}_p$. Reference to the Second Law of Thermodynamics (SLOT) in Chapter 3 shows that this means $\hat{S}_b = \hat{S}_a$, where $\hat{S}_i$ is the specific entropy of state i. This part of the cycle is isentropic, which is important in determining **state b**. A Carnot cycle would operate identically to drive the pump, as can be seen in Fig. 9.2, where the two cycles are graphed on a temperature (T)–entropy diagram (S).

The high pressure at **state b** is required so that the (compressed) liquid water can be forced through the pipes that circulate within the furnace that make up the boiler. The boiler operates at constant pressure for the Rankine cycle and heat at a rate of $\dot{Q}_b$ is supplied to boil the water. This is the heat flow supplied by burning coal or from the Sun. A Carnot cycle operates differently for this part of the process and heat transfer operates at constant temperature. This difference, constant pressure boiling of water rather than constant temperature, is the reason why the two cycles look so dissimilar in the T–S diagram. The only way the Carnot cycle can have constant temperature boiling of water is for the line b′-c′ to be within the vapor–liquid envelope. If it is not, then the temperature will rise as energy is supplied to either liquid water or water vapor (steam). As may be familiar to the reader, when water boils it occurs at constant temperature for an equilibrium process and all energy supplied to the water goes to achieving the phase transition. Before the phase transition, liquid water will increase in temperature while after the completion of the phase transition the vapor will increase in temperature. The line b-b′-c′-c in the Rankine cycle is boiling of water at constant pressure and so this line is an *isobar* which ends at **state c**, which is *superheated vapor* (compare Figures 9.1 and 9.2).

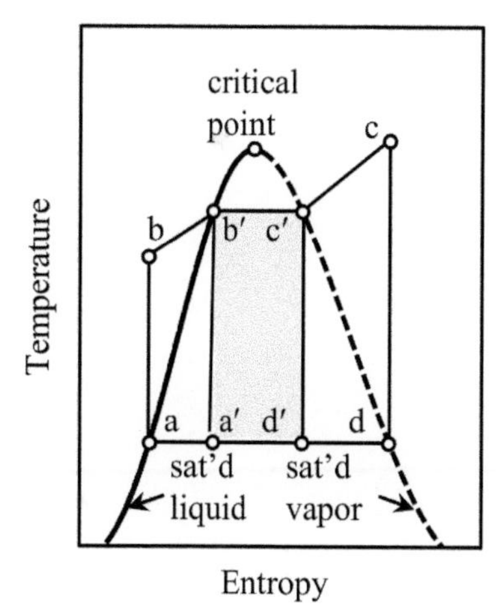

Fig. 9.2 Comparison of the ideal Rankine and Carnot cycles that operate between the same temperatures on a temperature–entropy diagram, whose shape is similar to that for water. The Rankine cycle operates in the cycle a-b-c-d and the Carnot cycle in the cycle a′-b′-c′-d′.

The high pressure, superheated vapor is forced through a turbine that makes electricity at a rate $\dot{W}_t$ along the *isentropic* (reversible, adiabatic) line c-d. The line c′-d′ for the Carnot cycle is a similar line in that it is isentropic too, however, it ends in the vapor–liquid envelope. This is undesirable since the **state d** should not have any liquid droplets in it which could damage the turbine blades should they hit them. So ideally, **state d** should be right at the saturated vapor line; in reality, the design usually has it slightly to the right of the line to ensure no water droplets are formed should the cycle vary in conditions during normal operation. In addition to **state d′** having water droplets which could damage the rotating blades, **state a′** in the Carnot cycle is a vapor–liquid mixture which is very difficult to pump, it is much easier to pump pure, liquid water. So, one can ascertain that the Carnot cycle is useful as a thermodynamic teaching tool, yet, its utility is negligible for power cycles due to its inherent engineering defects.

The final part of the process is condensation of the water vapor to liquid and to achieve **state a**. This occurs at constant pressure for the Rankine cycle and constant temperature for the Carnot cycle. If **state d** and **state a** are saturated vapor and liquid, respectively, then both cycles follow the same path since as long as the pressure is kept constant in the Rankine cycle it will be *isothermal* too. The heat extracted at a rate $\dot{Q}_c$ by the condenser is used to condense water vapor to liquid. Again, for a conservative design, **state d** may very well be a superheated vapor just to the right of the vapor envelope and **state a** may be compressed liquid to the left of the liquid envelope. If this is the case the line d-a will look similar to c-b.

Now it is possible to analyze the Rankine cycle and to do this the First and Second Laws of Thermodynamics, the FLOT and SLOT, are written below, since they will be required throughout this chapter.

$$\dot{m}\left[\hat{H}_2 - \hat{H}_1\right] = \dot{Q} + \dot{W}_m \quad \text{FLOT} \tag{3.12}$$

$$\dot{m}\left[\hat{S}_2 - \hat{S}_1\right] \geq \frac{\dot{Q}}{T} \quad \text{SLOT} \tag{3.13}$$

The change in kinetic and potential energies in the FLOT have been ignored. They can contribute, especially changes in kinetic energy, however, for the purposes of this chapter they will be ignored as are frictional energy losses in the pumping of liquids long distances in a pipe. Before delving into the various components that constitute the power cycle let's perform a FLOT analysis of the entire cycle to arrive at

$$0 = \dot{W}_p - \dot{W}_t + \dot{Q}_b - \dot{Q}_c \,[=]\, \mathrm{W} \tag{9.1}$$

where the signs of the heat and work terms have been explicitly written and the symbol [=] means 'has dimensions of.' The net work is defined as $\dot{W}_{net} = \dot{W}_t - \dot{W}_p$ and, as the definition implies, is the work obtained from the cycle ignoring the other loss terms (see Table 9.1) and is equivalent to the electrical output. The efficiency of the Rankine cycle η_R is written as

$$\eta_R = \frac{\dot{W}_{net}}{\dot{Q}_b} \tag{9.2}$$

representing the rate of energy one obtains from the cycle to that which is supplied. As mentioned above, a coal-fired power plant has an efficiency of about 40%. To understand the scale of these power plants and the coal that must be supplied to create electricity consider this example.

Example 9.1

Determine how much coal is required to operate the coal-fired power plant in Table 9.1 and the flow rate of water required to condense the water vapor coming from the turbine. Assume the cooling water comes from a nearby river and it can be returned from the power plant if it increases in temperature at most by 10 °C.

The power plant is large with a capacity of 500 MW and an efficiency of 39%, so, the rate of primary energy supply, from coal, should be $\dot{Q}_b$ = 500 MW/0.39 = 1.28 GW. The energy content of coal is approximately 2×10^4 kJ/kg making 64 kg/s or 5.5×10^6 kg/day required. A rail car can hold about 100 tonnes (10^5 kg) making 55 cars required per day. Each car is approximately 15 m long meaning that the delivery train would be 825 m long, or almost a kilometer, for each day! If $\dot{Q}_b$ were to be gathered from the Sun a rather large area would have to be used. Assuming the irradiance is 500 W/m^2 then an area of 1.6×1.6 km^2 would have to be used, assuming 100% collection efficiency.

The cooling water requirement is $\dot{Q}_c = 0.525 \times 1.28$ GW = 672 MW. This is the rate of energy given to the river water pumped to the power plant and subsequently returned to the river. One can write the FLOT for the river water going to and from the condenser as $\dot{m}_w C_{pw} \Delta T = \dot{Q}_c$, where $\dot{m}_w$ is the river water flow rate to/from the condenser, C_{pw}, the heat capacity of water, and ΔT, the water temperature rise which must be at most 10 °C. Rearranging this equation and substituting in numbers gives $\dot{m}_w = 1.6 \times 10^4$ kg/s.

The Susquehanna River, a large river in the east of the United States, has a flow rate of about 800 m^3/s or 8×10^6 kg/s. Injecting this much energy into the river will produce a temperature rise of 0.02 °C, which is negligible. However, care should be taken to ensure that there is little local temperature rise, since this will reduce oxygen in the water and promote conditions for algal blooms and invasive species propagation. Brunner Island in the river hosts three units that produce 1.49 GW, so, the temperature rise is expected to be three times this amount, still negligible. Yet, cooling towers have been installed on the island to reduce thermal discharge to the river, obviously local heating effects are severe or the company would not have done this. These challenges will be the same for a solar powered unit since this process also requires a heat sink.

The details of the Rankine power cycle can now be considered and an analysis of each process unit operation is shown in Table 9.2.[2] As explained in the side note the pump work is best obtained by

$$\dot{W}_p = \dot{m}\hat{V}_a [P_b - P_a] = \frac{\dot{m}}{\rho_a} [P_b - P_a] \tag{9.3}$$

where $\hat{V}_a$ is the specific volume of water in **state a**, P_a and P_b, pressures for **state a** and **state b**, respectively, and ρ_a, the water density at

[2]The best way to determine the enthalpy difference $\hat{H}_b - \hat{H}_a$ for water flowing through the pump is to use the thermodynamic relation: $Td\hat{S} = d\hat{H} - \hat{V}dP$. This relation is obtained from the FLOT for a closed system written in differential form: $dU = dQ + dW$. The SLOT for a reversible process in a closed system is: $dQ = TdS$, while we know that P-V work is defined as $dW = -PdV$. The definition of enthalpy is $H = U + PV$, and combining these equations results in $Td\hat{S} = d\hat{H} - \hat{V}dP$. Since the pump is a reversible process then $d\hat{S} \equiv 0$ and one can integrate the equation to find: $\hat{H}_b - \hat{H}_a = \dot{W}_p/\dot{m} = \int_a^b \hat{V}dP \approx \hat{V}_a[P_b - P_a]$. The approximation is a good one since liquid water is fairly incompressible for the conditions in a power plant. Also, see the discussion in Section 8.3.

Table 9.2 Conditions and equations governing the components to the Rankine cycle.

Component	Process line	FLOT	SLOT
Pump	Rev. & Adia. (a-b)	$\dot{m}\left[\hat{H}_b - \hat{H}_a\right] = \dot{W}_p$	$\dot{m}\left[\hat{S}_b - \hat{S}_a\right] = 0$
Boiler	Const. P (b-c)	$\dot{m}\left[\hat{H}_c - \hat{H}_b\right] = \dot{Q}_b$	not useful
Turbine	Rev. & Adia. (c-d)	$\dot{m}\left[\hat{H}_c - \hat{H}_d\right] = \dot{W}_t$	$\dot{m}\left[\hat{S}_d - \hat{S}_c\right] = 0$
Condenser	Const. P (d-a)	$\dot{m}\left[\hat{H}_d - \hat{H}_a\right] = \dot{Q}_c$	not useful

state a. One can use thermodynamic tables for compressed liquids to find $\hat{H}_a$ and $\hat{H}_b$, yet, eqn (9.3) is accurate enough for most conditions.

The process design is typically performed by stating the pressure level in the condenser and boiler and the power output, really, the electricity required. Or the boiler may operate at a given temperature and by knowing this and the condenser pressure together with the electrical output the entire process can be designed.

Clearly, the design entails finding the enthalpy of each state, which can be obtained through a variety of software packages. There are also tables of data as well as graphs. Appendix B has a pressure–enthalpy diagram which is very useful for visualizing the process as shown in Fig. 9.1 and will be used here.

Example 9.2

There are several small solar thermal energy generated electricity (STEGE) power plants in the Mojave Desert in California, United States, one of which is called SEGS VI (Solar Electric Generating System VI), which has generated 30 MW of electricity continuously since 1989. A simplified Rankine cycle for this power plant is that the steam at the exit from the boiler is at 375 °C and a pressure of 10 MPa. The pressure at the exit from the turbine is 10 kPa, this is absolute pressure and so is below atmospheric pressure and is a vacuum, while the exit state from the condenser will be saturated liquid. Determine the working fluid mass flow rate, the amount of heat rejected to the condenser, the heat required to make steam in the boiler and the efficiency of the process.

The FLOT and SLOT equations in Table 9.2 will be used to analyze this process, which is shown in Fig. 9.3. The place to start is at **state c**, which is superheated steam at a temperature of 375 °C and a pressure of 10 MPa, allowing one to determine the enthalpy and entropy: $\hat{H}_c = 3010$ kJ/kg and $\hat{S}_c = 6.078$ kJ/kg-K, respectively. Since the turbine operates isentropically **state d** will have $\hat{S}_d = 6.078$ kJ/kg-K and $P_d = 10$ kPa.

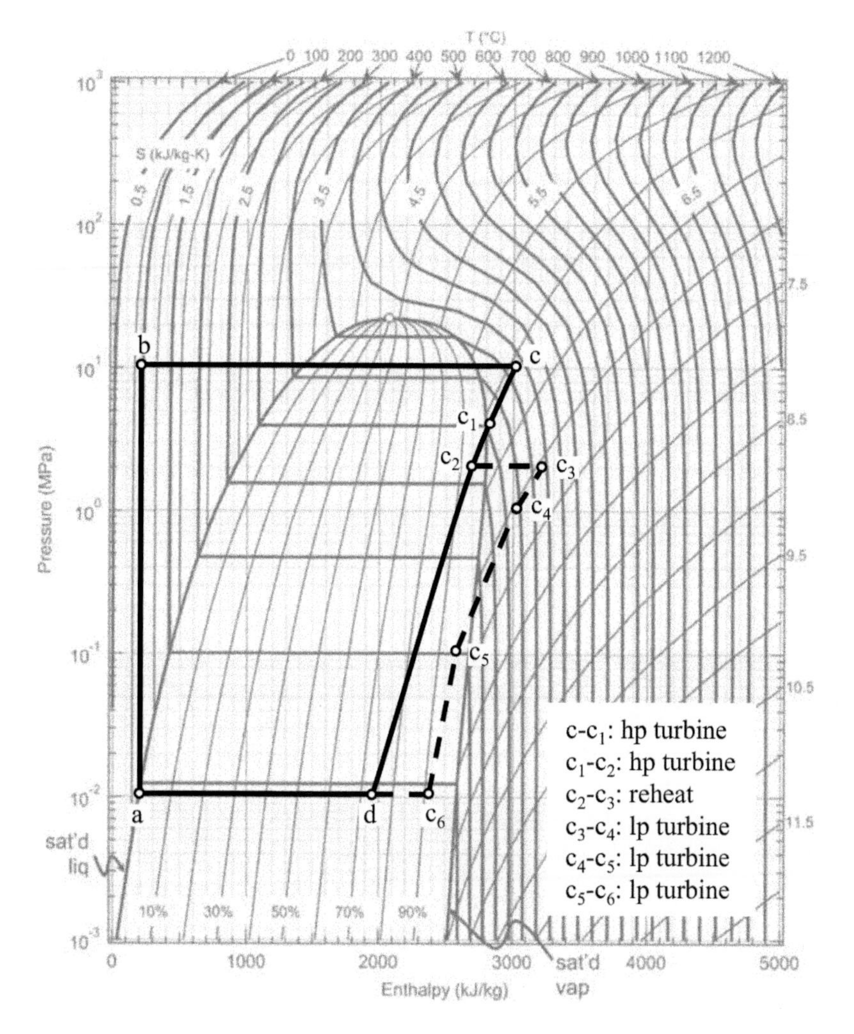

Fig. 9.3 Rankine cycle for the solar thermal energy generated electricity power plant. The simplified cycle is a-b-c-d-a, while the cycle which is similar to the SEGS VI power plant is a-b-c-c_1-c_2-c_3-c_4-c_5-c_6-a. In the real power plant both high pressure (hp) and low pressure (lp) turbines are used, as well as a reheat where the low pressure steam is reheated back to 375 °C, as shown in the table within the figure. The pressure–enthalpy diagram is from Appendix B.

Reference to the figure shows that **state d** will be a mixture of liquid and vapor, which is undesirable, since as previously mentioned the liquid droplets can ruin the turbine. We will use this as the exit state in this example for simplicity, a representation of the cycle which more closely resembles the power cycle is also given in Fig. 9.3. It involves reheat cycles and high and low pressure turbines. Although **states $\mathbf{c_2}$, $\mathbf{c_5}$** and $\mathbf{c_6}$ also appear to be within the vapor-liquid envelope, they are most likely not. This is due to the fact that a real turbine will not be reversible (nor adiabatic), which requires the entropy of the exit state to increase. This tends to put these states outside the envelope thereby avoiding formation of liquid droplets and is most easily visualized on a T–S diagram.

Now to determine the enthalpy of **state d** we need to find the fraction of vapor which is determined by

$$\hat{S}_d = [1 - x_d]\hat{S}_l + x_d\hat{S}_v \tag{9.4}$$

where x_d is the mass fraction of vapor (often called the quality), $\hat{S}_l$, the enthalpy of saturated liquid and $\hat{S}_v$, the enthalpy of saturated vapor. This is a simple mixing rule for vapor and liquid which is very accurate. At a pressure of 10 kPa one can determine $\hat{S}_l$ = 0.6492 kJ/kg-K and $\hat{S}_v$ = 8.1501 kJ/kg-K, making x_d = 0.724. Now we can also determine $\hat{H}_l$ = 191.8 kJ/kg and $\hat{H}_v$ = 2585 kJ/kg and, by using a similar mixing rule to that in eqn (9.4) for the enthalpy, one finds $\hat{H}_d$ = 1925 kJ/kg. The values for the saturated entropy and vapor were determined with steam tables (or one can do this with convenient software that is available) which is an accurate way of finding their value. However, to the accuracy of the design calculations applied in this chapter one can estimate them from the pressure–enthalpy diagram for water in the appendix that is reproduced in Fig. 9.3.

The above enthalpy values allow the work (electricity) per unit mass flow rate generated by the turbine to be calculated,

$$\frac{\dot{W}_t}{\dot{m}} = \hat{H}_c - \hat{H}_d = 1085\,\frac{\text{kJ}}{\text{kg}}$$

State a should be determined now to find the heat load on the condenser since **state d** has already been determined. There is no pressure drop through the condenser (this is an assumption) making P_a = 10 kPa and since it is saturated liquid, the enthalpy is $\hat{H}_a$ = 191.8 kJ/kg and $\hat{V}_a = 0.00101$ m^3/kg. Before considering other parts of the cycle it is best to determine the power required to drive the pump, which will in turn allow the net work to be found and the mass flow rate of the working fluid $\dot{m}$. Reference to eqn (9.3) and Table 9.2 yields

$$\frac{\dot{W}_p}{\dot{m}} = \hat{H}_b - \hat{H}_a = \hat{V}_a[P_b - P_a] = 1\frac{\text{kJ}}{\text{kg}}$$

and the enthalpy at **state b** can be determined *via* $\hat{H}_b = \hat{H}_a + \hat{V}_a[P_b - P_a]$ = 192.8 kJ/kg.

The net work per unit mass flow rate $\dot{W}_{net}/\dot{m}$ is $\dot{W}_t/\dot{m} - \dot{W}_p/\dot{m}$ = 1084 kJ/kg, so, the mass flow rate of the working fluid can be determined,

$$\dot{m} = \frac{\dot{W}_{net}}{\dot{W}_{net}/\dot{m}} = \frac{30\,\text{MW}}{1084\,\text{kJ/kg}} = 27.7\,\frac{\text{kg}}{\text{s}}$$

The condenser heat load $\dot{Q}_c$ is

$$\dot{Q}_c = \dot{m}\left[\hat{H}_d - \hat{H}_a\right] = 48.0 \text{ MW}$$

while the heat load on the boiler $\dot{Q}_b$ is

$$\dot{Q}_b = \dot{m}\left[\hat{H}_c - \hat{H}_b\right] = 78.0 \text{ MW}$$

Finally, the efficiency of the process η_R can be determined from eqn (9.2)

$$\eta_R = \frac{\dot{W}_{net}}{\dot{Q}_b} = \frac{30.0 \text{ MW}}{78.0 \text{ MW}} = 38.5\%$$

One can ascertain that the Rankine power cycle is a reasonably efficient cycle, with almost 40% of the heat source energy content being converted into electricity,[3] however, the waste of energy is apparent when the condenser heat load is determined. Unfortunately, this cannot be avoided.

[3] Compare this to an automobile which translates only 14–26% of the fuel energy content to movement.

Finally, the heat load on the boiler, 78.0 MW, is the amount of power that must be supplied by the Sun since this Rankine cycle is representative of that for the SEGS VI power plant in the Mojave Desert. Assuming there is 1000 W/m^2 of insolation available, an area of at least 78,000 m^2, or in terms of linear dimensions about 280×280 m^2, is needed. In reality, approximately twice this will be required since 1000 W/m^2 is a substantial power density and in fact the insolation is gathered over 188,000 m^2 ($\approx 430 \times 430$ m^2) at SEGS VI, which is approximately twice the above estimate.

STEGE is typically performed by raising the temperature of a heat transfer fluid with the Sun's energy and having a thermal reservoir to dampen out times when the Sun is obscured by clouds. This can be done by heating molten salt stored on-site that can be used at will to boil water; other technologies are also available. Variance in the high pressure steam properties (**state c**) would put strain on turbine operation if the quality of the steam entering it were to significantly change and the heat transfer fluid system is designed to minimize this. Furthermore, the heat transfer fluid reduces the pressures obtained in the field since it does not boil, which is desirable.

The heat transfer fluid is placed in thermal contact with water through a heat exchanger to subsequently generate steam in another part of the plant. This causes some process inefficiency and an effort does exist to directly generate steam to eliminate this deleterious effect. The heat transfer fluid is considered integral to STEGE and will be used in the designs discussed within this chapter as will become clear later. Before that is considered the manner in which solar radiation is concentrated is discussed.

9.2 Parabolic reflectors to concentrate insolation

In order to use solar energy to produce steam and hence power a Rankine cycle to generate electricity, one must concentrate the solar radiation. The concentrated insolation is absorbed by a pipe that has the heat transfer fluid flowing through it. It was demonstrated in the previous chapter that a flat plate solar energy collector can heat water fairly efficiently to order 50 °C, however, much higher temperatures, approaching 400 °C, are required for electricity generation. Here the use of parabolic reflectors is considered for electricity production.

Previously we found that concentrating the sunlight can produce a large temperature rise over a short length of pipe in a solar concentrator, in Example 3.6. The maximum concentration ratio for a linear collector, a pipe positioned in a mirrored device of some sort that concentrates the insolation, is

$$C_{R,i} = \frac{1}{\sin(\theta_{sub})} \quad \text{(1-Dimension)} \tag{3.35}$$

where $C_{R,i}$ is the maximum, ideal concentration ratio and θ_{sub} is the angle subtended between the Earth and Sun (0.264°). Using the value of the angle in the above equation one finds the maximum concentration ratio is 216. In practice, the concentration ratio, C_R, is of order 20–30 in power systems and these will be the values used here.

It is possible to estimate how C_R affects the maximum temperature the absorber can obtain. To do this we start with the definition of the extraterrestrial irradiance discussed in Chapter 2,

$$P_{ext} = \sin(\theta_S)^2 \sigma_S T_S^4 \equiv \frac{\sigma_S T_S^4}{C_{R,i2}} \,[=]\, \frac{\text{W}}{\text{m}^2} \tag{2.4}$$

where σ_S is Stefan's constant (5.670×10^{-8} W/m^2-K^4) and T_S is the Sun's temperature when modeled as a black body radiator (5793 K). Here $C_{R,i2}$ is the ideal concentration ratio in two dimensions which will yield the greatest absorber temperature, see eqn (3.34). Basically, this concentration ratio is equivalent to taking all the power that the Sun emits and focusing it on the absorber, and is equal to $C_{R,i2} = 1/\sin(\theta_S)^2 = 46,700$, as discussed in Chapter 3.

The maximum absorber temperature $T_{abs,m}$ is found by assuming the absorber loses heat through radiation only and is in equilibrium with the Sun emitting P_{ext} irradiance; attenuation by the atmosphere is ignored. This analysis is similar to that performed in the previous chapter when the stagnation temperature was found and is a balance of heat gain from insolation to heat loss from radiative heat transfer, one can write

$$\frac{\sigma_S T_S^4}{C_{R,i2}} A_{ap} = \sigma_S T_{abs,m}^4 A \,[=]\, \text{W}$$

where the absorber has been assumed to be equal to a perfect black body radiator and the aperture area A_{ap} and absorber area A multiply each term for generality. The ratio of these two areas is the concentration ratio, $C_R \equiv A_{ap}/A$, and the above equation can be rearranged to

$$T_{abs,m} = \sqrt[4]{\frac{C_R}{C_{R,i2}}}\, T_S \tag{9.5}$$

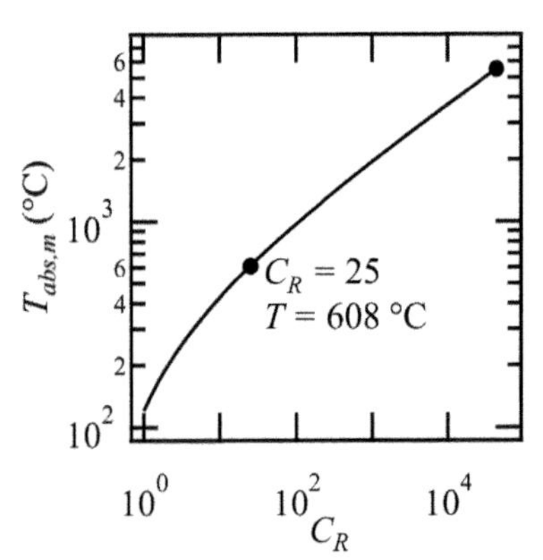

Fig. 9.4 Maximum absorber temperature versus concentration ratio.

The result of this model for the maximum absorber temperature is shown in Fig. 9.4. Convective heat transfer has been ignored, which will tend to reduce the temperature, and the emissivity for the absorber was allowed to be one or is a perfect black body. If the absorber is assumed to be a selective surface with an emissivity of 0.05–0.10, the fourth root of this number will be taken and the temperature for a concentration ratio of 1.0 will be approximately 500 °C, which is substantial, rather than 120 °C. But, as mentioned above, convective heat transfer will reduce this value, as will attenuation of the irradiance by the atmosphere.

However, as the absorber becomes hotter, at very high concentration ratios, its behavior will tend towards a black body. Thus, this calculation is most accurate near the highest concentration ratios, yet, the values for the lower concentration ratios are of the correct order and the maximum temperature rise for $C_R = 25$ is of order 600 °C. This is the concentration ratio used in the most mature STEGE plants.

The details of the reflector used to concentrate the insolation have not been considered yet. There is a vast literature on this topic and references which compare the various types of reflectors can be found at the end of this chapter. So, rather than an exhaustive study of the various types of reflectors, a simple one is considered, the parabolic reflector. The equation for a parabola in Cartesian coordinates is $y = x^2/[4f]$, where x and y are the coordinates and f is the focal distance. Polar coordinates are more suitable for analysis and with reference to Fig. 9.5 one can write the equation for a parabola as

$$r = \frac{2f}{1 + \cos(\phi)} \tag{9.6}$$

where r and ϕ are the radial and angular coordinates, respectively. The maximum radial position and angle, r_m and ϕ_m, respectively, are shown in the figure and are important in considering the amount of insolation trapped by the parabolic reflector. These details will not be considered here as the references at the end of the chapter adequately describe this aspect of parabolic reflectors. So, we consider the parabolic reflector as 100% efficient. The parabolic reflector's linear axis, of length L, is usually located in the north–south direction and tracking is employed by tilting it around that axis to follow the Sun through the day.

It is possible to use a truncated cylinder as a 'parabolic' reflector in some circumstances. Reference to Fig. 9.5 shows a circle whose center has been displaced from the origin along the y-axis by the radius R. The equation for the circle is $x^2 + [y - R]^2 = R^2$ which can be rearranged to $y = R - R\sqrt{1 - x^2/R^2}$. If $x \ll R$ then a Maclaurin series expansion can be performed to arrive at $y \approx R - R[1 - x^2/2R] = x^2/2R$. This is the equation for a parabola with $f = R/2$. In some cases a truncated cylinder may be easier to manufacture and as long as $W \ll R$ it is adequate as a concentrator of insolation. Most parabolic reflectors used in practice, though, have W = 5.8 m and f = 1.8 m, meaning that if a truncated cylinder were used $R \equiv 3.6$ m which invalidates the approximation. So, a true parabolic profile must be followed for systems used in the field.

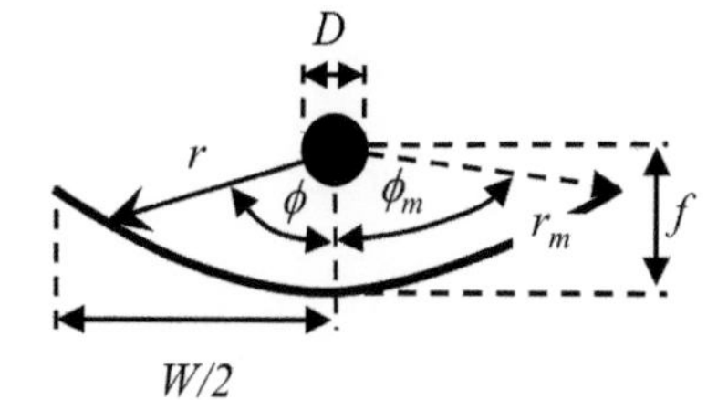

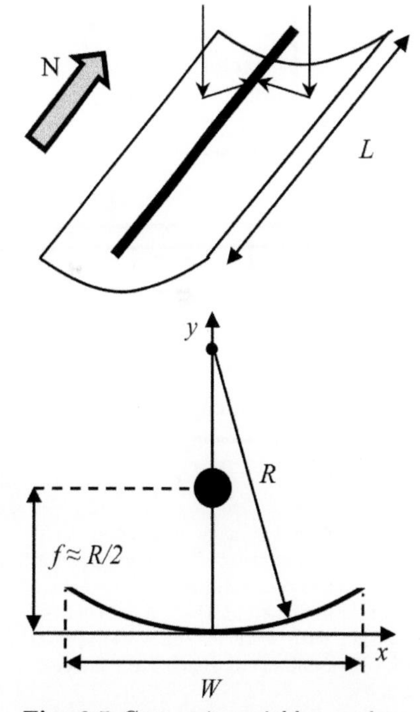

Fig. 9.5 Geometric variables used to define a parabolic reflector. The upper figure shows a parabolic reflector with a cylindrical absorber placed at the focal point. The middle figure shows a perspective of the parabolic reflector which has a length L and is normally located north to south. The lower figure shows a truncated cylinder that can approximate a parabolic trough for $R \gg W$.

9.3 The basic process

The basic parabolic trough STEGE process has a large area of parabolic reflectors that concentrate insolation to absorber pipes through which a heat transfer fluid (typically) flows to receive the energy. The pipes have a selective surface and are surrounded by a clear glass concentric tube to minimize heat transfer to the surroundings, similar to the cover used in a flat plate solar energy collector.

The heat transfer fluid can be a synthetic oil, like Therminol VP-1, whose properties are given in Appendix B, that can withstand the high temperatures required to establish an efficient process. Yet, the development of new heat transfer fluids is an active area of research and in some cases molten salts are being considered. The preferred properties of these materials mean that they have a vastly reduced vapor pressure as a result of a high boiling temperature that does not limit the process. If water were used, high pressure steam would be directly generated in the absorber pipes placing some engineering constraints on the system.

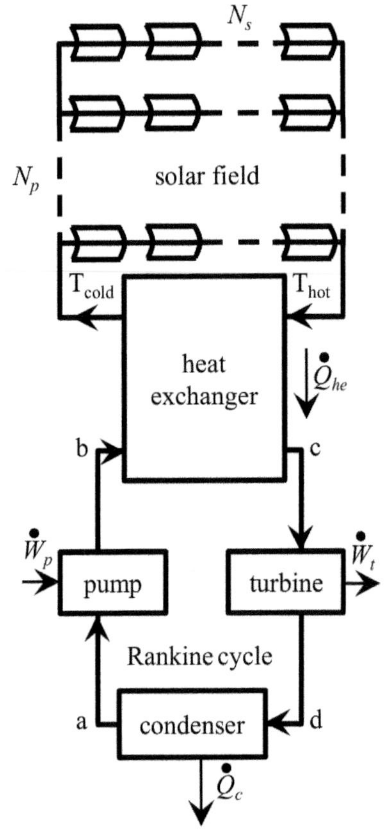

Fig. 9.6 Schematic of a solar thermal energy generated electricity plant, where the solar field is used to heat a heat transfer fluid that is in thermal contact with water as the working fluid of a Rankine cycle to generate the electricity.

A schematic of the process is shown in Fig. 9.6, demonstrating the two separate parts: the solar field which recirculates the heat transfer fluid and the Rankine cycle which has water as the working fluid to generate electricity. The solar field has N_s parabolic reflector units in series, called a *loop* as will be evident below, and there are N_p of these series systems in parallel for a total of $N_u = N_s \times N_p$ units.

The units placed in series will be used to heat the fluid up to its maximum temperature (*i.e.* from T_{cold} to T_{hot}) while the units in parallel are required to increase the overall heat content that can be supplied to the Rankine cycle. The total rate of heat transfer is $\dot{Q}_{he}$, which occurs within the heat exchanger, and is required to heat the water in the Rankine cycle. Of course, $\dot{Q}_{he} = \dot{Q}_b$ for the Rankine cycle, discussed above. The schematic has the two streams operating *counter-currently* in the heat exchanger, meaning that they travel in opposite directions. This mode of operation is fairly efficient in promoting heat transfer since the overall thermal driving force, the temperature difference between the two fluids, remains larger than for *co-current* operation. Co-current operation is used in cases where it is desired to reduce the average *hot* fluid temperature. This is useful if this fluid is temperature sensitive, however, a larger heat transfer area is required that will increase the cost. Suffice to say that the heat exchanger will be taken as a known quantity here and that it operates at 100% efficiency, allowing all the energy garnered within the solar field to be transferred to the water in the Rankine cycle.

The design of this plant begins with knowledge on the amount of electricity that one wants to generate $\dot{W}_{net}$ which will be 30 MW for the example given in this section, a value for some operational units and enough power for a small town or village. Then other details of the cycle are assumed, as discussed below in Example 9.2, to ultimately produce $\dot{Q}_{he} = \dot{Q}_b$.

Modeling of the solar field is similar to that for the flat plate solar energy collector. The *operating line* comes from the FLOT while the *design line* is determined through detailed heat transfer calculations for the fluid flowing in the pipes. Of course, the *operating point* is at the intersection of the two lines. There is a difference between the design discussed here to generate electricity and the flat plate water heater. In the previous chapter we typically had two or three units that were used to heat water, so, the area was previously assumed and all that was to be determined was the water temperature rise and ultimately how long it would take to heat up the hot water reservoir. Here the rate of heat generation is predetermined ($\dot{Q}_{he} = \dot{Q}_b$), placing emphasis on making sure there is enough energy supplied to the Rankine cycle. This changes the design procedure somewhat as will become clear below.

First consider the operating line and we write the FLOT for **one** parabolic collector unit as

$$\dot{m}_0 \left[\hat{H}_{out} - \hat{H}_{in}\right] = \left[C_R \|a\tau\| P_D(0) - \dot{q}_{out}\right] A \quad \text{operating line} \tag{9.7}$$

where $\dot{m}_0$ is the mass flow rate of heat transfer fluid through the pipe in the parabolic collector unit, H_{in} and H_{out}, the enthalpy in and out of the unit, respectively, C_R, the concentration ratio of insolation, $\|a\tau\|$, the calculated absorbance–transmittance product, $P_D(0)$, the insolation falling on the aperture whose plane is assumed to be normal to the direct beam, $\dot{q}_{out}$, heat loss out of the unit and A, the area of the absorber pipe. The area A is equal to πDL, where D is the absorber pipe outside diameter and L its length (see Figures 9.5 and 9.7).

The definition of the parabolic unit should be made clear. Within the STEGE literature there are two terms/acronyms typically used: Solar Collector Assembly (SCA) and Heat Collection Element (HCE). The SCA is the parabolic mirror mounted on a frame, together with controllers to rotate the parabolic trough through the day. The HCE is the absorber pipe mounted at the focal line of the parabolic reflector. The SCA can be up to 100 m long, have apertures (W) of almost 6 m and are huge structures. On the other hand, the HCE is of order 4 m long with an absorber pipe, which has a selective surface, with a diameter of 70 mm. Although long, their diameter is small, compared to the SCA's aperture, indicative of the amount of insolation that will be concentrated on them. The HCEs are joined together and mounted at the focal line in the massive SCA structures. Here the *unit* will be considered the SCA and HCE together and they are both assumed to be 4 m long, subsequently joined in series. The schematic for the Solar Electric Generating System Number 6 or SEGS VI located in the Mojave Desert is shown in Fig. 9.8, which has 9600 of these (equivalent) units.

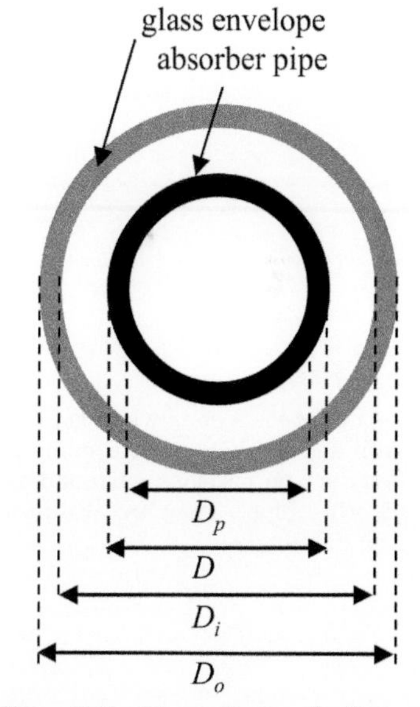

Fig. 9.7 Dimensions of the absorber tube and the surrounding glass envelope.

The definition of the concentration ratio C_R should be made clear and is given by

$$C_R = \frac{A_{ap}}{A} = \frac{W \times L}{\pi D \times L} = \frac{W}{\pi D} \tag{9.8}$$

In some cases it appears as if C_R is defined as W/D, which will increase it to about 80 for the systems listed above rather than approximately 25. The definition should be made clear at all times.

The heat capacity for the heat transfer fluid C_{pf} will be assumed to be constant over the temperature used in the field,[4] T_{cold} and T_{hot}, so

[4] The heat capacity and all other heat transfer fluid properties will be determined at the mean temperature, $T_m \equiv [T_{hot} + T_{cold}]/2$.

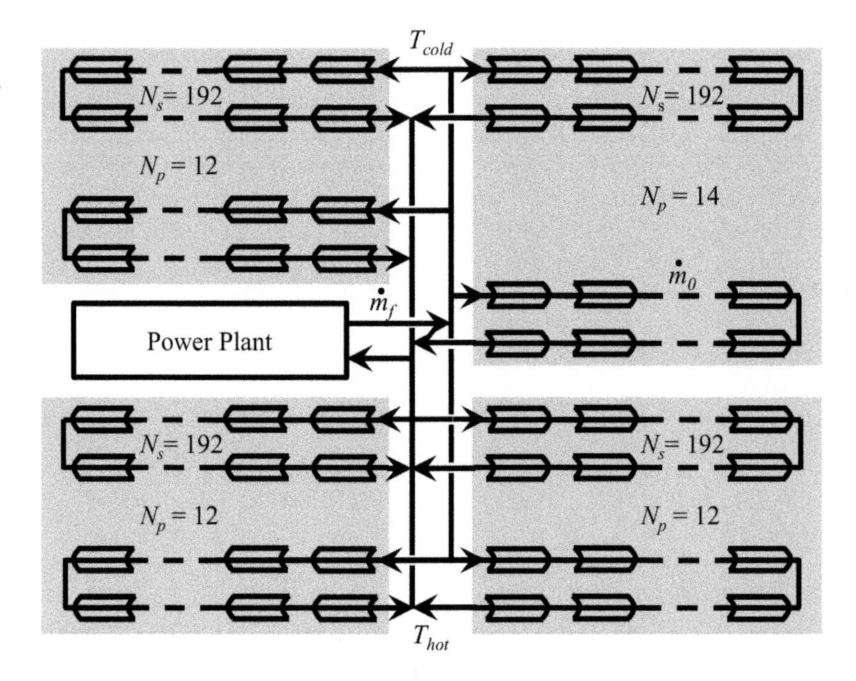

Fig. 9.8 An idealized version of the SEGS VI. There are N_s = 192 units in series, each unit is 4 m long, which makes up a loop. There are $N_p = 50$ loops that operate in parallel making the total number of units $N_s \times N_p = 9600$. The aperture W is 5.0 m and so the total aperture area is 192,000 m^2 which is slightly greater than the area used in reality, 188,000 m^2. The electrical power generated by this power plant is 30 MW. The total mass flow rate of the heat transfer fluid is $\dot{m}_f$ and the mass flow rate in each loop is $\dot{m}_0 = \dot{m}_f/N_p$.

the operating line (eqn (9.7)) for the entire solar field can be written in two forms:

$$\dot{Q}_{he} = \dot{m}_f C_{pf} \Delta T \tag{9.9a}$$

and

$$\dot{Q}_{he} = N_p N_s \left[C_R a P_D(0) - \dot{q}_{out} \right] A \tag{9.9b}$$

where $\Delta T \equiv T_{hot} - T_{cold}$ and $\dot{m}_f$ is the total mass flow rate of the heat transfer fluid. The absorber pipe was assumed to **not** have a surrounding glass envelope, so we can concentrate on the design rather than be involved with detailed heat transfer calculations through the envelope.[5] Thus, the amount of radiation absorbed is $C_R a P_D(0)$ and is similar to the approach taken in the previous chapter. Note that

[5] Note *envelope* is a noun while *envelop* is a verb and so the correct spelling is the first for the envelope surrounding the absorber pipe. They are also pronounced differently...

$$\dot{m}_0 = \frac{\dot{m}_f}{N_p} \tag{9.10}$$

which is the mass flow rate in each loop of N_s units.

The inherent assumption we will make in using eqn (9.9b) is that the absorber pipe temperature will be constant along its length. This sets $\dot{q}_{out}$ (see eqn (9.12) below) and everything in the equation is known except N_s, after N_p is determined. The proper, detailed way to determine the operating point of the system is to use eqn (9.7) as the operating line together with eqn (9.18), shown below, as the design line on each 4 m long unit and calculate the absorber pipe temperature as well as the heat transfer fluid temperature rise on that unit. The temperature rise is added to the entry temperature of the first unit and is taken as the entry temperature to the second unit and the operating point again determined. This process is repeated until the heat transfer fluid reaches the required temperature and one can calculate N_s (once N_p is known).

The procedure advocated here is conservative by assuming a value for the absorber pipe temperature (T_p) and will be justified later. The temperature T_p is set to be the maximum at which the heat transfer fluid can be used, its degradation temperature. This is typically of order 400 °C at least for contemporary organic oils. The temperature rise is set so that T_{cold} is not too low or the viscosity of the fluid will become too large and will increase the pressure drop in the loop. Operational STEGEs have ΔT equal to 100 °C thereby making T_{cold} = 300 °C since $T_{hot} \approx$ 400 °C. Having a smaller ΔT will reduce the pressure drop as well as subsequent pumping costs. If ΔT = 200 °C rather than 100 °C then pumping costs will increase by 10–20%, which is significant.

The conservative nature in assuming T_p = 400 °C is that the radiant heat loss will be larger than if the pipe temperature is allowed to gradually increase along the loop. This will occur as more solar energy is absorbed by the pipe and given to the heat transfer fluid. The error in the constant T_p assumption is only 10–15%, which is acceptable to the level of the design we wish to accomplish here (see Example 9.4).

A temperature rise in the solar field of 100 °C will be assumed and sets the heat transfer fluid mass flow rate

$$\dot{m}_f = \frac{\dot{Q}_{he}}{C_{pf}\Delta T} \tag{9.11}$$

after using eqn (9.9a). Remember $\dot{Q}_{he}$ is known since this must be the heat supply to run the Rankine cycle.

As mentioned above, the absorber pipe temperature T_p will be as large as possible, a temperature just at the organic oil's degradation temperature, and for Therminol VP-1 this is 400 °C. Other commercial products have similar degradation temperatures and T_p should be assumed accordingly. Knowing this temperature allows the heat transfer rate from the absorber pipe to the atmosphere, $\dot{q}_{out}$, to be calculated from

$$\dot{q}_{out} = h_{pa}\left[T_p - T_a\right] + e_p\sigma_S\left[T_p^4 - T_{sky}^4\right] \tag{9.12}$$

since the convective and radiative heat transfer rates operate in parallel. Here h_{pa} is the convective heat transfer coefficient for heat transfer from the absorber pipe to the air, T_a, the air temperature, e_p, the (selective surface) absorber pipe emissivity and T_{sky}, the effective sky temperature for radiative heat transfer calculations (see note 2 in Section 8.1).

The convective heat transfer coefficient h_{pa} can be estimated with a standard correlation for flow of air past a cylinder,

$$Nu = 0.35 + 0.56Re^{0.52} \tag{9.13}$$

where Nu and Re are the Nusselt and Reynolds number, respectively, given by $Nu = h_{pa}D/k_{air}$ and $Re = \rho_{air}v_wD/\mu_{air}$. The parameters are: k_{air} is the thermal conductivity of air, ρ_{air}, the density of air, μ_{air}, the viscosity of air and v_w, the wind velocity. As we have done before, the physical constants for air will be determined at the mean temperature $T_m = [T_p + T_a]/2$. Since the mean temperature is so high for this system, a glass envelope is not used, the correlations given in Appendix B are not valid and the ones in Table 9.3 should be used.

Table 9.3 Correlations for the physical properties of air at higher temperatures, valid for 100 °C $\leq T \leq$ 400 °C, in all cases the temperature T is given in °C .

Property	Equation
k_{air} (W/m-K)	2.49×10^{-2} $+6.74 \times 10^{-5}T$
ρ_{air} (kg/m^3)	$1.17 - 2.50 \times 10^{-3}T$ $+2.19 \times 10^{-6}T^2$
μ_{air} (Pa-s)	1.87×10^{-4} $+3.43 \times 10^{-8}T$

Now one can determine N_s since everything is known in eqn (9.9b), except N_s and N_p, and will be found from

$$N_s = \frac{\dot{Q}_{he}}{N_p \left[C_R a P_D(0) - \dot{q}_{out}\right] A} \tag{9.14}$$

Of course the value of N_p is one parameter that is not easily assumed, however, its value is determined by the amount of energy required by the Rankine power cycle and economic and pressure drop considerations. If it is too small then the flow rate in each loop is large, as is the pressure drop, which increases pumping costs. The pressure drop ΔP can be determined from the definition of the friction factor f given in eqn (8.14) to be

$$\Delta P = \frac{32}{\pi^2} \frac{\dot{m}_0^2 N_s L}{\rho D_p^2} f \tag{9.15}$$

where D_p is the inner diameter of the absorber pipe. The flow will be found to be turbulent in the absorber pipe and so the friction factor is fairly insensitive to the Reynolds number, see eqn (8.15b), and of order 3×10^{-2} for typical conditions within a loop. So, one can determine the variables to which the pressure drop is most sensitive by using eqns (9.10), (9.11), (9.14) and (9.15) to arrive at

$$\Delta P \sim \frac{\dot{Q}_{he}^3}{N_p^3 \Delta T^2} \tag{9.16}$$

where ~ means ‘scales as,’ see note 3 in Section 7.1. The dependence of the pressure drop on N_p is large, as it is with $\dot{Q}_{he}$, while it is less sensitive to ΔT. If N_p is halved from 50, the value used for SEGS VI, the pressure drop is expected to increase by a factor of 8, quite large. In fact, a detailed calculation shows it will increase by a factor of 6.7, demonstrating that the simple relation in eqn (9.16) is a reasonable one. Note, increasing the heat load on the solar field $\dot{Q}_{he}$ similarly increases the pressure drop and doubling it also increases it by a factor of 6.7, again substantial. Thus, one must perform a true optimization for the process; an increase in N_p will reduce ΔP and the associated operational costs versus an increase in N_p that will increase the capital cost of the power plant. This type of design is beyond the scope of this monograph.

Here the value of N_p will be determined heuristically rather than having to perform a complete power plant design that would also have to include economic considerations. The SEGS VI STEGE has 1.67 loops/MW while more recent STEGEs in the Mojave Desert have 1.78 loops/MW and 1.85 loops/MW for the SEGS VIII and SEGS IX, respectively. Given these values one can assume a value of 1.75 loops/MW or

$$N_p \approx 1.75\,\dot{W}_{net}(\mathrm{MW}) \tag{9.17}$$

Note N_s will be the same for all-sized power plants since this will set the heat transfer fluid temperature rise, all other variables being equal.

The reader may notice that no consideration of heat transfer from the absorber pipe wall to the flowing heat transfer fluid has been mentioned. This was quite central to the design of the flat plate solar energy collector. Consider the *design line* for flow of the fluid through the pipe which was given in eqn (8.17) and is written in the present variables as

$$\frac{T_{hot} - T_{cold}}{T_p - T_{cold}} = 1 - \exp\left(-\frac{h_f N_s A_p}{\dot{m}_0 C_{pf}}\right) \quad \text{design line} \tag{9.18}$$

where h_f is the heat transfer coefficient for energy exchange between the absorber pipe inner wall and the flowing fluid and A_p is the inner heat transfer area of the absorber pipe $\pi D_p L$. The coefficient is determined with the same correlation used in the previous chapter (eqn (8.13))

$$Nu = \frac{\frac{1}{2}f[Re - 1000]Pr}{1.07 + 12.7\sqrt{\frac{1}{2}f}\left[Pr^{2/3} - 1\right]}\left[\frac{\mu(T_m)}{\mu(T_p)}\right]^{0.11} \tag{8.13}$$

since we will find that flow is turbulent in the absorber pipes. Here $T_m = [T_{hot} + T_{cold}]/2$ and the fluid viscosity is determined at both T_m and T_p, the mean and pipe wall temperatures, respectively. All the physical constants will be determined at T_m.[6]

The key factor is the ratio of parameters within the exponential term in eqn (9.18). It is the ratio of heat transfer through convection $h_f N_s A_p \Delta T$ to the sensible heat rise of the heat transfer fluid $\dot{m}_0 C_{pf} \Delta T$. If this ratio is a large number then heat transfer is not a limiting factor and will be denoted as the *Absorber Number* or *Ab*

$$Ab \equiv \frac{h_f N_s A_p}{\dot{m}_0 C_{pf}} \tag{9.19}$$

The absorber number is related to the *Stanton number St* often used in heat transfer problems and they are related through

$$Ab = \frac{\pi D_p L}{\frac{\pi}{4} D_p^2} St = 4\frac{L}{D_p} St \tag{9.20}$$

In most cases *Ab* will be much greater than 10 for a loop, making the exponential term essentially zero, implying $T_p = T_{hot}$. Remember the absorber pipe temperature T_p was designed, by using the correct number of units in series, N_s, to be below the degradation temperature of the heat transfer fluid. Thus, it is the length of the loop that determines the heat transfer fluid's temperature and it is not limited by heat transfer. One should always check eqn (9.18), however, it appears that for most industrial power plants heat transfer is not a rate limiting step.

[6] The value of Nu can be increased by the heat transfer fluid absorbing radiation emitted by the absorber pipe. Since the pipe will have a wall temperature of ≈ 400 °C it will emit radiation centered around a wavelength of ≈ 4000 nm by using the Wien displacement law, eqn (2.7). This is in the infrared region and if the absorption coefficient α of the heat transfer fluid is large for this wavelength then the effective heat transfer coefficient can increase. Two dimensionless parameters are important: αD_p and $k\alpha/4n^2\sigma_S T_p^3$, where k and n are the heat transfer fluid's thermal conductivity and refractive index, respectively, and σ_S is the Stefan-Boltzmann constant. The larger the first parameter ($\gtrsim 10$) and the smaller the second ($\lesssim 0.1$) then Nu is increased above that for pure convective heat transfer. For the conditions used in SEGS VI then $\alpha \gtrsim 1$ cm^{-1} and $T_p \gtrsim 400$ °C for this effect to be important. Therminol VP-1 is a commonly used heat transfer fluid, however, its absorption coefficient is not known for ascertaining whether this is a large effect.

Example 9.3

Given the 30 MW Rankine cycle in Example 9.2, design the solar field required to generate the steam in the power plant. By design it is meant to find $\dot{m}_f$, N_p and N_s. Use the standard conditions in Table 9.4 and the heuristic for N_p given in eqn (9.17). This solar field will be close to the size of the SEGS VI STEGE in the Mojave Desert.

The first parameter to be determined is N_p. The heuristic given in eqn (9.17) is used to find, to the nearest integer and rounding up, $N_p = 53$. Now N_s can be found from eqn (9.14),

$$N_s = \frac{\dot{Q}_{he}}{N_p \left[C_R a P_D(0) - \dot{q}_{out} \right] A_p} \tag{9.14}$$

however, $\dot{q}_{out}$ is not known *a priori.* This is found from

$$\dot{q}_{out} = h_{pa} \left[T_p - T_a \right] + e_p \sigma_S \left[T_p^4 - T_{sky}^4 \right] \tag{9.12}$$

since T_p will be assumed to be equal to 400 °C. Everything is known in this equation except h_{pa}, the heat transfer coefficient between the absorber pipe and the surrounding air. This can be determined from eqn (9.13),

$$Nu = 0.35 + 0.56 Re^{0.52} \tag{9.13}$$

The mean temperature of the pipe and air is 212.5 °C and care must be taken to use the correlations in Table 9.3 for the physical properties of air. Since the wind velocity is 3 m/s one finds $Re = 5960$ and $Nu = 51.8$, making $h_{pa} = 29.0$ W/m^2-K, which is quite large. Inserting numbers in eqn (9.12), the rate of heat transfer from the system is found to be $\dot{q}_{out} = 1.26 \times 10^4$ W/m^2-K.

One can determine N_s from eqn (9.14). Before this is done the net rate of heat transfer to the fluid is calculated by $C_R a P_D(0) - \dot{q}_{out}$ to be 8.99×10^3 W/m^2. So, even though the amount of heat coming into the system is large, $C_R a P_D(0) = 2.16 \times 10^4$ W/m^2, there is a large rate of heat transfer out of the system, resulting in a relatively small net rate of heat transfer to the heat transfer fluid. This is obviously due to the fact that an envelope is not used around the absorber pipe and is a similar circumstance to the flat plate solar hot water heater without a cover.

Putting in all the numbers results in $N_s = 186$, close to that used in SEGS VI, which has approximately 200 in series. The total number of units is $N_u = N_s \times N_p = 9858$ and since the aperture area of each unit is 20 m^2 the total aperture of the solar field is 1.97×10^5 m^2, while that for SEGS VI is 1.88×10^5 m^2.

Table 9.4 Standard conditions used in designing a solar thermal energy generated electricity power plant in this chapter. See Figures 9.5, 9.6 and 9.7 for definitions of some variables. The insolation was assumed to be 1000 W/m^2, similar to the AM1.5G spectrum, allowing ready comparison to solar photovoltaic devices and unless otherwise stated this is direct beam insolation. The *parabolic trough unit aperture* is the aperture of the *solar collector assembly* or SCA which is the device that holds the parabolic reflector and tilts the reflector to have the aperture normal to the direct beam. Each SCA holds a number of *heat collection elements* or HCEs which are the absorber tubes and are called the *parabolic trough unit length* in the table.

Insolation (B_{ter})	1000 W/m^2
Heat (energy) required ($\dot{Q}_b$)	78 MW
Air temperature (T_a)	25 °C
Wind velocity (v_w)	3 m/s
Rankine cycle	
Pump inlet water pressure (P_a)	10 kPa
Water state at pump entrance	Saturated liquid
Heat exchanger exit temperature (T_c)	375 °C
Heat exchanger exit pressure (P_c)	10 MPa
Solar field	
Heat transfer fluid temperature rise (ΔT)	100 °C
Absorber pipe temperature (T_p)	400 °C
Temperature difference ($\delta T \equiv T_p - T_{hot}$)	5 °C
Concentration ratio (C_R)	22.7
Focal length (f)	1.49 m
Parabolic trough unit aperture (W)	5 m
Parabolic trough unit length (L)	4 m
Absorber pipe inner diameter (D_p)	66 mm
Absorber pipe outer diameter (D)	70 mm
Envelope inner diameter (D_i)	109 mm
Envelope outer diameter (D_o)	115 mm
Absorber pipe short wavelength absorbance (a_p)	0.95
Absorber pipe long wavelength emissivity (e_p)	0.15
Envelope absorptivity (α_e)	0.04 cm^{-1}
Envelope emissivity (e_e)	0.1
Heat transfer fluid	Therminol VP-1

One should determine if heat transfer is a limiting process in a loop. The mass flow rate of the heat transfer fluid, $\dot{m}_f$, from eqn (9.11) is found first, which is written as $\dot{m}_f = \dot{Q}_{he}/C_{pf}\Delta T$. The heat load on the heat exchanger (or $\dot{Q}_b$) is 78.0 MW from Example 9.2, while the temperature rise for the fluid ΔT will be ≈ 100 °C. All that needs to be determined is the the heat capacity for Therminol VP-1, which is given in Appendix B as a function of temperature.

The exit temperature from a given loop is T_{hot} which is δT below the absorber pipe temperature T_p. This takes into account any heat losses, such as to the SCA structure itself, that are not accounted for, as well as reduction in the heat transfer due to fouling the inside of the absorber pipe. Thus, the mean temperature is $T_m = [T_{hot} + T_{cold}]/2 = 345$ °C and C_{pf} is 2465 J/kg-K. Combining the above results in $\dot{m}_f = 316.5$ kg/s, quite large!

This allows $\dot{m}_0$ to be determined from eqn (9.10), which is 5.971 kg/s. The Reynolds number can be found for the heat transfer fluid ($Re = [4/\pi] \times [\dot{m}_0/\mu D_p]$) after calculating the viscosity μ at T_m for Therminol VP-1 to be 6.13×10^5 Pa-s. The friction factor is calculated with eqn (8.15b) to be 2.82×10^{-3}. Now the Nusselt number Nu can be determined with eqn (8.13) and is equal to 2.25×10^3, quite large too!

The heat transfer coefficient h_f is now calculated to be 2.97×10^3 W/m^2-K after the thermal conductivity for the fluid is determined at T_m (0.0873 W/m-K). This makes the absorber number Ab equal to 31.2 for the 186 units in series. Clearly, based on eqn (9.18) heat transfer is not a limiting factor.

Although the absorber pipe temperature is assumed constant to be at 400 °C this is not a poor assumption and greatly simplifies the design. The next example is given to gauge the goodness of this assumption and a more detailed calculation is given.

Example 9.4

Determine how accurate the assumption of a constant absorber pipe temperature is in Example 9.3.

Demonstrating this is not conceptually difficult, one must apply eqn (9.9b) on each unit, starting with the first where the cold heat transfer fluid enters. This equation is re-written as

$$\dot{Q}_0 \equiv \frac{\dot{Q}_{he}}{N_p} = [C_R a P_D(0) - \dot{q}_{out}]\, A \quad \text{operating line} \tag{9.21}$$

for a single 4 m long unit. The definition of the rate of heat transfer required for each loop $\dot{Q}_0$ is evident from the equation. The rate of heat transfer out of the unit is given by eqn (9.12) and is

$$\dot{q}_{out} = h_{pa}\left[T_p - T_a\right] + e_p \sigma_S \left[T_p^4 - T_{sky}^4\right] \tag{9.12}$$

The design line for a unit is given in eqn (9.18) and is re-written as

$$\frac{T_{out} - T_{in}}{T_p - T_{in}} = 1 - \exp\left(-\frac{h_f A}{\dot{m}_0 C_{pf}}\right) \quad \text{design line} \tag{9.22}$$

where T_{in} and T_{out} are the inlet and outlet temperatures to/from the unit. The initial unit's inlet temperature will be T_{cold} and the final unit's outlet temperature will be T_{hot}. The heat transfer fluid's physical properties will be determined at the mean temperature for the given unit, $T_m = [T_{in} + T_{out}]/2$. The heat transfer coefficient will be found from the Nusselt number correlation given by eqn (8.13).

Everything is known in the above equations except T_p and T_{out} for each unit. One starts with the first unit and then continues to the next and so-forth. The result of this calculation is given in Fig. 9.9 assuming each unit to be 4 m long. The cold heat transfer fluid enters the first unit at 295.0 °C and the absorber pipe achieves a temperature of 300.2 °C while there is a 0.81 °C fluid temperature rise to make T_{out} = 295.8 °C. The second unit now has an inlet temperature of 295.8 °C and so on. The upper graph in Fig. 9.9 shows this calculation, which requires about 166 units to have T_{out} = 400 °C.

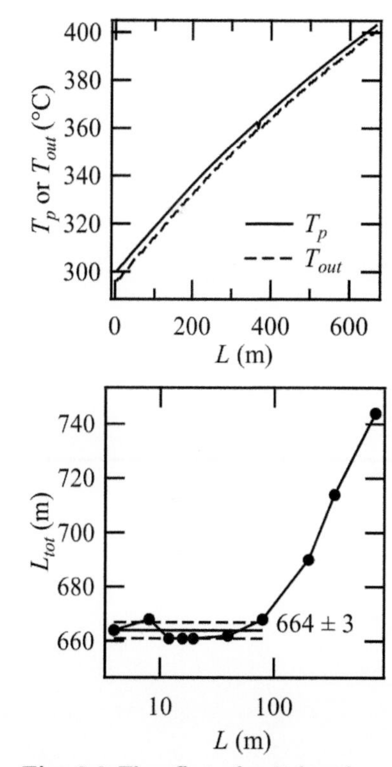

Fig. 9.9 The effect of unit length on the performance of STEGE. The upper figure shows the absorber pipe temperature T_p and the outlet temperature T_{out} of each 4 m-long unit as a function of position, in terms of length, L down the loop. The effect of unit length on the total length L_{tot} of the loop for the heat transfer fluid to reach 400 °C .

The lower graph in the figure shows the effect of unit length on the total loop length to achieve a final outlet temperature of 400 °C. Various unit lengths L were assumed to discretize the system more and more coarsely until only one unit of length 744 m was assumed, which is the loop length in the previous example. Of course for this unit length the absorber pipe temperature was also 400 °C for its entire length.

Remarkably, assuming unit lengths of 4–80 m made little difference and a total loop length of 664 ± 3 m was required. As mentioned above, having only one unit in the loop makes the overall heat loss from the system greater, requiring a longer loop length. The heat transfer rate out of the system $\dot{q}_{out}$ was 1.26×10^4 W/m^2 using a single unit with a length of 744 m in the loop, compared to 1.02×10^4 W/m^2 if 166 × 4 m long units were placed in series and the absorber pipe temperature was allowed to gradually increase. Interestingly, the result of this calculation is that it is only 12% longer than that from the more detailed calculation making this a reasonable approximation.

These examples demonstrate an approximate way to design a solar thermal energy generated electricity system. However, to keep the analysis as simple as possible, the effect of an envelope around the absorber pipe was not considered. This complicates the analysis in the same manner that a cover did for the flat plate solar hot water heater, an additional criterion for equilibrium must be established. Most power plants have envelopes and the effect of this additional element in the STEGE is discussed in the next section.

9.4 The effect of an envelope

When a cover was placed on a flat plate solar energy collector it was found that the device became more thermally efficient despite the fact that the cover admitted less insolation. The gauge of the insolation reaching and absorbed by the absorber plate was $\|a\tau\|$, which is a complicated calculation requiring detailed knowledge of the cover and absorber material properties. In spite this, consideration of the product was given in Chapter 8; now the analysis becomes more challenging with a curved surface such as the envelope. In addition, the reflection optics, that is the reflection of insolation from the parabolic mirror to the absorber pipe, must also be considered to find an effective $\|a\tau\|$. Much has been done on this topic and can be found in the references at the end of this chapter. Here we will write $\|a\tau\|$ as $a_p\tau_e$, where a_p is the absorbance of the absorber pipe and τ_e is the transmittance of the envelope ($\exp(-\alpha_e d)$, where α_e is the envelope absorptivity and d is the envelope thickness, $[D_o - D_i]/2$). One can always use this product as an empirical correction factor and a value of 0.7 may be appropriate after all the various factors are considered. Reference to Table 9.4 shows that $a_p\tau_e = 0.94$, which will overestimate the performance of the absorber pipe system. Regardless, this value will be used and results derived through its use can, of course, be rectified by using other values, such as 0.7 mentioned above.

An envelope modifies the rate of heat transfer out of the system and is very effective at this. There will be series heat transfer as heat is lost by the absorber pipe to the envelope then from the envelope to the surrounding atmosphere. In addition there will be parallel heat transfer processes occurring through radiant and conductive and convective heat transfer from the absorber pipe and envelope.

The absorber pipe will lose heat to the surrounding glass envelope through radiative and conductive or convective heat transfer. Which mechanism dominates depends on the pressure in the annulus, as will become clear below. Similarly to what was done in the previous chapter the rate of heat transfer for a unit of length L from the pipe to the envelope $\dot{Q}_{pe}$ is written

$$\dot{Q}_{pe} = \frac{2\pi k_{pe}}{\ln\left(\dfrac{D_i}{D}\right)}[T_p - T_e]L + \frac{\sigma_S}{\dfrac{1}{e_p} + \dfrac{1-e_e}{e_e}\dfrac{D}{D_i}}\left[T_p^4 - T_e^4\right]A\,[=]\,\mathrm{W} \quad (9.23)$$

where k_{pe} is an effective thermal conductivity for the gas in the annulus between the absorber pipe and the glass envelope, D_i, the inner diameter of the glass envelope, T_e, the envelope temperature (the inner and outer part of the envelope are considered to be at the same temperature) and e_e, the emissivity of the envelope. The first term on the right-hand side of the equation is simple conductive heat transfer through an annulus, while the second term is radiative heat transfer across the same annulus where the area A is that for the absorber pipe πDL (compare to eqn

(8.31a) for radiative heat transfer between two parallel, flat plates). This radiative heat transfer term will not be derived here and can be found in most undergraduate heat transfer texts.

The effective thermal conductivity is a complicated function of pressure. At very low pressure, say below 100 Pa, one has free molecular conduction and

$$\frac{k}{k_{pe}} = 1 + b\lambda\left[\frac{1}{D} + \frac{1}{D_i}\right]\ln\left(\frac{D_i}{D}\right) \quad P \leq 100 \text{ Pa} \tag{9.24}$$

where k, is the gas thermal conductivity at standard temperature and pressure (*i.e.* 0 °C and 1 atm.), b, an interaction coefficient and λ, the mean free path between molecular collisions in the gas, which was mentioned in note 3 in Section 5.1. Parameters required for eqn (9.24) and below are given in Table 9.5. The remarkable fact is that for air, and most gases in general, k, as well as other physical properties, are not a large function of pressure. They are a function of temperature (except C_p to some degree) and so the pressure dependence of k_{pe} is solely through the pressure dependence of λ.

The mean free path is written here as (assuming ideal gas behavior)

$$\lambda = \frac{k_B T_m}{\sqrt{2}\pi d^2 P} \tag{9.25}$$

with k_B being Boltzmann's constant, T_m, the mean temperature in the annulus ($T_m = [T_p + T_e]/2$), d, the molecular (collision) diameter of the gas molecule and P, pressure in the annulus.

The interaction coefficient is given by

$$b = \frac{[2-a]\times[9\gamma - 5]}{2a[\gamma + 1]} \tag{9.26}$$

where a is the thermal accommodation coefficient[7] and γ is the ratio of the constant pressure to constant volume heat capacity. An ideal gas would have $\gamma = 5/3$ while the ratio for air is 1.4.

Although one may expect air to be the gas to leak into the annulus, it is in fact Hydrogen which can cause difficulties. Hydrogen permeates through the absorber pipe after formation through degradation of the heat transfer fluid. Since Hydrogen has a thermal conductivity approximately 6–7 times greater than air, even a small pressure of pure Hydrogen can affect performance. Thus, the molecular diameter, accommodation coefficient, interaction coefficient and heat capacity ratio are given in Table 9.5 for both air and Hydrogen. Hydrogen gas can be taken out of the annulus by evacuation or the use of a 'getter' material placed within the annulus to absorb Hydrogen.

The limiting value of k_{pe} at higher pressures is k that occurs when λ decreases in value, and this happens at a reasonably low pressure; for air this is at approximately 30 Pa, which is quite small. Increasing the pressure even further results in natural convection to increase k_{pe}, similarly to that seen in the flat plate solar hot water heater, and the Rayleigh number Ra becomes important to determine k_{pe}

Table 9.5 Parameters used to calculate the thermal conductivity of air and Hydrogen at low pressures. The accommodation coefficient a depends on the surface and its roughness so the maximum value of 1 is used. The heat capacity ratio is slightly smaller than 1.4 for both gases, however, since a is not known, an approximate value for γ is given.

Property	Air	Hydrogen
d (nm)	0.353	0.240
a	1	1
b	1.57	1.57
γ	1.4	1.4

[7]The thermal accommodation coefficient is frequently written as $a = [T_r - T_i]/[T_s - T_i]$, where T_r is the temperature of the reflected gas molecule, T_i, the temperature of the incident gas molecule and T_s, the temperature of the solid. Since temperature is proportional to energy one can use energy rather than temperature in the definition. A theoretical value for a is due to Boule in 1914 (Ann. der Physik **44**(1) (1914) 145), $a = 2\mu/[1+\mu]^2$, where μ is the ratio of the gas to surface material's molecular weight. A more accurate prediction is obtained if 2 is replaced by 2.4 in the Boule relation and is valid for clean, smooth surfaces. Surface roughness and cleanliness affect a by making it lower; a value of one is conservative and will overestimate the rate of heat transfer.

$$\frac{k_{pe}}{k} = 0.386\left[\frac{Pr}{0.861 + Pr}Ra^*\right]^{1/4} \quad \text{for} \quad 10^2 \lesssim Ra^* \lesssim 10^7 \tag{9.27a}$$

with Ra^* given by

$$Ra^* \equiv \frac{\ln(D_i/D)^4}{L_{pe}^3\left[D^{-3/5} + D_i^{-3/5}\right]^5}Ra \tag{9.27b}$$

where Ra was defined in eqn (8.34) and is written in present variables as

$$Ra = \frac{g\chi\rho^2 C_p}{k\mu}\Delta T_{pe}L_{pe}^3 \tag{9.27c}$$

where g is the gravitational constant, χ, the thermal expansion coefficient, ρ, the density, C_p, the heat capacity, k, the thermal conductivity at the temperature of interest, μ, the viscosity, ΔT_{pe}, the temperature difference between the pipe and envelope (in our case this will be $T_p - T_e$) and L_{pe}, the gap between the pipe envelope ($[D_i - D]/2$). As before all gas properties will be determined at the mean temperature $[T_p + T_e]/2$. We note here that k_{pe} in this regime of natural convection is not strictly correct, the assumption of using a thermal conductivity rather than a heat transfer coefficient is certainly not conceptually true. Yet, the changeover from conduction to convection necessitates the use of this nomenclature, where k_{pe} should be referred to as an effective thermal conductivity, at least when natural convection effects occur.

The Prandtl number Pr for air can be assumed to be constant over the temperature range 200 °C–400 °C and equal to 0.68, while the other variables such as k, C_p, χ and μ are not a (large) function of pressure and are only that of temperature. One can determine χ with the ideal gas law ($\chi = 1/T$) as well as $\rho = PM/RT$, where M is the molecular weight and R, the gas constant. The other variables can be determined with the relation in Appendix B or in Table 9.3.

The effective thermal conductivity will be taken as the maximum of either eqn (9.24) or (9.27) so

$$k_{pe} = \text{MAX}(\text{eqn (9.24)}, \text{eqn (9.27)}) \tag{9.28}$$

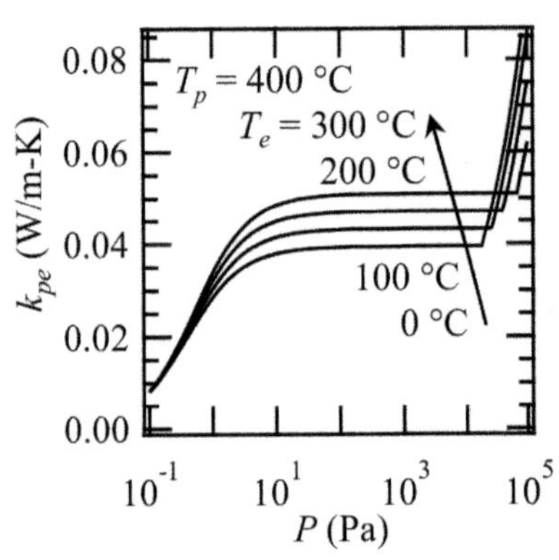

Fig. 9.10 The effective thermal conductivity k_{pe} for the standard conditions listed in Table 9.4 as a function of air pressure in the annulus. The absorber pipe temperature T_p was assumed to be 400 °C while the envelope temperature was assumed to have values from 0 °C to 300 °C.

The result in determining k_{pe} for the standard conditions listed in Table 9.4 is shown in Fig. 9.10. The air pressure in the annular region between the absorber pipe and envelope was varied between 10^{-1} and 10^5 Pa. The amazing property of gases is that the thermal conductivity is relatively constant over an extended pressure range. Molecular conduction dominates until pressures of order 10^4 Pa are approached. Thereafter, depending on the temperature driving force, natural convection can occur to promote more efficient heat transfer. Obviously this is undesirable.

Now, just as with the flat plate solar energy collector, the rate of energy transfer from the system can be determined. The heat transfer rate from the absorber pipe to the envelope, $\dot{Q}_{pe}$, is given in eqn (9.23), and must equal that from the envelope to the surrounding atmosphere, $\dot{Q}_{ea}$, given by eqn (9.12), with the variables updated. This is because they occur in series. Also, each rate, *i.e.* $\dot{Q}_{pe}$ and $\dot{Q}_{ea}$, has two heat transfer rates that occur in parallel too. So, we can write the rate of heat transfer out of the system as

$$\begin{aligned}\dot{Q}_{out} = \dot{Q}_{pe} &= \frac{2\pi k_{pe}}{\ln\left(\dfrac{D_i}{D}\right)}\left[T_p - T_e\right]L + \frac{\sigma_S}{\dfrac{1}{e_p} + \dfrac{1-e_e}{e_e}\dfrac{D}{D_i}}\left[T_p^4 - T_e^4\right]A \\ &= \dot{Q}_{ea} = h_{ea}\left[T_e - T_a\right]A_e + e_e\sigma_S\left[T_e^4 - T_{sky}^4\right]A_e\end{aligned} \quad (9.29)$$

where the subscript ea represents envelope-to-atmosphere, A is the absorber pipe area noted above and A_e is the envelope area $\pi D_o L$. The heat transfer coefficient h_{ea} is determined with eqn (9.13), as with the bare absorber pipe above. Note how $\dot{Q}_{out}$ must be considered, rather than $\dot{q}_{out}$, since the area of the absorber pipe is different to that of the envelope, in addition to the manner in which k_{pe} is defined.

The absorber pipe temperature T_p will be assumed to be constant for the entire system and the surrounding air T_a and sky T_{sky} temperatures will be known, thus, eqns (9.29) are used to find the envelope temperature T_e. As mentioned above, the envelope material, glass, represents little thermal resistance relative to other components and so the assumption of the temperature being constant across the envelope thickness introduces little error.

The operating line for the 4 m-long unit can be written, after reference to eqns (9.7) and (9.21), as

$$\dot{Q}_0 \equiv \frac{\dot{Q}_{he}}{N_p} = N_s\left[C_R\|a\tau\|P_D(0)A - \dot{Q}_{out}\right] \quad \text{operating line} \quad (9.30)$$

which is similar to eqn (9.21), except that $\|a\tau\| \approx a_p\tau_e$ is included.

In addition to the absorbance–transmission product introducing some complexity to the analysis, heat transfer from the system is also more complicated than from the bare absorber pipe, as can be seen in eqn (9.29). The following example is included to estimate the rate of heat transfer from the absorber pipe when an envelope is included, before consideration of a full power plant, including an envelope surrounding the absorber pipe, is examined.

Example 9.5

Determine the rate of heat transfer from the absorber pipe of the standard system given in Table 9.4 when an envelope is included. Assume the air pressure in the annulus is 5 Pa (≈ 0.04 torr).

Firstly, the absorber pipe temperature is assumed to be equal to 400 °C and to not vary along the length of an entire loop. The surrounding air temperature is 25 °C and so one must use eqn (9.29) and change T_e until $\dot{Q}_{pe}$ is equal to $\dot{Q}_{ea}$, which is easily accomplished with a numerical equation solver. Results for manually changing T_e are shown in Table 9.6 and the envelope temperature is found to be 60.1 °C.

Some intermediate results required to find $\dot{Q}_{out}$, which is 1521 W, are that the thermal conductivity of air at atmospheric pressure at the mean temperature in the annulus (T_m = 230.1 °C) is 0.04041 W/m-K. Since the air pressure is 5 Pa in the annulus one determines that k_{pe} = 0.03881 W/m-K, which is approximately 4% lower. The manufacturer's specification for the annulus pressure in the SEGS VI STEGE is apparently 0.013 Pa or 10^{-4} torr. Should this be the case, then k_{pe} = 0.002396 W/m-K and much smaller. This reduces $\dot{Q}_{out}$ to 830.0 W and T_e = 44.1 °C, which is low enough that one could touch it without damage to a finger! It is amazing that having a glass envelope which is only about 20 mm from an absorber pipe at a temperature of 400 °C could be touched. In addition, the rate of heat transfer has also been reduced by almost a factor of almost two, should the high vacuum be used relative to a pressure of 5 Pa.

Table 9.6 Solution to eqn (9.29) to allow $\dot{Q}_{pe} = \dot{Q}_{ea}$ for a unit of length L = 4 m by changing T_e.

T_e (°C)	$\dot{Q}_{pe}$ (W)	$\dot{Q}_{ea}$ (W)
100	1426	3252
75	1487	2165
50	1543	1085
60.1	1521	1521

The steady - state heat loss without an envelope was 7905 W for the 4 m-long absorber pipe. However, including the envelope reduces the heat loss to 1521 W, a more than a five times reduction in energy loss! An envelope certainly reduces the heat loss and is worth its cost.

Calculation of k_{pe} shown in Fig. 9.10 shows a large increase in the effective heat transfer coefficient at larger pressures due to natural convection. Since this is such a large pressure, ≈ 10^4 Pa, this effect does not have to be considered since the unit would most likely be at a condition that is not operational. However, if k_{pe} is doubled, as is the size of the natural convection effect, then $\dot{Q}_{pe}$ increases to 2208 W and T_e = 76.0 °C. Both have considerable increases in their value and significantly affect the power plant design. For example, the total aperture area increases by almost 5% after the full design is completed (two more units are required in a loop to have the heat transfer fluid achieve the operational temperature).

A final comment about the air pressure assumed in the annulus. The manufacturer's specification for the annulus pressure is 0.013 Pa or 10^{-4} torr. However, there is evidence that the pressure is above this, as the envelope temperature is greater than expected (see Price et al., 'Field Survey of Parabolic Trough Receiver Thermal Performance', NREL/CP-550-39459). This was determined by measuring the envelope temperature relative to the heat transfer fluid temperature (see Exercise 9.9). Furthermore, there was some evidence that Hydrogen gas was in some of the annuli which tremendously increased the rate of heat transfer and the envelope temperature. So, the standard conditions assumed for the STEGE in this chapter have a pressure of 5 Pa which will produce a conservative estimate of the power plant design.

The design of the solar field having a glass envelope surrounding the absorber pipe is similar to the bare absorber pipe considered in the previous section. Here eqn (9.14) is re-written with the absorbance–transmittance product $\|a\tau\|$ written as $a_p\tau_e$

$$N_s = \frac{\dot{Q}_{he}}{N_p\left[C_R a_p \tau_e P_D(0) A - \dot{Q}_{out}\right]} \tag{9.31}$$

where $\dot{Q}_{out}$ must be determined according to Example 9.5. All the variables in the equation have been defined above, however, the definition of A is reinforced to be πDL, where D and L are the absorber pipe *unit* outer diameter (70 mm) and length (4 m), respectively.

A drawback in assuming that the absorber pipe has a constant temperature along its total length can be seen upon inspection of the above equation. It diverges when $C_R a_p \tau_e P_D(0) A = \dot{Q}_{out}$. This will occur if $P_D(0)$ becomes too small and there is more heat transfer out of the system than into it. Of course, this is the result of our simplified model and in reality an infinite number of units in series will not have to be used, yet, N_s is expected to increase rapidly when $P_D(0)$ becomes too low. For example, if the heat collection element has only the bare absorber pipe and no envelope then the heat transfer fluid will have increased in temperature by only about 15 °C in 500 m for the standard conditions if the insolation is reduced to 500 W/m^2. Thus, low values of $P_D(0)$ do indeed affect performance and certainly one should have a system design that includes an envelope around the absorber pipe.

This effect is represented graphically in Fig. 9.11 (see Example 9.6 below). The absorber pipe temperature T_p was assumed to be constant along the absorber pipe length at 400 °C and the loop was sized so the heat transfer fluid achieved a temperature of 395 °C. It is clear that if an envelope is not used then the aperture area A_{ap} is given by

$$A_{ap} \equiv N_s N_p W L \tag{9.32}$$

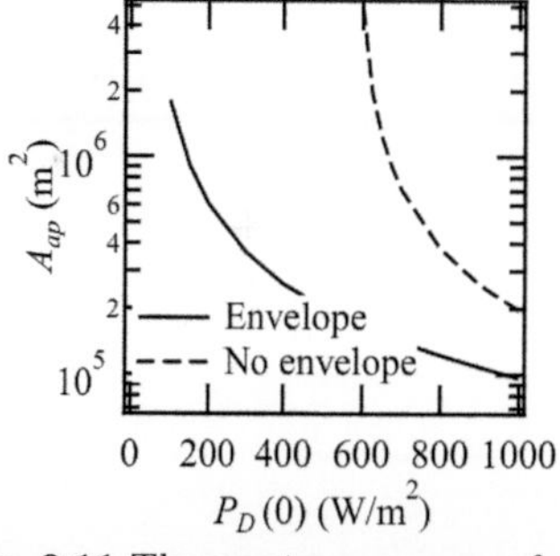

Fig. 9.11 The aperture area as a function of insolation for the standard system given in Table 9.4. Both an envelope and no envelope were assumed for the absorber pipe.

Here W is the width of the parabolic trough (see Fig. 9.5) and assumed to be 5 m for the standard conditions, so, each unit has an aperture area of 20 m^2. The aperture area diverges at a relatively small value of $P_D(0)$ and is due to the very high heat transfer rate from the system. This is only due to the assumption of a constant absorber pipe temperature, as mentioned above. If the more detailed calculation is performed the area does not uniquely diverge, it does go to very large values though.

The calculations required to produce Fig. 9.11 are straightforward after the heat transfer rate out of the system $\dot{Q}_{out}$ is determined as described in Example 9.5. If $\dot{Q}_{out}$ is ignored though one can find the minimum aperture area $A_{ap,min}$ by rearranging eqn (9.31),

$$A_{ap,min} = \frac{\dot{Q}_{he}}{a_p \tau_e P_D(0)} \tag{9.33}$$

which is a reasonable equation to begin the design of the system especially when an envelope is used.

Efficiency has not been discussed and can be determined with an equation similar to eqn (9.2). In this case, though, we have to include all the energy harvested by the Sun and so it is given by

$$\eta_{sf} = \frac{\dot{Q}_{he}}{P_D(0)A_{ap}} \tag{9.34}$$

where η_{sf} is the efficiency of the solar field. The maximum value of η_{sf} is $a_p\tau_e$ and is limited only by the optical properties of the system. However, the efficiency of the Rankine cycle has not been taken into account for an overall efficiency. Equations (9.2) and (9.34) are combined to yield the overall system efficiency η_{sf-R},

$$\eta_{sf-R} \equiv \eta_{sf} \times \eta_R = \frac{\dot{W}_{net}}{P_D(0)A_{ap}} \tag{9.35}$$

Example 9.6

Using the result given in Example 9.5 determine the number of units in a loop, the total aperture area and solar field and overall efficiencies for a solar field designed to deliver 30 MW of net power. In this case use an envelope and the standard conditions given in Table 9.4. Assume that the air pressure in the annulus is 5 Pa (≈ 0.04 torr).

Firstly, the absorber pipe temperature is assumed to be equal to 400 °C and not to vary along the length of a loop. The number of parallel loops is determined from eqn (9.17), so one has $N_p = 53$ for 30 MW of power.

Now, the rate of heat loss to the surroundings was found in Example 9.5 to be $\dot{Q}_{out} = 8.319 \times 10^2$ W while the amount coming in $C_R a_p \tau_e P_D(0)A = 1.696 \times 10^4$ W, allowing one to calculate N_s with eqn (9.31) to be 91 and so the total number of units is $N_u = N_s \times N_p = 4823$, which is approximately half the number required if no envelope were used ($N_u = 9858$). Clearly the envelope makes a large impact on the performance. Since each unit has an aperture of 20 m^2 the total aperture area is 9.631×10^4 m^2.

This aperture area can be compared to the minimum using eqn (9.33) to find $A_{ap,min} = 9.184 \times 10^4$ m^2, which is only 5% less than the more rigorously determined area. This is due to the fact that the envelope significantly decreases the rate of heat transfer from the absorber pipe. Thus, with an envelope in place one can reasonably estimate the aperture area due to the reduced rate of heat transfer out of the system.

As mentioned earlier in this chapter, the SEGS VI power plant has an aperture area of 1.88×10^5 m^2, which is approximately twice as large as we have calculated. Reference to Fig. 9.11 shows that the aperture area depends on the amount of radiation received by the absorber, as expected. The standard conditions have $P_D(0) = 1000$ W/m^2 so this technology can be compared to photovoltaic devices that are tested with an AM1.5G spectrum whose total power is equal to this. However, if $P_D(0)$ is reduced to 500 W/m^2 then the area is increased to 2.032×10^5 m^2 and nearer to that actually used. It could be that the engineers who designed SEGS VI actually designed the system for a lower solar power density and so our result for the lower $P_D(0)$ may represent the true design value. Finally, $A_{ap,min}$ is still only 10% lower than the more rigorously calculated value for this lower insolation level, again suggesting that for quick estimates eqn (9.33) is good enough, especially when an envelope is used.

The efficiencies can be calculated with eqns (9.34) and (9.35) to be $\eta_{sf} = 80.99\%$ and $\eta_{sf-R} = 31.20\%$. These are too high compared to data from the SEGS VI power plant, which are approximately half of these values (depending on the time of year). Again, using a lower apparent absorbance for the absorber pipe can lower the efficiency and lower values are determined. Furthermore, we have not included other heat transfer terms which can be important, such as heat loss to the structure that holds the heat collection element (*i.e.* the absorber pipe) which acts as a thermal sink, clearly undesirable and which will reduce the efficiency.

Finally we discuss heat transfer effects since the number of units in series N_s is 91 rather than 186. Since N_s is reduced by effectively a factor of two, will heat transfer be a limiting factor for the envelope case? The answer is no, since the absorber number *Ab* is still quite large and equal to 15.2. Thus, heat transfer is certainly not a limiting factor and one merely requires the loop to be long enough to absorb enough insolation to increase the temperature of the heat transfer fluid to the operating level.

9.5 Conclusion

Making electric power from the Sun was considered in this chapter. This is not direct generation of electricity as with a photovoltaic module, rather it is production of electricity using a contemporary Rankine cycle with energy supplied by the Sun. A difference to the flat plate solar energy collector is required and that difference is concentration of the solar energy. In order to manufacture the higher temperature needed to operate the Rankine cycle, concentration optics are needed. This imposes extra engineering and operational costs on this technology.

However, the cost of the fuel is nothing for solar thermal energy generated electricity, so, it should have an operational advantage over a contemporary coal-fired power plant. The challenge is that the temperatures that can be achieved are lower than when using coal as an energy source and so the efficiency of solar thermal energy generated electricity is less. In addition, inclusion of a heat transfer fluid, to lower the effect of fluctuations in solar radiation, adds additional cost to the maintenance and capital expenditures. However, depending on one's goals this technology can certainly be a viable option.

No mention has been made of the condenser required to operate the

Rankine cycle. At the beginning of this chapter there was a discussion about the immense amount of energy that is rejected to the environment, see Table 9.1. This is certainly a challenge to the operation of these power plants and water used in traditional condensers or even make-up water required when cooling towers are used can limit where these plants are built. Water should be available. Many solar thermal power plants use air cooled condensers though. Although these are less efficient than a water cooled condenser they do eliminate the water requirement. So, the efficiency of these power plants will further suffer, yet, as mentioned above, the fuel is free and minimal carbon dioxide emissions occur.

9.6 General references

D. Barlev, R. Vidu and P. Stroeve, 'Innovation in concentrated solar power,' Solar Energy Mater. & Solar Cells **95** (2011) 2703

J.A. Duffie and W.A. Beckman, 'Solar energy of thermal processes,' John Wiley and Sons, 3rd Edition (2006)

R. Forristall, 'Heat transfer analysis and modeling of a parabolic trough solar receiver implemented in engineering equation solver,' National Renewable Energy Laboratory (2003) NREL/TP-550-34169

P. Gleckman, J. O'Gallagher and R. Winston, 'Concentration of sunlight to solar-surface levels using non-imaging optics,' Nature **339** (1989) 198

F.P. Incropera and D.P. DeWitt, 'Fundamentals of heat transfer,' John Wiley and Sons (1981). This book has many heat transfer correlations in it, including that for k_{pe}.

S.A. Kalogirou, 'Solar thermal collectors and applications,' Prog. Energy Comb. Sci. **30** (2004) 231

S.A. Kalogirou, 'Solar energy engineering, Processes and systems,' Elsevier (2009)

A. Rabl, 'Active solar collectors and their applications,' Oxford University Press (1985)

Exercises

(9.1) Solar thermal energy generated electricity is produced by operating the solar field at 400 °C; estimate the maximum possible efficiency for this power plant.

(9.2) Show that the area enclosed by a-b-c-d-a in Fig. 9.1 is the net work for a Rankine cycle. Use the First and Second Laws of Thermodynamics and indicate any assumptions made.

(9.3) Confirm the equations given in Table 9.2 assuming an operational, ideal Rankine cycle.

(9.4) Generally the higher the pressure and temperature of the steam entering the turbine of a Rankine cycle the more efficient the cycle becomes. In fact, as materials of construction become better there are now supercritical Rankine cycles, meaning that the pressure and temperature are above the critical point of water, approximately 373 °C and 22 MPa. Change the temperature and pressure entering the turbine in Example 9.2 to 525 °C and 28 MPa and calculate the efficiency η_R; does it increase or decrease? Also determine the work required to drive the pump, the heat load on the boiler and condenser and the mass flow rate on the working fluid.

(9.5) Determine the values of ϕ_m and r_m for the parabolic mirror used in the standard conditions listed in Table 9.4.

(9.6) Assuming you can measure W and d for the parabolic mirror in the figure below; calculate the focal length f. The distance d is that from the very bottom of the parabola to the rim.

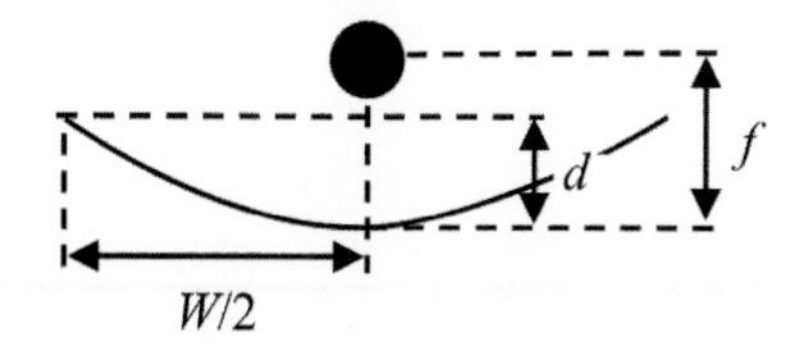

(9.7) The U.S. Department of Energy's Office of Energy Efficiency and Renewable Energy gives the following equation to calculate the pumping cost within a pipe C in \$US

$$C = 8.14 \times 10^{16} f \frac{\left[\frac{\dot{m}}{\rho}(\mathrm{m}^3/\mathrm{s})\right]^3 L(\mathrm{m})}{D(\mathrm{mm})^5} \times \frac{t(\mathrm{h})\chi(\$\mathrm{US/kW\text{-}h})}{\eta_p(\%)}$$

where f is the friction factor, $\dot{m}$, the mass flow rate in the pipe, ρ, the liquid density, L, the pipe length, D, the inner pipe diameter, t the operating time (there are approximately 8760 h in a year), χ, the cost of electricity in \$US/kW-h and η_p, the pump efficiency in % (*i.e.* use 75 and not 0.75). Determine the yearly pumping cost for one loop and the entire solar field in Example 9.3, assuming that the pump is 75% efficient and that electricity costs \$US 0.1/kW-h.

(9.8) The heat transfer rate across the envelope thickness $\dot{q}_e$ is given by

$$\dot{q}_e = \frac{2\pi k_e}{\ln(\frac{D_o}{D_i})}[T_i - T_o]L$$

where k_e is the thermal conductivity of the envelope material (glass), D_o and D_i are the envelope's outer and inner diameters, T_o and T_i are the envelope's outer and inner temperatures and L the length under consideration. Assume the thermal conductivity of the glass envelope is given by

$$k(\mathrm{W/m\text{-}K}) = 1.141 + 1.086 \times 10^{-3} T(^\circ\mathrm{C})$$

estimate the temperature drop across the envelope thickness for the conditions given in Example 9.5.

(9.9) Price et al. ('Field Survey of Parabolic Trough Receiver Thermal Performance', NREL/CP-550-39459) have studied the performance of the SEGS VI solar thermal energy generating electricity power plant and measured the envelope temperature T_e relative to the ambient temperature T_a, compared to the heat transfer fluid temperature T_{hot} relative again to T_a. The results are shown in the figure below.

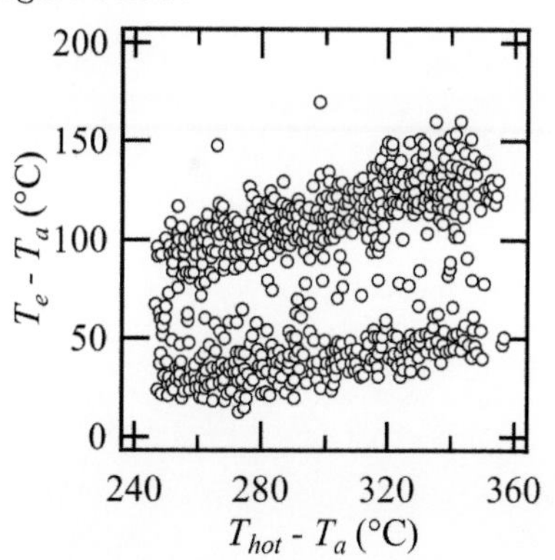

There appear to be two bands of data. The lower band is reported to be working according to manufacturer's specifications with an air pressure of 0.013 Pa (or 10^{-4} torr) while the upper band appears to have hydrogen gas in the annulus. Determine if these are reasonable conclusions and estimate the air and hydrogen gas pressures for the lower and upper bands of data. Assume that the standard conditions in Table 9.4 apply. The thermal conductivity of Hydrogen at atmospheric pressure is given by

$$k_e \mathrm{W/m\text{-}K}) = 0.1744 + 3.977 \times 10^{-4} T(^\circ\mathrm{C})$$

and should be accurate enough for the conditions in the annulus of an absorber pipe. The parameters for calculating k for Hydrogen at smaller pressures can be found in Table 9.5.

(9.10) Some have proposed to use an organic liquid in the solar field for direct use in the power cycle, omitting the need for a heat transfer fluid and the extra cost. This is due to their lower boiling temperature and associated pressures allowing more ready use in the solar field. These plants will typically be used to produce a smaller amount of energy, on the order of 1 MW. Assuming that this net power is to be produced, use n-Pentane as the working fluid in a Rankine cycle (the pressure–enthalpy diagram is given in Appendix B) working between the pressures of 0.1 MPa and 4 MPa with a temperature exiting the solar field of 280 °C. Determine the cycle efficiency, work required to drive the pump, the heat load on the solar field and condenser and the mass flow rate of the working fluid.

(9.11) Determine the aperture area (in m^2) for a solar field with n-Pentane as the working fluid for direct use in a power cycle (see Appendix B for a pressure–enthalpy diagram of Pentane). This means that n-Pentane will be heated in the solar field then passed through a Rankine cycle rather than a heat exchanger. Organic liquids are being considered for small power plants, to be used in the so-called Organic Rankine Cycle. Use the same pressures and temperatures as those given in Exercise 9.10 and assume the absorber pipe and concentrator are the same as those in Table 9.4 and a glass envelope is used. The insolation level is reduced from the standard conditions to 500 W/m^2.

(9.12) Repeat Example 9.6 assuming annular pressures in the heat collection element from 10^{-1} to 10^5 Pa using the standard conditions given in Table 9.4.

Useful numbers, constants and relations

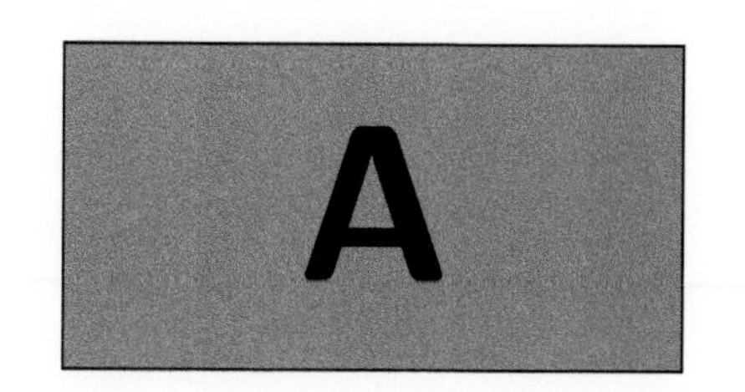

Table A.1 Numerical prefixes

Prefix	Symbol	Value	Prefix	Symbol	Value
Yotta	Y	10^{24}	Yocto	y	10^{-24}
Zetta	Z	10^{21}	Zepto	z	10^{-21}
Exa	E	10^{18}	Atto	a	10^{-18}
Peta*	P	10^{15}	Femto	f	10^{-15}
Tera	T	10^{12}	Pico	p	10^{-12}
Giga	G	10^{9}	Nano	n	10^{-9}
Mega	M	10^{6}	Micro	μ	10^{-6}
Kilo	k	10^{3}	Milli	m	10^{-3}
Hecto	h	10^{2}	Centi	c	10^{-2}
Deca	da	10^{1}	Deci	d	10^{-1}

*The prefix *Quadrillion* is used in the USA energy industry to represent 10^{15}, such as Quadrillion BTUs or Quads.

Table A.2 Energy and power conversion factors

$\frac{6.242 \times 10^{18}\,\text{eV}}{\text{J}}$	$\frac{4.184\,\text{J}}{\text{cal}}$	$\frac{1055\,\text{J}}{\text{BTU}}$
$\frac{10^{15}\,\text{BTU}}{\text{Quad}}$	$\frac{1.055 \times 10^{18}\,\text{J}}{\text{Quad}}$	$\frac{745.7\,\text{W}}{\text{hp}}$

Table A.3 Useful constants

Constant	Value	Units
c	2.998×10^{8}	m/s
g	9.807	m/s^2
k_B	1.381×10^{-23}	J/K
	8.617×10^{-5}	eV/K
h	6.626×10^{-34}	J-s
	4.136×10^{-15}	eV-s
k_B/h	2.084×10^{10}	1/s-K
N_A	6.022×10^{23}	#/mol
R	8.315	J/mol-K = Pa-m^3/mol-K
T_S	5793 (AM0)	K
	5359 (AM1.5G)	K
σ_S	5.670×10^{-8}	W/m^2-K^4
θ_{sub}	0.265	deg
ϵ_0	8.854×10^{-12}	C/V-m
	Electron properties	
m_e	9.109×10^{-31}	kg
q	1.602×10^{-19}	C

Table A.4 Useful relations

Relation	Page reference
$E = \dfrac{hc}{\lambda} = \dfrac{1240\,\text{eV}}{\lambda(\text{nm})}$	18
$g_s = \pi \sin(\theta_{sub})^2 = 6.720 \times 10^{-5}$	18
$\lambda T_S\|_{max} \approx \dfrac{hc}{5k_B} = 2.9 \times 10^6$ nm-K	20
$\dfrac{2g_s k_b^3}{h^3c^2} = \dfrac{1.353 \times 10^{10}}{\text{s-m}^2\text{-K}^3}$	56
$V_{th} = \dfrac{k_BT}{q} = 25.86$ mV (at 300 K)	112

Physical properties

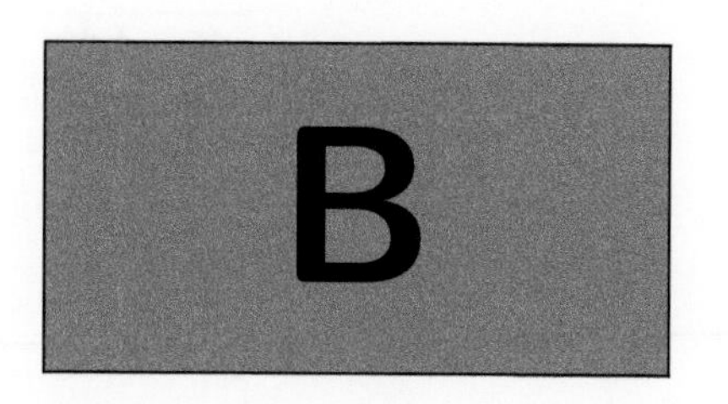

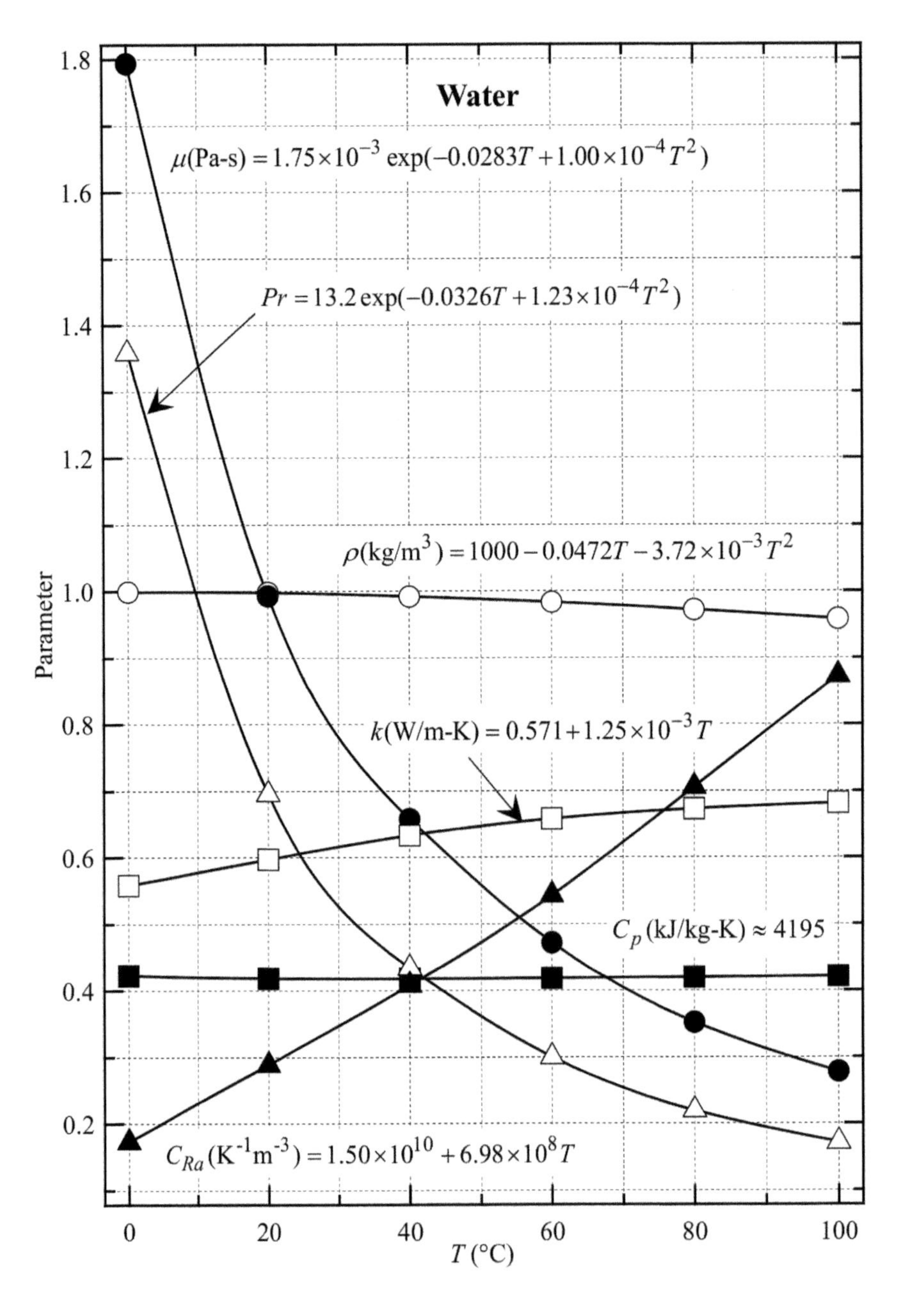

Fig. B.1 Various properties of liquid water including: viscosity (μ, •), density (ρ, ∘), heat capacity (C_p, ▪), thermal conductivity (k, ▫), a group of parameters used in the calculation of the Rayleigh number ($Ra = [g\chi\rho^2 C_p/k\mu]\Delta T L^3 \equiv C_{Ra}\Delta T L^3$, see the text for the definition of the parameters used to determine C_{Ra}, ▴) and the Prandtl number (Pr, ▵). The dimensions for each parameter are given in the equation and their magnitude on the graph can be ascertained by putting 0 °C in the equation. All the equations in the graph have the temperature T in °C. Data are from J. R. Welty et al., 'Fundamentals of momentum, heat and mass transfer,' Third Edition, John Wiley & Sons (1984).

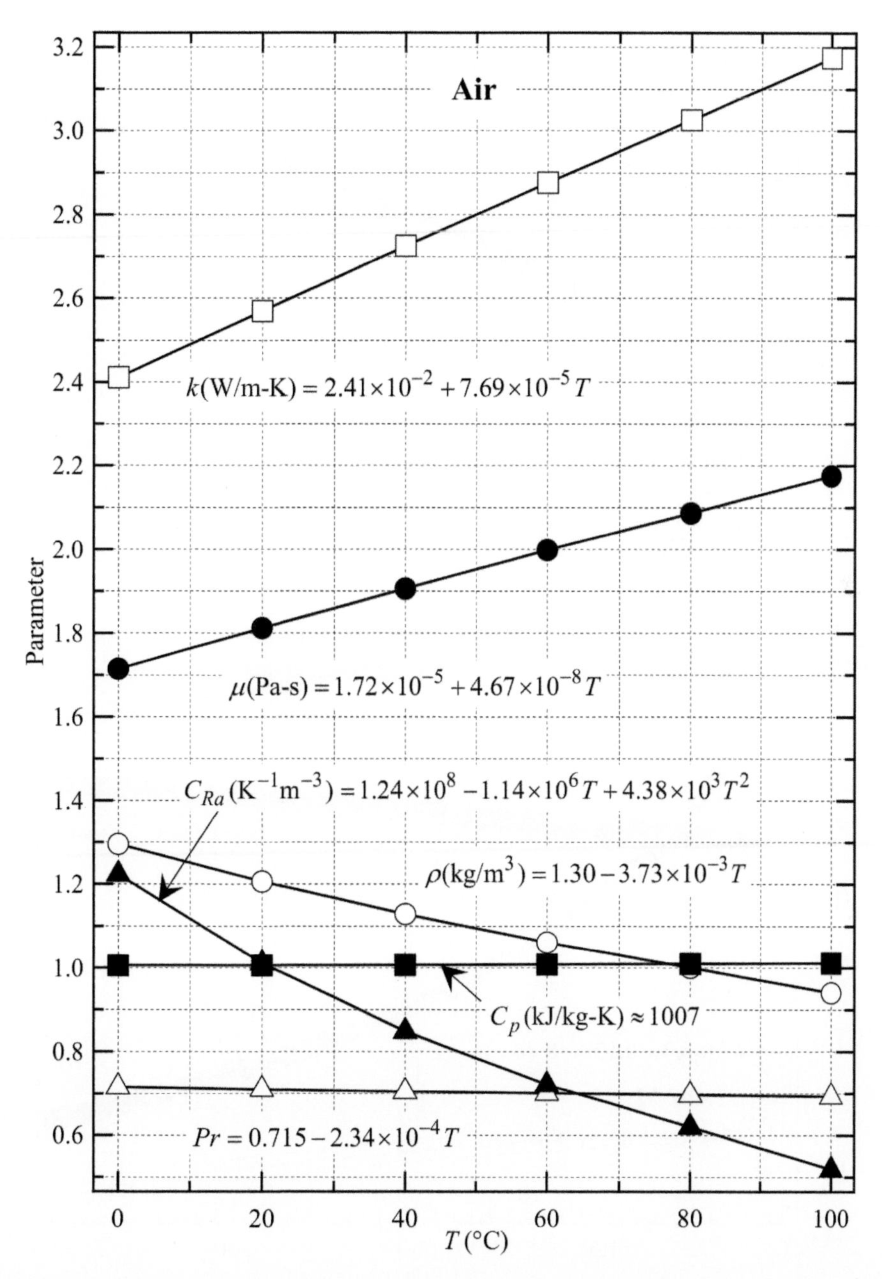

Fig. B.2 Various properties of air including: viscosity (μ, •), density (ρ, ∘), heat capacity (C_p, ▪), thermal conductivity (k, ▫), a group of parameters used in the calculation of the Rayleigh number ($Ra = [g\chi\rho^2 C_p/k\mu]\Delta T L^3 \equiv C_{Ra}\Delta T L^3$, see the text for the definition of the parameters used to determine C_{Ra}, ▴) and the Prandtl number (Pr, ▵). The dimensions for each parameter are given in the equation and their magnitude on the graph can be ascertained by putting 0 °C in the equation. All the equations in the graph have the temperature T in °C. Data are from J. R. Welty et al., 'Fundamentals of momentum, heat and mass transfer,' Third Edition, John Wiley & Sons (1984).

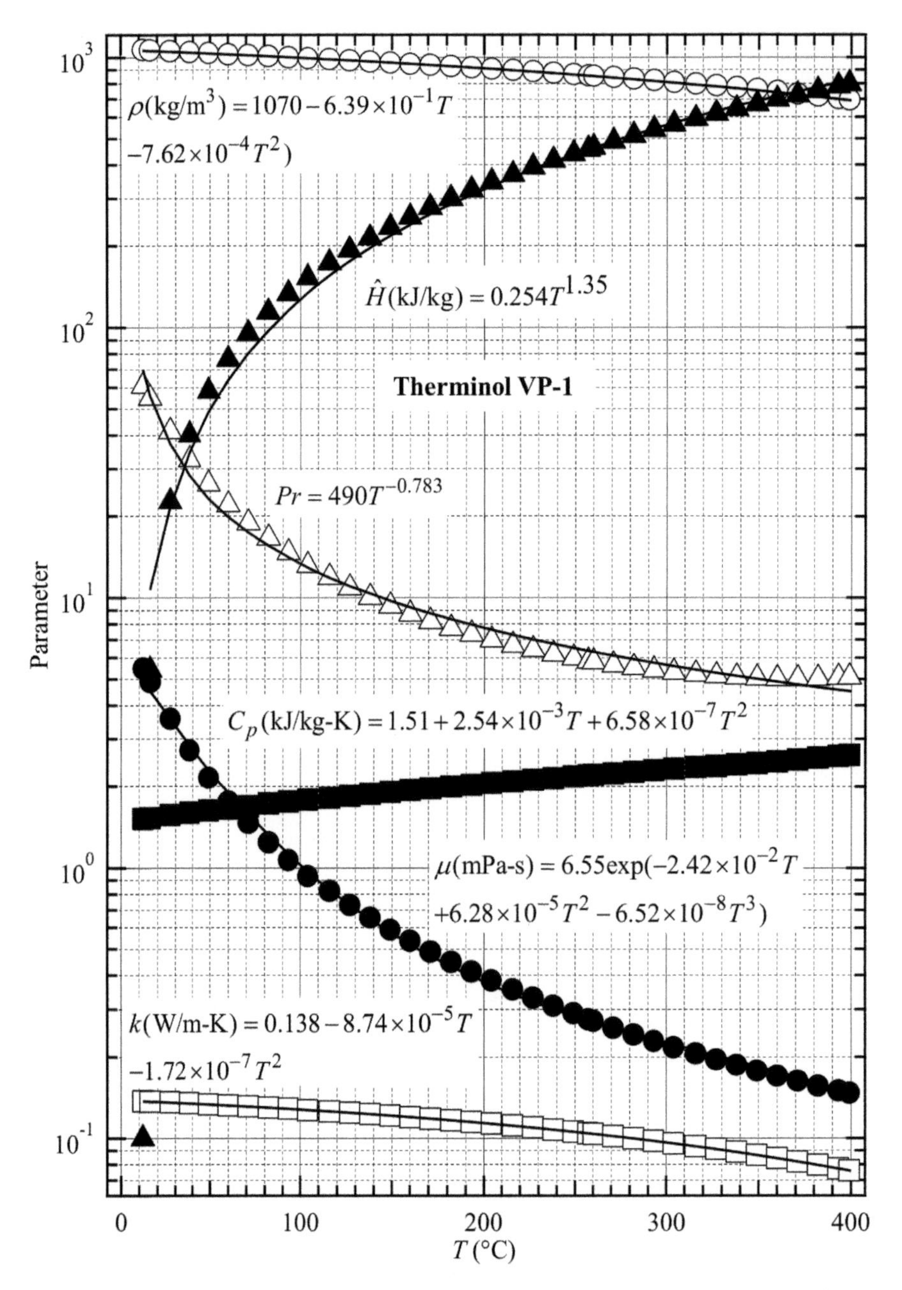

Fig. B.3 Various properties of Therminol VP-1 including: viscosity (μ, •), density (ρ, ◦), heat capacity (C_p, ▪), thermal conductivity (k, ▫), specific enthalpy ($\hat{H}$, ▴) and the Prandtl number (Pr, ▵). The dimensions for each parameter are given in the equation and the temperature T is in °C. The symbols are data given by Solutia Inc. while the lines represent the correlations to the data. Data are from the Dow Chemical Company.

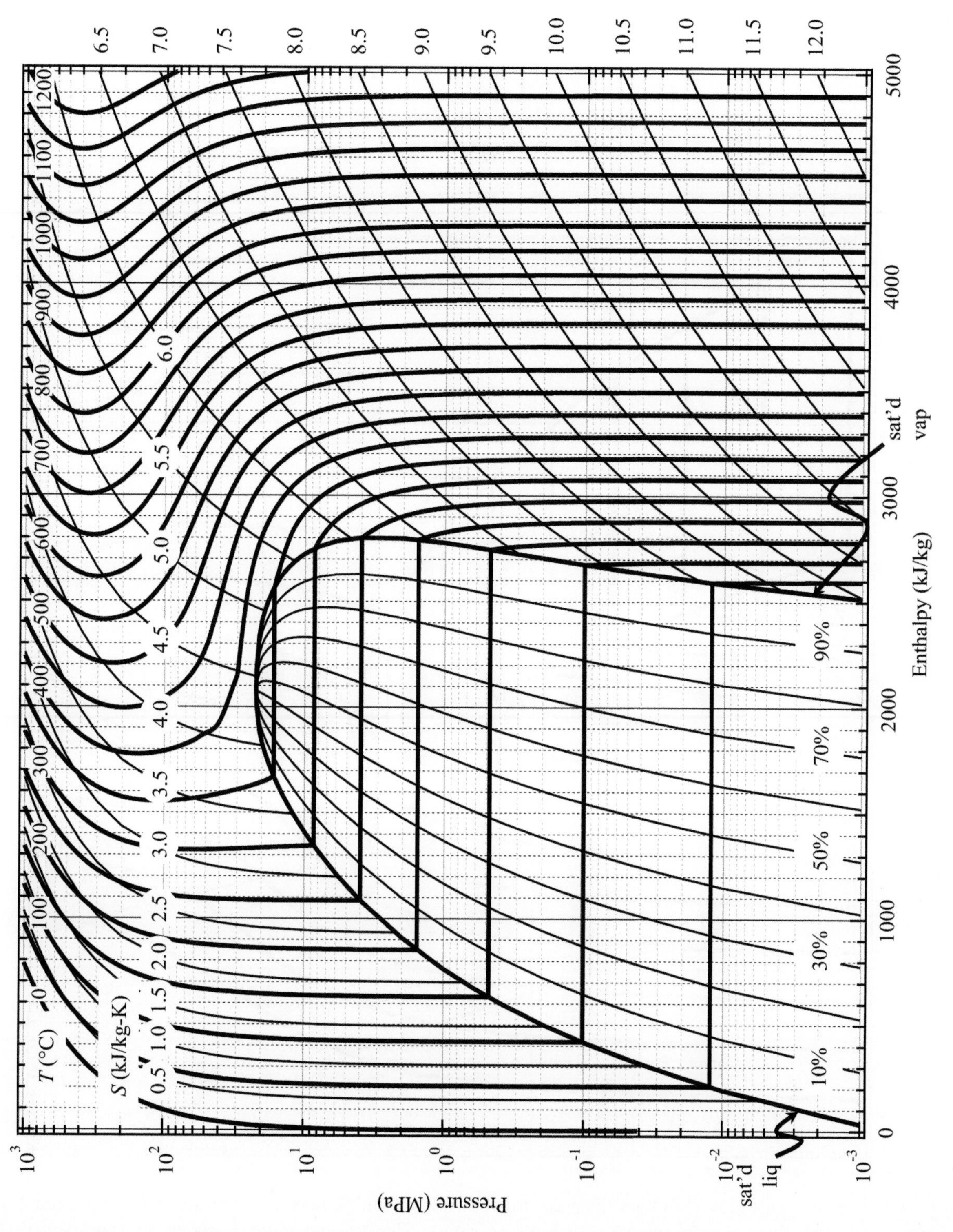

Fig. B.4 Pressure–enthalpy (Mollier) diagram for water. Thick lines are isotherms (T, constant temperature) and thin lines, isentropes (S, constant entropy). The lines labeled %10, *etc.* are lines of constant quality which is a constant amount of gaseous water (vapor). The critical point is shown at the apex of the vapor–liquid envelope. Data are from International Association for the Properties of Water and Steam.

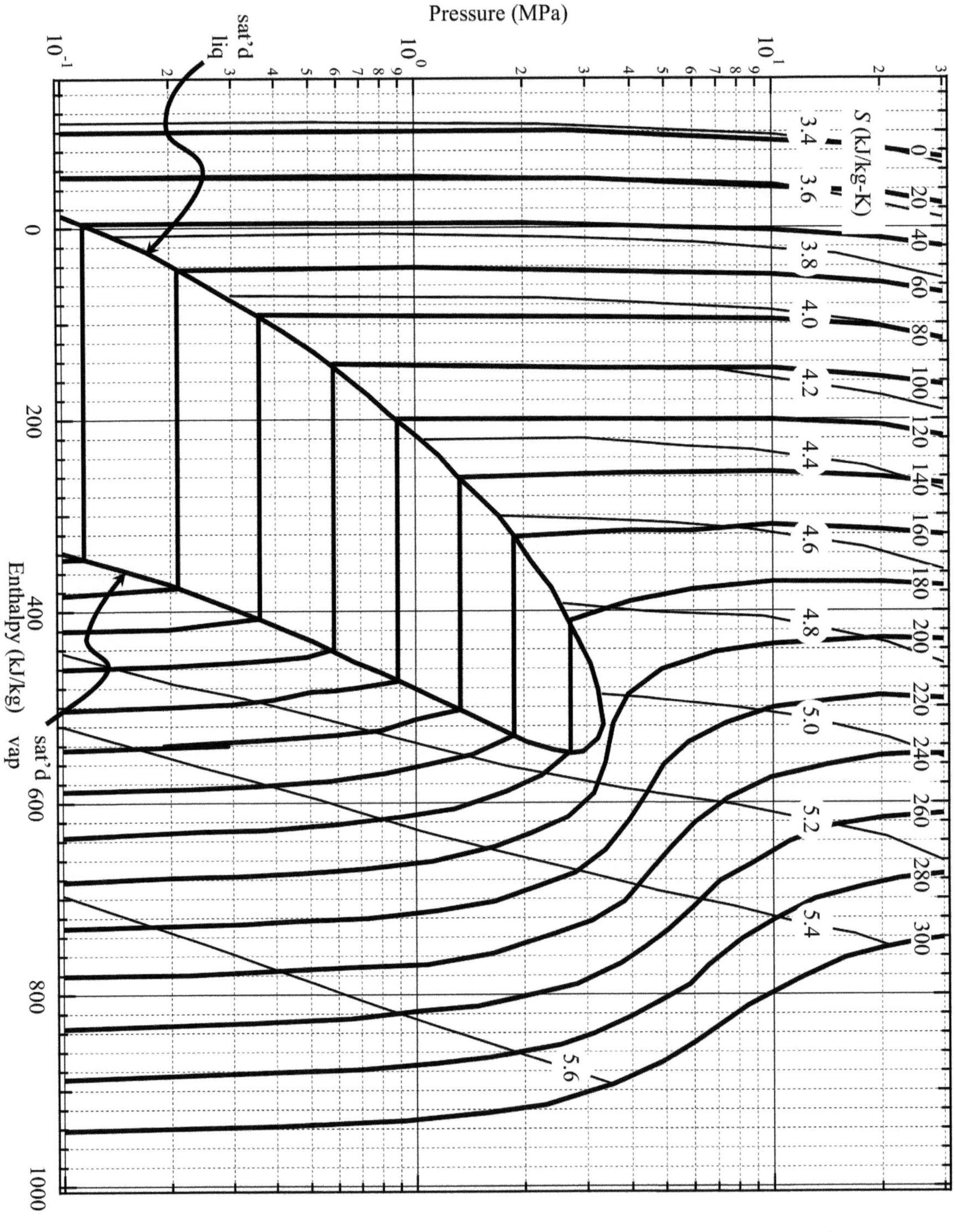

Fig. B.5 Pressure–enthalpy (Mollier) diagram for n-Pentane. Thick lines are isotherms (T, constant temperature) and thin lines, isentropes (S, constant entropy). The critical point is shown at the apex of the vapor–liquid envelope. Data are from Starling *et al.*, 'Geothermal binary-cycle working-fluid properties information,' DOE Report # DOE/ID/01719–4.

Dimensionless numbers and momentum and heat transfer correlations

Before there were computers we let nature be the *computer*. This was performed by using dimensionless numbers and developing correlations between them, particularly when designing complicated flow or heat transfer processes. The essence of the procedure was to take the differential equation applicable to the process at hand and make it dimensionless. The variables were all made dimensionless with process parameters such that they were all of order one in dimensionless form. How this allows nature to be the computer will become evident below.

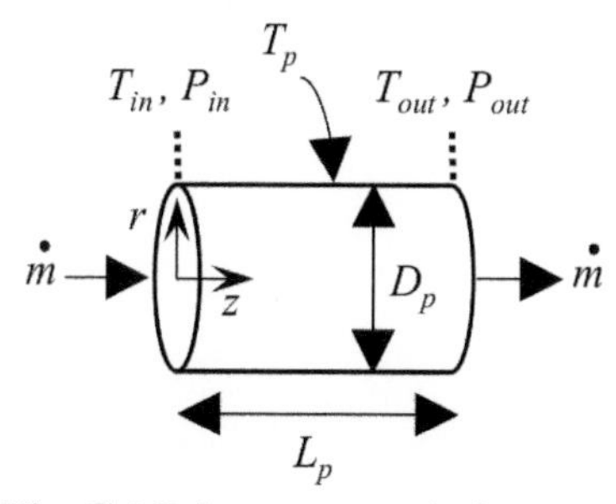

Fig. C.1 Relevant geometrical parameters for heat transfer from a pipe wall at temperature T_p to a fluid flowing within it.

Consider the velocity gradient $\partial v_z/\partial r$ for pressure driven flow of water in a pipe, see Fig. C.1. The appropriate velocity component of interest is that in the z-direction, v_z for cylindrical coordinates. The radial position is given by r. We know there is a gradient since the fluid velocity at the wall will (normally) be zero and gradually rise to a maximum at the center-line ($r = 0$). In fact, for a Newtonian fluid of constant viscosity the velocity profile is a parabola.

By convention, the radial position r is made dimensionless with the pipe diameter D_p, while v_z is made dimensionless with the average velocity in the pipe, V_w (the subscript w stands for water since we are primarily concerned with water here). The average velocity is the ratio of the volumetric flow rate to the pipe area and determined from the water mass flow rate $\dot{m}$ by

$$V_w = \frac{\dot{m}}{\rho_w \times \frac{\pi}{4} D_p^2} [=] \frac{\mathrm{m}}{\mathrm{s}} \tag{C.1}$$

where ρ_w is the density of water and the symbol [=] means 'has dimensions of.' One can now make the velocity gradient dimensionless to arrive at

$$\frac{\partial v_z}{\partial r} = \frac{V_w}{D_p}\frac{\partial \hat{v}_z}{\partial \hat{r}} \quad \text{with,} \quad \hat{v}_z = v_z/V_w \text{ and, } \hat{r} = r/D_p$$

where the ^ indicates a dimensionless number. In the above equation all dimensionless numbers are of order one, despite the value of D_p or V_w. The key to dimensional analysis is this, all the variables and derivatives of variables should be of order one and, as the reader will see below, the physics will be confined to the dimensionless numbers made from the numerical pre-factors.

The steady-state momentum equation for pressure driven flow of a Newtonian fluid in a pipe at constant temperature is given by

$$\frac{\partial P}{\partial z} = \mu_w \frac{1}{r}\frac{\partial}{\partial r}\left(r\frac{\partial v_z}{\partial r}\right) \tag{C.2}$$

where P is the pressure at any given axial position z and μ_w is the viscosity of water. Physically this equation states that the pressure drop or gradient along the pipe is caused by momentum transfer within the fluid (*i.e.* dissipation) from the center-line to the pipe wall.

If the upstream and downstream pressures, P_{in} and P_{out}, respectively, are kept constant one expects $\partial P/\partial z$ to also be constant along the pipe axis since there is no physical reason for it to vary other than linearly as long as the fluid temperature is constant (we have not begun to discuss heat transfer yet which will make the viscosity vary along the pipe length!). From a more mathematical point of view, since we are concerned with steady-state flow, pressure is expected to only be a function of z and the velocity only a function of r; since one side of the equation in a function of z and the other side r they must be equal to a constant. Now, one can discern why v_z is a parabolic function in r since its second derivative must be a constant. Regardless, let's make eqn (C.2) dimensionless to arrive at

$$\frac{\Delta P}{L_p}\frac{\partial \hat{P}}{\partial \hat{z}} = \frac{\mu_w V_w}{D_p^2}\frac{1}{\hat{r}}\frac{\partial}{\partial \hat{r}}\left(\hat{r}\frac{\partial \hat{v}_z}{\partial \hat{r}}\right)$$

where $\Delta P \equiv P_{in} - P_{out}$ and has been used to make the pressure dimensionless while z was made dimensionless with the pipe length L_p, see Fig. C.1. Because of convention and many years (well over a hundred) of studying flow in a pipe, the dimensionless equation is written

$$f\frac{\partial \hat{P}}{\partial \hat{z}} = \frac{1}{Re}\frac{1}{\hat{r}}\frac{\partial}{\partial \hat{r}}\left(\hat{r}\frac{\partial \hat{v}_z}{\partial \hat{r}}\right) \text{ with, } f \equiv \frac{\Delta P}{2\rho_w V_w^2}\frac{D_p}{L_p} \text{ and, } Re \equiv \frac{\rho_w V_w D_p}{\mu_w} \tag{C.3}$$

where f and Re are the friction factor and Reynolds number, respectively. The factor of 2 in the friction factor is used to define the Fanning friction factor, while the Moody friction factor has a value of $1/2$ rather than 2; we use the Fanning friction factor here.

So, one might do a series of experiments and measure the pressure drop as a function of mass flow rate for a variety of pipes with various lengths and diameters. Equation (C.3) can be used to show that if you plot f as a function of Re all the data should collapse onto one master curve; as is found. Although one can solve eqn (C.2) analytically for laminar flow it is not possible to do so for turbulent flow and this is where the dimensionless numbers are of use and how nature becomes a computer.[1] The f–Re correlation should collapse all the data for both laminar and turbulent flows allowing the designer to find the pressure drop for his or her circumstance.

[1] There has been some progress in understanding turbulent flow to determine the structure within it, which seems like an oxymoron. So, the statement that eqn (C.2) can only be solved for laminar flow is not exactly correct.

Now consider heat transfer by having a cold fluid entering a heated pipe, as one may encounter in a flat plate solar energy collector. The equation of energy with constant transport properties (*i.e.* the heat capacity C_{pw} and thermal conductivity k_w of water are assumed to be constant) for steady-state, non-isothermal flow in a pipe, is

$$\rho_w C_{pw} v_z \frac{\partial T}{\partial z} = k_w \left[\frac{1}{r} \frac{\partial}{\partial r} \left(r \frac{\partial T}{\partial r} \right) + \frac{\partial T}{\partial z} \right]$$

where T is temperature. This equation can be written in terms of dimensionless variables as

$$\frac{\rho_w C_{pw} V_w}{L_p} \hat{v}_z \frac{\partial \hat{T}}{\partial \hat{z}} = k_w \left[\frac{1}{D_p^2 \hat{r}} \frac{\partial}{\partial \hat{r}} \left(\hat{r} \frac{\partial \hat{T}}{\partial \hat{r}} \right) + \frac{1}{L_p^2} \frac{\partial \hat{T}}{\partial \hat{z}} \right]$$

where the dimensionless temperature $\hat{T}$ was written as $[T - T_{in}]/[T_{out} - T_{in}]$; one can subtract T_{in} from T in the derivatives since it is a constant and will naturally drop out when the derivative is taken. The subtraction was done so $\hat{T}$ is of order one at all times. This equation can be divided by μ_w on both sides and re-written as

$$Re\, \hat{v}_z \frac{\partial \hat{T}}{\partial \hat{z}} = \frac{1}{Pr} \left[\frac{L_p}{D_p \hat{r}} \frac{\partial}{\partial \hat{r}} \left(\hat{r} \frac{\partial \hat{T}}{\partial \hat{r}} \right) + \frac{D_p}{L_p} \frac{\partial \hat{T}}{\partial \hat{z}} \right] \text{ with, } Pr = \frac{C_{pw} \mu_w}{k_w} \qquad \text{(C.4)}$$

Equation (C.4) can be used to demonstrate the importance of the Reynolds and Prandtl (Pr) numbers in heat transfer and any correlation must be developed with them in mind since $\hat{T}$ will be a function of them. This equation can be solved for laminar flow, however, we need to develop a correlation that relates the heat input from the hot pipe wall to the cold fluid under any flow condition, laminar or turbulent.[2]

[2] This being said eqn (C.4) is an adventure in applied mathematics, even in laminar flow.

Equation (C.4) will allow calculation of the temperature rise for a cold fluid flowing in a hot pipe. Yet, details of the heat transfer from the hot pipe to the fluid have not been discussed. Any time there is an interface between a liquid or gas and a solid there is not good thermal contact between the two materials, at least not as good as within a solid, homogenous material alone.

Consider simple heat transfer from the heated pipe to liquid water, this is usually written as

$$\dot{q}_w = h_w [T_p - T_\infty] [=] \frac{\text{W}}{\text{m}^2} \qquad \text{(C.5)}$$

similar to that written in Section 8.1 and shown in eqn (8.10). This is Newton's law of cooling and relates the rate of heat transfer per unit area to the difference between the pipe temperature T_p (see Fig. C.1) and the temperature far from it, T_∞ (this is an ambiguous temperature and is included to relate this heat transfer relation to thermal conduction, suffice to say it is 'far' from the pipe wall). The heat transfer coefficient h_w can only be found by correlations and is not solely a material property. It is dependent on the material's properties as well as process conditions.

The type of heat transfer considered in eqn (C.4) is only conduction from the pipe wall to the water, via knowledge of the thermal conductivity k_w. This can be written

$$\dot{q}_w = k_w \left.\frac{\partial(T - T_\infty)}{\partial x}\right|_{x=0} \equiv \frac{k_w}{D_p} \left.\frac{\partial(T - T_\infty)}{\partial \hat{x}}\right|_{\hat{x}=0} \tag{C.6}$$

by applying Fourier's law right at the pipe wall. The temperature of the fluid just at the pipe wall T is unknown. The second equation results by making the distance x dimensionless with the pipe diameter D_p. Of course, this is an imprecise discussion since radial coordinates are not being used; the mathematical details are simplified to promote the gist of the discussion.

These two rates of heat transfer should be equal when describing the same system and using eqns (C.5) and (C.6) one arrives at

$$Nu \equiv \frac{h_w D_p}{k_w} = \frac{\left.\frac{\partial(T - T_\infty)}{\partial \hat{x}}\right|_{\hat{x}=0}}{T_p - T_\infty} \tag{C.7}$$

So, the Nusselt number represents a dimensionless temperature gradient. The utility of the Nusselt number is this: assuming one could solve eqn (C.4) the boundary condition of water having a temperature T_p at the pipe wall would have to be used. One does not know the temperature of the water at this position since heat transfer from the pipe at temperature T_p to the water in contact with it is unknown. There will be a finite temperature drop at the interface. The experimentally determined Nusselt number corrects for this.

Thus, correlations must be determined experimentally and the Nu correlated with Re and Pr. This would be accomplished by measuring the temperature rise $T_{out} - T_{in}$ for a given pipe maintained at T_p, allowing one to determine $\dot{q}_w$, the rate of heat transfer per unit area of pipe, *via*

$$\dot{q}_w = \dot{m} C_{pw} [T_{out} - T_{in}] \tag{C.8}$$

The heat capacity of water C_{pw} will be determined at the mean temperature, see eqn (8.11). This is just the sensible heat rise of the water. A similar equation exists in eqn (8.10), written in slightly different form here,

$$\dot{q}_w = h_w \Delta T_{lm} \tag{C.9}$$

since this is the rate of heat transfer per unit area based on the log mean temperature difference. Equating eqns C.8 and C.9 allows one to determine h_w,

$$h_w = \frac{\dot{m} C_{pw} [T_{out} - T_{in}]}{\Delta T_{lm}}$$

One would calculate h_w from this equation and then determine Nu and a correlation with Re and Pr is made.

A simple correlation is the Dittus-Boelter correlation which is of limited utility,

$$Nu = 0.023Re^{0.8}Pr^n$$

where $n = 0.3$ if the fluid is cooled and $n = 0.4$ if the fluid is heated. Equation (8.13) is more complicated, and accurate, as it was developed over many years and includes the friction factor f.

Below is a summary of the dimensionless numbers used in fluid flow and heat transfer in this monograph. Also included is the absorber number, introduced in eqn (9.19), that is a useful grouping of variables and is a modified Stanton number. The dimensionless numbers are formulated as a ratio of *forces* like the ratio of inertial to viscous, for Re, or rates.

Table C.1 Dimensionless numbers used in fluid flow and heat transfer in this monograph.

Dimensionless number	Definition	Interpretation
Absorber number*	$Ab = \dfrac{hN_sA_p}{\dot{m}_0C_p}$	Ratio of convective heat transfer to sensible temperature rise
Friction factor (Fanning)	$f = \dfrac{\Delta P}{2\rho V^2}\dfrac{D_p}{L_p}$	Dimensionless pressure drop for flow in a pipe
Nusselt number	$Nu = \dfrac{hD}{k}$	Ratio of convective to conductive heat transfer
Prandtl number	$Pr = \dfrac{C_p\mu}{k}$	Ratio of momentum to thermal diffusivities
Reynolds number	$Re = \dfrac{\rho VD}{\mu}$	Ratio of inertial to viscous forces

*As discussed in Section 9.3 and shown in eqn (9.20), the absorber number is related to the Stanton number through, $Ab = 4\dfrac{L}{D_p}St$, where $Ab = \dfrac{hN_s\pi D_pL_p}{\dot{m}_0C_p}$, A_p is the internal pipe area.

List of symbols

a Absorbance [=] -
Acceleration [=] m/s^2
Lattice constant or parameter [=] m
Length of one-dimensional box [=] m
Thermal accommodation coefficient [=] -

A Absorber pipe area based on outside diameter [=] m^2
Area [=] m^2

A^* Group of constants [=] $J^{-1/2}$-m^{-1}

A_{ap} Aperture area [=] m^2

$A(\beta)$ Albedo irradiance striking a device at the angle β [=] W/m^2

AM Air mass [=] -

b Interaction coefficient [=] -

$B(\beta)$ Beam irradiance striking a device at the angle β [=] W/m^2

B_{ter} Terrestrial beam irradiance [=] W/m^2

c Speed of light = 2.998×10^8 m/s

C_c Correction factor for flow in a chimney [=] -

C_p Heat capacity at constant pressure [=] J/kg-K

C_R Concentration ratio [=] -

$C_{R,i}$ Ideal or maximum concentration ratio in 1-dimension [=] -

$C_{R,2i}$ Ideal or maximum concentration ratio in 2-dimensions [=] -

C_v Heat capacity at constant volume [=] J/kg-K

C_w Correction factor for flow in a chimney [=] -

d Distance between the Sun and Earth [=] 1.5×10^8 km

D Diameter of the Sun = 1.4×10^6 km
Outside diameter of absorber pipe [=] m

D_i Inside diameter of absorber pipe envelope [=] m

D_o Outside diameter of absorber pipe envelope [=] m

D_p Inside diameter of absorber pipe [=] m

$D(\beta)$ Diffuse irradiance striking a device at the angle β [=] W/m^2

e Emissivity [=] -

E Energy [=] J

$\dot{E}$ Rate of energy transfer [=] W

E_{a-S}	Specular exchange factor between the absorber and Sun [=] -
E_c	Energy at the bottom of the conduction band [=] J or eV
E_f	Final energy level [=] J
E_g	Band gap energy [=] J or eV
E_i	Initial energy level [=] J
E_n	Energy of the nth energy level [=] J
E_p	Phonon energy exchange [=] J
E_U	Urbach tail energy parameter [=] J
E_v	Energy at the top of the valence band [=] J or eV
$F(E)$	Probability of having a fermion at energy E [=] -
FF	Fill factor [=] -
FF_0	Fill factor under ideal conditions [=] -
FF_s	Fill factor correcting for series resistance [=] -
FF_{sh}	Fill factor correcting for shunt resistance [=] -
g	Gravitational acceleration constant = 9.807 m/s^2
g_s	Geometric factor = $g_S = \pi \sin(\theta_S) = 6.720 \times 10^{-5}$
$G(E, x)$	Excited state generation rate for energy E and position x [=] #/s-m^3-J
$G_L(E)$	Excited state generation rate for thickness L [=] #/s-m^2-J
G_{LE}	Total excited state generation rate [=] #/s-m^2
h	Height [=] m
	Planck's constant = 6.626×10^{-34} J-s = 4.136×10^{-15} eV-s
	Minimum atmospheric thickness [=] m
	Heat transfer coefficient [=] W/m^2-K
$\hbar$	Dirac or reduced Planck constant = $h/2\pi = 1.055 \times 10^{-34}$ J-s
H	Chimney height [=] m
$\hat{H}$	Specific enthalpy [=] J/kg
i	Imaginary number = $\sqrt{-1}$
I	Energetic irradiance [=] #/s-m^2
J	Current density [=] A/m^2
J_0	Dark current density in a photovoltaic device [=] A/m^2
J_{01}	Dark current density with quasi-neutral region recombination [=] A/m^2
J_{02}	Dark current density with depletion region recombination [=] A/m^2
J_L	Light generated current density [=] A/m^2
$\hat{J}_L$	Dimensionless light generated current density [=] -
J_m	Current density at the maximum power point [=] A/m^2
$\hat{J}_m$	Dimensionless maximum power current density = J_m/J_0 [=] -
J_{m0}	Ideal maximum power point current density [=] A/m^2

J_{max}	Maximum current density from a photovoltaic device [=] A/m^2
J_{sc}	Short circuit current density [=] A/m^2
k	Extinction coefficient [=] -
	Thermal conductivity [=] W/m-K
k_B	Boltzmann's constant = 1.380×10^{-23} J/K = 8.617×10^{-5} eV/K
K	Group of variables = $15 \sin(\theta_S)^2 \sigma_S T_S^4/\pi^4$ [=] W/m^2
K_J	Group of variables = $2\pi q \sin(\theta_S)^2 [k_B T_S]^3/[h^3c^2]$ [=] A/m^2
K_N	Group of variables = $2\pi \sin(\theta_S)^2 [k_B T]^3/[h^3c^2]$
K_T	Clearness index [=] -
L	Latitude [=] deg or rad
	Length [=] m
	Spectral irradiance [=] W/m^2-nm
L_{act}	Active region thickness [=] m
L_{bb}	Spectral irradiance from a black body [=] W/m^2
L_e	Diffusion length for electrons [=] m
L_{gb}	Spectral irradiance from a gray body [=] W/m^2
L_h	Diffusion length for holes [=] m
L_H	House length scale [=] m
L_n	Depletion region length in n-type material [=] m
L_p	Depletion region length in p-type material [=] m
m	Mass [=] kg
$\dot{m}$	Mass flow rate [=] kg/s
m_e	Mass of an electron = 9.109×10^{-31} kg
m_e^*	Effective mass of an electron [=] kg
m_h^*	Effective mass of a hole [=] kg
n	Day number (Jan 1, $n = 1$) [=] -
	Non-ideality index of a diode [=] -
	Number of electrons per unit volume [=] $\#/m^3$
	Photon number rate [=] #/s-sr-J-m^2
	Principal quantum number [=] -
	Whole number [=] -
n_i	Number of intrinsic carriers [=] $\#/m^3$
n_n	Number density of electrons in n-type material [=] $\#/m^3$
n_p	Number density of electrons in p-type material [=] $\#/m^3$
N_A	Avogadro's number [=] 6.022×10^{23} #/mol
	Number density of acceptor atoms [=] $\#/m^3$
N_A^+	Number density of ionized acceptor atoms [=] $\#/m^3$
N_c	Effective density of states in the conduction band [=] $\#/m^3$
N_D	Number density of donor atoms [=] $\#/m^3$

N_D^+	Number density of ionized donor atoms [=] #/m^3
$N(E)$	Number of electrons per unit volume per unit energy [=] #/m^3-J
$N'(E)$	Number of holes per unit volume per unit energy [=] #/m^3-J
N_h	Number of hours between sunrise and sunset [=] h
N_L	Maximum number of photons absorbed in length L [=] #/s-m^2
N_{out}	Rate of electrons out of a photovoltaic device [=] #/s-m^2
N_p	Integrated photon number rate [=] #/s-J-m^2
N_v	Effective density of states in the valence band [=] #/m^3
Nu	Nusselt number [=] -
p	Momentum [=] kg-m/s
	Number of holes per unit volume [=] #/m^3
p_n	Number density of holes in n-type material [=] #/m^3
p_p	Number density of holes in p-type material [=] #/m^3
P	Pressure [=] Pa
	Irradiance [=] W/m^2
	Dimensionless potential in Kronig–Penney model [=] -
P_{bb}	Irradiance from a black body [=] W/m^2
$P_D(\beta)$	Total irradiance striking a device at the angle β [=] W/m^2
P_Δ	Useful irradiance between two energy levels [=] W/m^2
P_{ext}	Extraterrestrial irradiance [=] W/m^2
P_{gb}	Irradiance from a gray body [=] W/m^2
P_m	Maximum power density in a solar cell [=] W/m^2
P_{out}	Rate of power out of a photovoltaic device [=] W/m^2
P_{ter}	Terrestrial irradiance [=] W/m^2
Pr	Prandtl number [=] -
q	Charge of an electron = 1.602×10^{-19} C
$\dot{q}$	Rate of heat transfer per unit area [=] W/m^2
Q	Heat [=] J
Q_H	Heat expelled from high temperature reservoir [=] J
Q_L	Heat expelled to low temperature reservoir [=] J
$\dot{Q}$	Rate of heat transfer [=] W
$\dot{Q}_b$	Rate of heat transfer in the boiler [=] W
$\dot{Q}_c$	Rate of heat transfer in the condenser [=] J/s = W
$\dot{Q}_{gen}$	Rate of heat generated by radiation absorption [=] W/m^2
$\dot{Q}_S$	Rate of heat transfer from the Sun [=] W
$\dot{Q}_u$	Rate of useful heat transfer [=] W
r	Radial position [=] m
r_s	Dimensionless series resistance [=] -
r_{sh}	Dimensionless shunt resistance [=] -

R_{ch} Characteristic resistance of a solar cell [=] Ohm
$\hat{R}_{ch}$ Dimensionless solar cell characteristic resistance [=] -
R_E Radius of the Earth = 6380 km
Ra Rayleigh number [=] -
Ra^* Modified Rayleigh number [=] -
Re Reynolds number [=] -
S Entropy [=] J/K
$\hat{S}$ Specific entropy [=] J/kg-K
t_{cor} Time correction [=] h
t_{day} Daylight saving time correction [=] h
t_{EOT} Equation of time value [=] h
t_{GMT} Greenwich mean time [=] h
t_L Local (clock) time [=] h
t_{LST} Local solar time [=] h
T Temperature [=] K
T_a Temperature of surrounding air [=] K
T_c Cover temperature [=] K
T_H Temperature of high temperature reservoir [=] K
T_L Temperature of low temperature reservoir [=] K
T_p Absorber plate temperature [=] K
T_S Black body temperature of the Sun [=] K
T_{sky} Sky temperature [=] K
T_{stag} Stagnation temperature [=] K
U Internal energy [=] J
$\hat{U}$ Specific internal energy [=] J/kg
v Velocity [=] m/s
v_w Wind velocity [=] m/s
V Voltage [=] V
Average velocity in a conduit [=] m/s
$\hat{V}$ Specific volume [=] m^3/kg
V_0 Potential in Kronig–Penney model [=] J
Potential [=] J
V_a Applied voltage (potential) [=] V
V_m Voltage at the maximum power point [=] V
$\hat{V}_m$ Dimensionless maximum power point voltage = V_m/V_{th} [=] -
V_{m0} Ideal maximum power point voltage [=] V
V_{max} Maximum voltage from a photovoltaic device [=] V
V_{oc} Open circuit voltage [=] V
$\hat{V}_{oc}$ Dimensionless open circuit voltage = V_{oc}/V_{th} [=] -

V_{th}	Thermal voltage $= k_B T/q$ = 25.86 mV at 300 K
V_w	Average velocity of water in a pipe [=] m/s
W	Aperture width [=] m
	Work [=] J
	Total depletion region length [=] m
$\dot{W}$	Rate of work [=] W
$\dot{W}_m$	Rate of mechanical work [=] W
$\dot{W}_{net}$	Net rate of work [=] W
$\dot{W}_p$	Rate of work to drive a pump [=] W
$\dot{W}_{PV}$	Rate of pressure–volume work [=] W
$\dot{W}_t$	Rate of work from a turbine [=] W
x	Atmospheric thickness between observer and Sun [=] m
$\bar{x}$	Photon mean free path [=] m
x_g	Dimensionless band gap energy $= E_g/k_B T_S$
x_i	Dimensionless energy of state $i = E_i/k_B T$ [=] -
$Z(E)$	Electrons' density of energy states per unit energy [=] #/m^3-J
$Z'(E)$	Holes' density of energy states per unit energy [=] #/m^3-J
Z_{eff}	Effective charge number [=] -

Greek Symbols

α	Absorption coefficient [=] m^{-1}
	Dimensionless energy [=] -
α_S	Solar angular altitude [=] deg or rad
$\alpha_{S,N}$	Solar angular altitude at solar noon [=] deg or rad
α_U	Urbach tail absorption coefficient [=] m^{-1}
β	Device tilt angle with the horizontal [=] deg or rad
	Dimensionless variable in Kronig–Penney model [=] -
β_{opt}	Optimum device tilt angle at solar noon [=] deg or rad
γ	Ratio of C_p to C_v [=] -
δ	Effective angular declination [=] rad or deg
δ_0	Angular declination $= 23°27'$ = 23.45°
Δ	Parameter in the equation of time [=] deg or rad
ϵ	Dielectric constant [=] -
	Molar absorption coefficient [=] L/mol-cm
ϵ_0	Permittivity of free space [=] 8.854×10^{-12} F/m
$\zeta(\bullet)$	Riemann zeta function [=] -
η	Efficiency [=] -
θ	Polar angle [=] deg or rad
θ_A	Apparent angle for albedo radiation transmission [=] deg or rad

θ_D	Apparent angle for diffuse radiation transmission [=] deg or rad
θ_S	Angle between area normal and solar beam [=] deg or rad
θ_{sub}	Angle subtended between the Sun and Earth = 0.265°
θ_Z	Solar zenith angle [=] deg or rad
κ	Wave (momentum) vector [=] m^{-1}
λ	Gas molecule mean free path [=] m
	Wavelength [=] m
Λ_{STM}	Local standard time meridian [=] deg or rad
μ	Reduced mass of an electron = $1.05m_e$ for Hydrogen
	Chemical potential [=] J or eV
	Viscosity [=] kg/m-s or Pa-s
μ_e	Mobility of an electron [=] m^2/s-V
μ_h	Mobility of a hole [=] m^2/s-V
μ_n	Chemical potential of n-type material [=] J
μ_p	Chemical potential of p-type material [=] J
ν	Frequency [=] cycles/s
$\tilde{\nu}_{ab}$	Wave number = ν/c [=] cm^{-1}
π	pi = 3.14159
ρ	Density [=] kg/m^3
	Reflectance [=] -
	Space charge density [=] C/m^3
ρ_A	Albedo reflection factor [=] -
σ	Absorption cross-section [=] m^2
σ_S	Stefan-Boltzmann constant = 5.670×10^{-8} W/m^2-K^4
ΣP_{ter}	Terrestrial irradiance summed over a given time period [=] J/m^2
τ	Transmittance or transmission coefficient [=] -
τ_{abs}	Transmission coefficient for absorption [=] -
τ_A	Transmission coefficient for albedo component [=] -
τ_B	Transmission coefficient for beam component [=] -
τ_D	Transmission coefficient for diffuse component [=] -
τ_{refl}	Transmission coefficient for reflection [=] -
τ_{scat}	Transmission coefficient for scattering [=] -
ϕ	Azimuthal angle [=] deg or rad
ψ	Position dependent wave function [=]
ψ	Potential [=] V
Ψ	Time and position dependent wave function [=] -
ψ_0	Potential developed at a p-n junction [=] V
χ	Isothermal compressibility [=] K^{-1}

ω	Frequency [=] rad/s
	Hour angle [=] deg or rad
ω_S	Sunset angle [=] deg or rad
Ω	Solid angle [=] sr

Other symbols and operations

$< \bullet >$	Operation yielding average value [=] -
$\lvert\lvert \bullet \rvert\rvert$	Operation to calculate absorbance-transmittance product [=] -
dy/dx	Derivative of y with respect to x [=] varies
Δ	Difference between two quantities [=] - varies
IQE	Internal quantum efficiency [=] -
$\parallel$	Superscript for the parallel polarized radiation component [=] -
$\perp$	Superscript for the perpendicular polarized radiation component [=] -
$\mathcal{P}$	Probability per unit volume per unit time that radiation is absorbed [=] m^{-3}-s^{-1}
$\partial y/\partial x$	Partial derivative of y with respect to x [=] varies
$\sum_i^j$	Summation from i to j [=] -
SSE	Sum of squares error [=] varies
$\square$	Superscript for either the $\perp$ or $\parallel$ polarized component [=] -
[=]	Means 'has dimensions of'

Index